# Surveys and Tutorials in the Applied Mathematical Sciences

## Volume 4

*Editors*
S.S. Antman, J.E. Marsden, L. Sirovich

# Surveys and Tutorials in the Applied Mathematical Sciences

Volume 4

*Editors*
S.S. Antman, J.E. Marsden, L. Sirovich

Mathematics is becoming increasingly interdisciplinary and developing stronger interactions with fields such as biology, the physical sciences, and engineering. The rapid pace and development of the research frontiers has raised the need for new kinds of publications: short, up-to-date, readable tutorials and surveys on topics covering the breadth of the applied mathematical sciences. The volumes in this series are written in a style accessible to researchers, professionals, and graduate students in the sciences and engineering. They can serve as introductions to recent and emerging subject areas and as advanced teaching aids at universities. In particular, this series provides an outlet for material less formally presented and more anticipatory of needs than finished texts or monographs, yet of immediate interest because of the novelty of their treatments of applications, or of the mathematics being developed in the context of exciting applications. The series will often serve as an intermediate stage of publication of materials which, through exposure here, will be further developed and refined to appear later in one of Springer's more formal series in applied mathematics.

Yalchin Efendiev  •  Thomas Y. Hou

# Multiscale Finite Element Methods

## Theory and Applications

 Springer

Yalchin Efendiev
Department of Mathematics
Texas A & M University
College Station, TX 77843
USA
efendiev@math.tamu.edu

Thomas Y. Hou
Applied and Computational Mathematics,
217-50
California Institute of Technology
Pasadena, CA 91125
USA
hou@acm.caltech.edu

*Editors:*

S.S. Antman
Department of Mathematics
*and*
Institute for Physical Science
and Technology
University of Maryland
College Park
MD 20742-4015
USA
ssa@math.umd.edu

J.E. Marsden
Control and Dynamical
System, 107-81
California Institute
of Technology
Pasadena, CA 91125
USA
marsden@cds.caltech.edu

L. Sirovich
Laboratory of Applied
Mathematics
Department of
Bio-Mathematical Sciences
Mount Sinai School of Medicine
New York, NY 10029-6574
USA
Lawrence.Sirovich@mssm.edu

ISBN 978-0-387-09495-3      ISBN 978-0-387-09496-0 (eBook)
DOI 10.1007/978-0-387-09496-0

Library of Congress Control Number: 2008943964

Mathematics Subject Classification (2000): 65N99, 76S05, 35B27

Printed on acid-free paper

springer.com

Dedicated to my parents, Rafik and Ziba,
my wife, Denise, and my son, William
Yalchin Efendiev


Dedicated to my parents, Sum-Hing and Sau-Ying,
my wife, Yu-Chung, and my children, George and Anthony
Thomas Y. Hou

# Preface

The aim of this monograph is to describe the main concepts and recent advances in multiscale finite element methods. This monograph is intended for the broader audience including engineers, applied scientists, and for those who are interested in multiscale simulations. The book is intended for graduate students in applied mathematics and those interested in multiscale computations. It combines a practical introduction, numerical results, and analysis of multiscale finite element methods. Due to the page limitation, the material has been condensed.

Each chapter of the book starts with an introduction and description of the proposed methods and motivating examples. Some new techniques are introduced using formal arguments that are justified later in the last chapter. Numerical examples demonstrating the significance of the proposed methods are presented in each chapter following the description of the methods. In the last chapter, we analyze a few representative cases with the objective of demonstrating the main error sources and the convergence of the proposed methods.

A brief outline of the book is as follows. The first chapter gives a general introduction to multiscale methods and an outline of each chapter. The second chapter discusses the main idea of the multiscale finite element method and its extensions. This chapter also gives an overview of multiscale finite element methods and other related methods. The third chapter discusses the extension of multiscale finite element methods to nonlinear problems. The fourth chapter focuses on multiscale methods that use limited global information. This is motivated by porous media applications where some type of nonlocal information is needed in upscaling as well as multiscale simulations. The fifth chapter of the book is devoted to applications of these methods. Finally, in the last chapter, we present analyses of some representative multiscale methods from Chapters 2, 3, and 4.

# Acknowledgments

We are grateful to J. E. Aarnes, C. C. Chu, P. Dostert, L. Durlofsky, V. Ginting, O. Iliev, L. Jiang, S. H. Lee, W. Luo, P. Popov, H. Tchelepi, and X. H. Wu for many helpful comments, discussions, and collaborations. The partial support of NSF and DOE is greatly appreciated.

College Station & Pasadena                                  *Yalchin Efendiev*
August 2008                                                  *Thomas Y. Hou*

# Contents

# 1

# Introduction

## 1.1 Challenges and motivation

A broad range of scientific and engineering problems involve multiple scales. Traditional approaches have been known to be valid for limited spatial and temporal scales. Multiple scales dominate simulation efforts wherever large disparities in spatial and temporal scales are encountered. Such disparities appear in virtually all areas of modern science and engineering, for example, composite materials, porous media, turbulent transport in high Reynolds number flows, and so on. A complete analysis of these problems is extremely difficult. For example, the difficulty in analyzing groundwater transport is mainly caused by the heterogeneity of subsurface formations spanning over many scales. This heterogeneity is often represented by the multiscale fluctuations in the permeability (hydraulic conductivity) of the media. For composite materials, the dispersed phases (particles or fibers), which may be randomly distributed in the matrix, give rise to fluctuations in the thermal or electrical conductivity or elastic property; moreover, the conductivity is usually discontinuous across the phase boundaries. In turbulent transport problems, the convective velocity field fluctuates randomly and contains many scales depending on the Reynolds number of the flow.

The direct numerical solution of multiple scale problems is difficult even with the advent of supercomputers. The major difficulty of direct solutions is the size of the computation. A tremendous amount of computer memory and CPU time are required, and this can easily exceed the limit of today's computing resources. The situation can be relieved to some degree by parallel computing; however, the size of the discrete problem is not reduced. Whenever one can afford to resolve all the small-scale features of a physical problem, direct solutions provide quantitative information of the physical processes at all scales. On the other hand, from an application perspective, it is often sufficient to predict the macroscopic properties of the multiscale systems. Therefore, it is desirable to develop a method that captures the small-scale effect on the large scales, but does not require resolving all the small-scale features.

Y. Efendiev, T.Y. Hou, *Multiscale Finite Element Methods: Theory and Applications*,
Surveys and Tutorials in the Applied Mathematical Sciences 4,
DOI 10.1007/978-0-387-09496-0_1, © Springer Science+Business Media LLC 2009

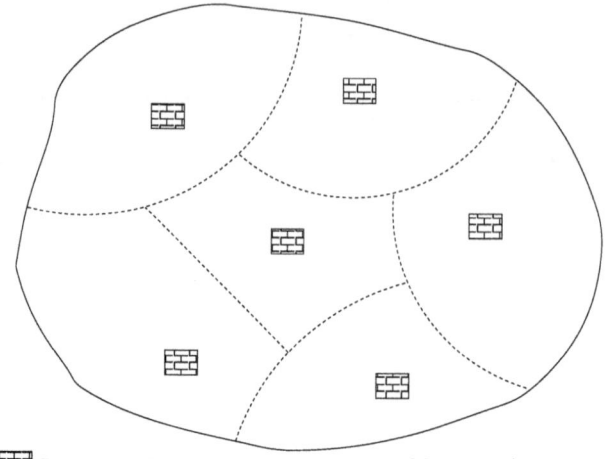

⊞ Representative Volume Element; ⋯⋯ Macroscopic region boundaries

**Fig. 1.1.** Schematic description of Representative Volume Element and macroscopic elements.

The methods discussed in this book attempt to capture the multiscale structure of the solution via localized basis functions. These basis functions contain essential multiscale information embedded in the solution and are coupled through a global formulation to provide a faithful approximation of the solution. Typically, we distinguish between two types of multiscale processes in this book. The first type has scale separation. In this case, the small-scale information is captured via local multiscale basis functions computed based only on the information within local regions (coarse-scale grid blocks). The other types of multiscale processes do not have apparent scale separation. For these processes, the information at different scales (e.g., nonlocal information) is used for constructing effective properties, such as multiscale basis functions. Next, we present a more in-depth discussion of scale issues that arise in multiscale simulations.

When dealing with multiscale processes, it is often the case that input information about processes or material properties is not available everywhere. For example, if one would like to study the fluid flows in a subsurface, then the subsurface properties at the pore scale are not available everywhere in the reservoir. Similarly, material properties of fine-grained composites are not often available everywhere. In this case, one can use Representative Volume Element (RVE) which contains essential information about the heterogeneities. For example, pore scale distribution in an extracted rock core can be regarded as representative information about the pore scale distribution over some macroscale region. Assuming that such information is available over the entire domain in macroscopic regions (see Figure 1.1 for illustration), one can perform upscaling (or averaging) and simulate a process over the entire region. Multiscale methods discussed in this book can easily handle such cases with scale separation, and the proposed basis functions can be computed only in RVE.

We would like to note that there are many methods (see next subsection) that can solve macroscopic equations given the information in RVE. These approaches can be divided into two groups: fine-to-coarse approaches and coarse-to-fine approaches. In fine-to-coarse approaches, the coarse-scale equations are not formulated explicitly and representative fine-scale information is carried out throughout the simulations. On the other hand, coarse-to-fine approaches assume the form of coarse-scale equations and the coarse-scale parameters are computed based on the calculations in RVE. These approaches share similarities. The methods discussed in this book belong to the class of fine-to-coarse approaches.

To illustrate the above discussion in a simple case, we consider a classical example of steady-state heterogeneous diffusion

$$\text{div}(k(x)\nabla p) = f(x). \tag{1.1}$$

Here $k(x)$ is a spatial field varying over multiple scales. It is possible that the full description (details) of $k(x)$ at the finest resolution is not available, and we can only access it in small portions of the domain. These small regions are RVE (see Figure 1.1), and one can attempt to simulate the macroscopic behavior of the material or subsurface processes based on RVE information. However, the latter assumes that the material has some type of scale separation because RVE information is sufficient to determine the macroscopic properties of the material.

In many other applications, the fine-scale description of the media is given or can be obtained everywhere based on prior information. This information is usually not precise and contains uncertainties. However, this information often contains some important features of the media. For example, in porous media applications, the subsurface properties typically contain some large-scale (nonlocal) features such as connected high-conductivity regions. In the example (1.1), $k(x)$ represents the permeability (or hydraulic conductivity). Modern geostatistical tools allow us to prescribe $k(x)$ at every grid block which is usually called the fine-scale grid block. Usually, the detailed subsurface model is built based on prior information. This information is a combination of fine-scale information coming from core samples and large-scale information coming from seismic data and macroscopic inversion techniques. The large-scale features typically provide the information about the connectivity of the porous media and can be quite complex, for example, tortuous long channels with small varying width or multiple connectivity information embedded to each other.

In Figure 1.2, we illustrate the multiscale nature of the conductivity in typical subsurface problems. Here, we illustrate that pore scale information is needed for understanding the conductivity of the core sample. However, it is also essential to understand the large-scale features of the media in order to build a comprehensive model of porous media. More complicated situations

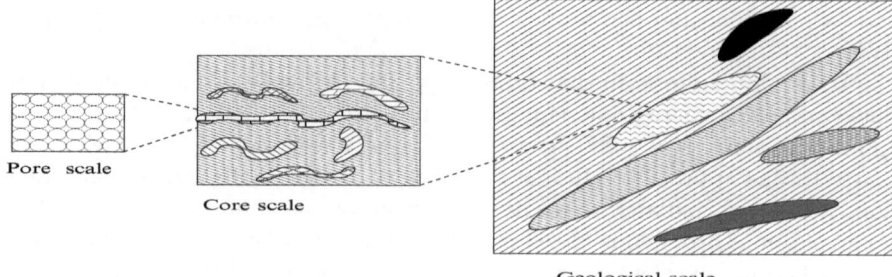

**Fig. 1.2.** Schematic description of various scales in porous media.

in geomodeling can occur. In Figure 1.3, geological variation over multiple scales is shown. Here one can observe faults (red lines in Figure 1.3(e)) with complicated geometry, thin but laterally extensive compaction bands that represent low-conductivity regions (blue lines in Figure 1.3(e); see also 1.3(d)) as well as other features at different scales. A blowup of the fault zone is shown in Figure 1.3(c). The fault rock is of low conductivity and the slip band sets consist of fractures that are filled (fully or partially) with cement. Pore-scale views of portions of a slip band set are shown in Figures 1.3(a) and 1.3(b)[1]. When simulating based only on RVE information as discussed before, the large-scale nonlocal information is disregarded, and this can lead to large errors. Thus, it is crucial to incorporate the multiscale structure of the solution at all scales that are important for simulations.

Materials with multiscale properties occur in many other applications. For example, composite material properties, similar to subsurface properties, can vary over many length scales. In Figure 1.4[2], fiber materials are depicted. Materials such as papers, filtration materials, and other engineered materials can have fibers of various sizes and geometry. As we see from Figure 1.4, the fibers can have complicated geometry and connectivity patterns. As in subsurface processes, the multiscale features of these materials at different scales are needed to perform reliable simulations.

Although small-scale features of the media are important, the large-scale connectivity can play a crucial role. When both fine- and coarse-scale information are combined, the resulting media properties have scale disparity and vary over many scales. In these problems, one cannot simply use RVE because there is no apparent scale separation. Moreover, the solution of (1.1) can be prohibitively expensive or unaffordable to compute. This situation is further

---

[1] We are grateful to L. Durlofsky for providing us the figures and the explanations. We would like to thank the authors (see the caption of Figure 1.3) for allowing us to use the figures in the book.

[2] Published with permission of Engineered Fibers Technology, LLC (www.EFTfibers.com).

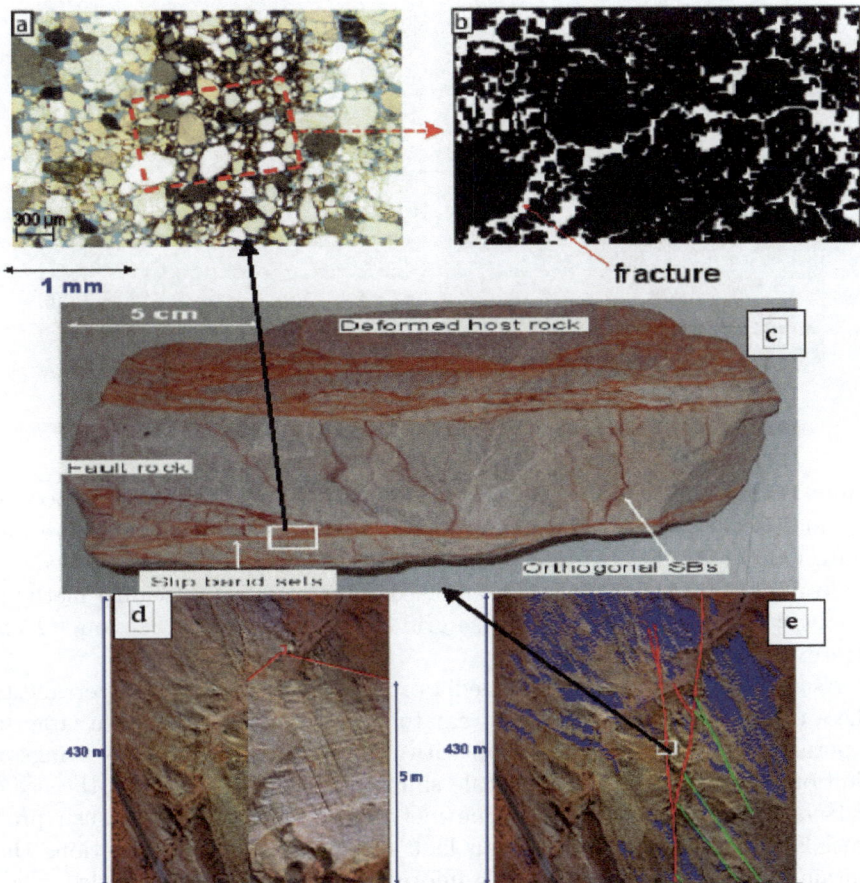

**Fig. 1.3.** Schematic description of hierarchy of heterogeneities in subsurface formations ((a) and (b) are from [20], (c) is from [19], (d) and (e) are courtesy of Kurt Sternlof).

complicated because of the fact that the flow equations (e.g., (1.1)) need to be solved many times for different source terms ($f(x)$ in (1.1)), mobilities ($\lambda(x)$ in (1.2)), and so on. For example, in a simplest situation, two-phase immiscible flow in heterogeneous porous media is described by

$$\mathrm{div}(\lambda(x)k(x)\nabla p) = f(x), \qquad (1.2)$$

where $\lambda(x)$, $f(x)$ are coarse-scale functions that vary dynamically (in time). As the physical processes become complicated due to additional physics arising in multiphase flows, it becomes impossible to simulate these processes without coarsening model equations. When performing simulations on a coarse grid, it

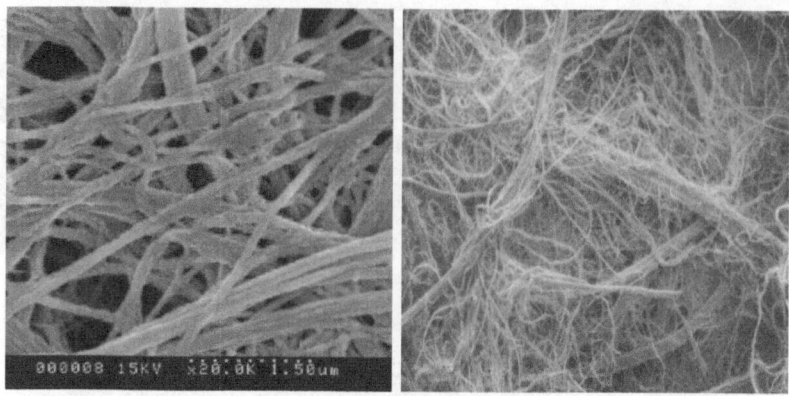

**Fig. 1.4.** Schematic description of fiber materials.

is important to preserve important multiscale features of physical processes. The multiscale methods considered in this book are intended for these purposes. Our multiscale methods compute effective properties of the media in the form of basis functions which are used, as in classical upscaling methods, to solve the processes on the coarse grid (see discussions in Section 1.2 and Figure 1.5 for an illustration).

As we mentioned earlier, the media properties often contain uncertainties. These uncertainties are usually parameterized and one deals with a large set of permeability fields (realizations) with a multiscale nature. This brings an additional challenge to the fine-scale simulations and necessitates the use of coarse-scale models. The multiscale methods are important for such problems. For these problems, one can look for multiscale basis functions that contain both spatio-temporal scale information and the uncertainties. These basis functions allow us to reduce the dimension of the problem and simulate realistic stochastic processes. We show that the multiscale finite element methods studied in this book can easily be generalized to take into account both multiscale features of the solution and the associated uncertainties.

## 1.2 Literature review

Many multiscale numerical methods have been developed and studied in the literature. In particular, many numerical methods have been developed with goals similar to ours. These include generalized finite element methods [33, 31, 30], wavelet-based numerical homogenization methods [56, 87, 84, 168], methods based on the homogenization theory (cf. [49, 95, 80]), equation-free computations (e.g., [166, 238, 224, 176, 242, 241]), variational multiscale methods [154, 59, 155, 209, 165], heterogeneous multiscale methods [97], matrix-dependent multigrid-based homogenization [168, 84], generalized $p$-FEM in homogenization [197, 198], mortar multiscale methods [228, 27, 226], upscaling methods (cf. [91, 199]), network methods [48, 44, 45, 46] and other methods [181, 180, 222, 223, 210, 81, 65, 63, 62, 124, 195]. The methods based on the homogenization theory have been successfully applied to determine the effective properties of heterogeneous materials. However, their range of applications is usually limited by restrictive assumptions on the media, such as scale separation and periodicity [43, 164].

Before we present a brief discussion about various multiscale methods, we would like to mention that multiscale finite element methods (MsFEMs) share similarities with upscaling methods. Upscaling procedures have been commonly applied and are effective in many cases. The main idea of upscaling techniques is to form coarse-scale equations with a prescribed analytical form that may differ from the underlying fine-scale equations. In multiscale methods, the fine-scale information is carried throughout the simulation and the coarse-scale equations are generally not expressed analytically but rather formed and solved numerically. For problems with scale separation, one can establish the equivalence between upscaling and multiscale methods (see Section 2.8).

The MsFEMs discussed in this book take their origin from a pioneering work of Babuška and Osborn [33, 31]. In this paper, the authors propose the use of multiscale basis functions for elliptic equations with a special multiscale coefficient which is the product of one-dimensional fields. This approach is extended in the work of Hou and Wu [145] to general heterogeneities. Hou and Wu [145] showed that boundary conditions for constructing basis functions are important for the accuracy of the method. They further proposed an oversampling technique to improve the subgrid capturing errors. Later on, the MsFEM of Hou and Wu were generalized to nonlinear problems in [104, 112]. In these papers, various global coupling approaches and subgrid capturing mechanisms are discussed.

There are a number of approaches with the purpose of forming a general framework for multiscale simulations. Among them are equation-free [166] and heterogeneous multiscale method (HMM) [97]. These approaches are intended for solving macroscopic equations based on the information in RVE and cover a wide range of applications. When applied to partial differential equations, MsFEMs are similar to these approaches. For such problems, multiscale basis functions presented in the book are approximated using the solutions in RVE. We note that for MsFEMs the local problems can be described by the set of equations different from the global equations. An important step in multiscale simulations is often to determine the form of the macroscopic equations and the variables upon which the basis functions depend. In many linear problems and problems with scale separation, these issues are well understood. Many general numerical approaches for multiscale simulations do not address the issues related to determining the variables on which the macroscopic quantities (e.g., multiscale basis functions) depend (see [176] where some of these issues are discussed).

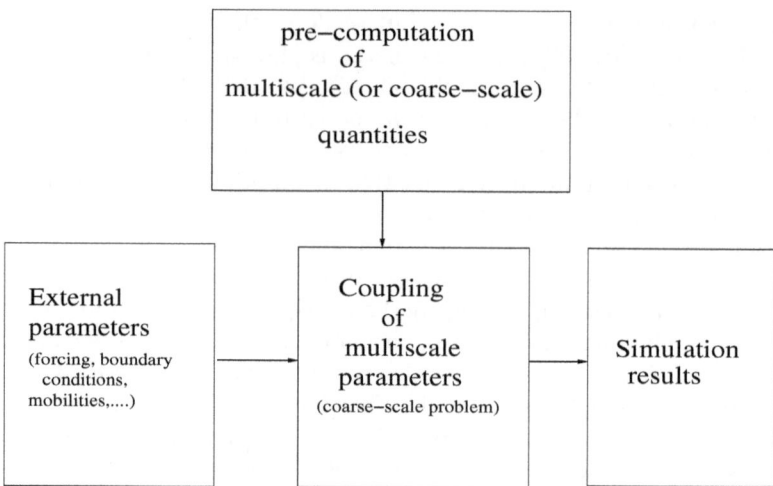

**Fig. 1.5.** A schematic illustration of upscaling concept.

MsFEMs also share similarities with variational multiscale methods [59, 154, 155]. In this approach, the solution of the multiscale problem is divided into resolved (coarse) and unresolved parts. The objective is to compute the resolved part via the unresolved part of the solution and then approximate the unresolved part of the solution. This is shown in the framework of linear equations. Typically, the approximation of the unresolved part of the solution requires some type of localization (e.g., [154, 24]). The localization leads to methods similar to the MsFEM. This is discussed in detail in Section 2.8.

There are also many other multiscale techniques with goals similar to ours. In particular, many methods share similarities in approximating the subgrid effects (the effects of the scales smaller than the coarse grid block size). There are a number of approaches that rely on techniques derived from homogenization theory (e.g., [198]). These methods are often restricted in terms of structure of multiscale coefficients. However, these approaches are more robust and accurate when the underlying multiscale structure satisfies the necessary constraints.

The multiscale methods considered in this book pre-compute the effective parameters that are repeatedly used for different sources and boundary conditions. In this regard, these methods can be classified as upscaling methods where upscaled parameters are pre-computed. An illustration for the concept of upscaling is presented in Figure 1.5. In multiscale approaches discussed in this book, one can re-use pre-computed quantities to form coarse-scale equations for different source terms, boundary conditions and so on. Moreover, adaptive and parallel computations can be carried out with these methods where one can downscale the computed coarse-scale solution in the regions of interest. These features of upscaling methods and MsFEMs are exploited in subsurface applications. The multiscale methods considered in this book differ from domain decomposition methods (e.g., [257]) where the local problems are solved many times. Domain decomposition methods are powerful techniques for solving multiphysics problems; however, the cost of iterations can be high, in particular, for multiscale problems. These iterations guarantee the convergence of domain decomposition methods under suitable assumptions. Multiscale methods with upscaling concepts in mind, on the other hand, attempt to find accurate subgrid capturing resolution and avoid the iterations. This may not be always possible, and for that reason, some type of hybrid methods with accurate subgrid modeling can be considered in the future (e.g., [93]).

One of the recent directions in multiscale simulations has been the use of some type of limited global information. The use of limited global information is not new in upscaling methods. The main idea of these approaches is to use some simplified surrogate models to extract important information about non-local multiscale behavior of physical processes. The surrogate models are typically solved off-line in a pre-computation step and their computations can be expensive. However, they allow us to compute effective parameters that will render a more accurate description of dynamic problems with varying source terms, boundary conditions, and so on. An example is two-phase immiscible flow in highly heterogeneous media. In [69, 103], single-phase flow information is used for accurate upscaling of two-phase flow and transport. In particular, the global single-phase flow equation is solved several times to compute upscaled permeabilities (conductivities) which are then used in the simulations of two-phase flow and transport on a coarse grid. These upscaling computations are performed off-line. Similar to upscaling methods using global information, multiscale finite element methods using limited global information are introduced in [1, 103, 218]. The work of [218] provides a theoretical foundation for upscaling using limited global information. These methods use limited global information to construct multiscale basis functions. Finally, we would like to mention that multiscale methods with limited global information share some similarities with reduced model techniques (e.g., [234]) where snapshots of the solution at previous times can be used to construct a reduced basis for approximating the solution.

There are many other multiscale methods in the literature that discuss bridging scales in various applications. In this book, we mostly focus on methods that are most relevant to MsFEMs. We note that a main feature of MsFEMs is the use of variational formulation at the coarse scale which allows us to couple multiscale basis functions. Fine-scale formulation of the problem that allows computing multiscale basis functions is not necessarily based on partial differential equations and can have a discrete formulation. In this regard, MsFEMs share conceptual similarities with some approaches that couple atomistic (discrete) and continuum effects. These approaches use a variational formulation at the coarse scale, but use a discrete atomistic description at finest scales (e.g., quasi-continuum method ([251, 252])). This method has been widely used in material science applications.

## 1.3 Overview of the content of the book

The purpose of this book is to review some recent advances in multiscale finite element methods and their applications. Here, the notion "multiscale finite element methods" refers to a number of methods, such as multiscale finite volume, mixed multiscale finite element method, and the like. The concept that unifies these methods is the coupling of oscillatory basis functions via various variational formulations. One of the main aspects of this coupling is the subgrid capturing errors that are extensively discussed in this book.

The book is laid out in a way that it is accessible to a broader audience. Each chapter is divided into the description of the numerical method and the computational results. At the end of each chapter, the section "Discussions" is presented. This section discusses extensions, existing methods, and other relevant research in this area. The analysis of the proposed methods is discussed in the last chapter. We have attempted to keep the book concise and therefore present convergence analysis only for a few representative cases. Some of the results are referred to earlier in the chapter to convey the convergence of the proposed methods.

The book is organized in the following way. In the second chapter, we review MsFEM for solving partial differential equations with multiscale solutions; see [145, 147, 146, 107, 71, 260, 14, 103]. The central goal of this approach is to obtain the large-scale solutions accurately and efficiently without resolving the small-scale details. The main idea is to construct finite element basis functions that capture the small-scale information within each element. The small-scale information is then brought to the large scales through the coupling of the global stiffness matrix. The basis functions are constructed from the leading-order homogeneous elliptic equation in each element. As a consequence, the basis functions are adapted to the local microstructure of the differential operator. We discuss various global coupling techniques and the computational issues associated with multiscale methods. Simple examples and pseudo-codes are presented. Issues such as performance and implementation of MsFEMs are discussed in Section 2.9. We present the comparison between the MsFEM and some other multiscale methods in Section 2.8. Some comments on generalizations of MsFEMs are presented in Section 2.4.

In Chapter 3, we discuss the extension of MsFEMs to nonlinear problems. Our aim is to show that one can naturally extend the multiscale methods to nonlinear problems by replacing the multiscale basis functions with multiscale maps. Indeed, because the underlying equations are nonlinear, the small-scale features of the problem do not form a linear space. We show that with this modification MsFEMs can be used for solving nonlinear partial differential equations. After presenting the methodology, some numerical examples are presented for solving nonlinear elliptic equations. The chapter also includes discussions on the extension of the method to nonlinear parabolic equations and multiphysics problems.

Multiscale methods discussed in Chapter 2 and Chapter 3 apply local calculations to determine basis functions. Although effective in many cases, global effects can be important for some problems. The importance of global information has been illustrated within the context of upscaling procedures as well as multiscale computations in recent investigations (e.g., [69, 143, 1, 103, 218]). These studies have shown that the use of limited global information in the calculation of the coarse-scale parameters (such as basis functions) can significantly improve the accuracy of the resulting coarse model. In the fourth chapter of the book, we describe the use of limited global information in multiscale simulations. The chapter starts with a motivation and a motivating numerical example which show that the accuracy of multiscale methods deteriorates for problems with strong nonlocal effects. We introduce the basic idea of multiscale methods using limited global information. These approaches are used if the problem is solved repeatedly for varying parameters but keeping the source of heterogeneities fixed. Typical problems of this type arise, for example, in porous media applications. Numerical examples both for structured and unstructured grids for mixed MsFEM and MsFVEM are presented in this chapter. In general, one can use simplified global information combined with local multiscale basis functions for accurate simulation purposes which is discussed in Section 4.4.

Chapter 5 is devoted to the applications of multiscale methods to multiphase flow and transport in highly heterogeneous porous media. We limit ourselves to a few applications. For two-phase flow and transport simulations, we consider the applications of multiscale methods to hyperbolic equations describing the dynamics of the phases and their coupling to multiscale methods for pressure equations in Section 5.2. In this section, the applications of nonlinear multiscale methods to hyperbolic equations are presented along with various subgrid treatment techniques for hyperbolic equations. We present an application of nonlinear MsFEMs to Richards' equation and fluid flows in highly deformable porous media in Sections 5.3 and 5.4. We include two short sections (contributed by J.E. Aarnes and S.H. Lee et al.) summarizing the applications of mixed MsFEM and multiscale finite volume (MsFV) to reservoir modeling and simulation in Sections 5.5 and 5.6. These sections discuss the use of MsFEMs in the simulations of multiphase flow and transport which include various additional physics arising in more realistic petroleum applications. The extension of MsFEMs to stochastic differential equations is described in Section 5.7. To handle the uncertainties in heterogeneous coefficients, we propose to use a few realizations of the permeability to generate multiscale basis functions for the ensemble. Uncertainty quantification in inverse problems using multiscale methods is also discussed in this section. The aim is to speedup uncertainty quantification in inverse problems using fast multiscale finite element methods as surrogate models. We finish the chapter with a discussion on other applications of multiscale finite element methods.

In Chapter 6, we present an analysis of MsFEMs discussed in Chapters 2, 3, and 4. Only some representative cases are studied in the book with the aim to demonstrate the main ideas and techniques used in the analysis of MsFEMs. We try to keep the presentation accessible to a broader audience and avoid some technical details in the presentation. Some basic analysis of MsFEMs for linear problems in a homogenization setting is presented in Section 6.1. We study the convergence of MsFEMs for problems with scale separation. Our analysis reveals sources of the resonance errors. Next, the convergence of MsFEM with oversampling is studied. We also present analysis of the mixed MsFEM for problems with scale separation in Section 6.1. In Section 6.2, we present the analyses of nonlinear MsFEMs. We restrict ourselves to a periodic case and refer to [113, 112] where the analysis for more general cases are studied. In Section 6.3, we study the convergence of MsFEMs with limited global information. The analysis is performed under the assumption that the solution can be smoothly approximated via known global fields.

Finally, the brief overview of linear and nonlinear homogenization theory is presented in Appendix B. We present the formal asymptotic expansion as well as some partial results on the convergence of these expansions. These results are used in the convergence analysis of MsFEMs presented in Chapter 6.

# 2

# Multiscale finite element methods for linear problems and overview

## 2.1 Summary

In this section, the main concept of multiscale finite element methods (Ms-FEM) is presented. We keep the presentation simple to make it accessible to a broader audience. Two main ingredients of MsFEMs are the global formulation of the method and the construction of basis functions. We discuss global formulations using various finite element, finite volume, and mixed finite element methods. As for multiscale basis functions, the subgrid capturing errors are discussed. We present simplified computations of basis functions for cases with scale separation. We also discuss the improvement of subgrid capturing errors via oversampling techniques. Finally, we present some representative numerical examples and discuss the computational cost of MsFEMs. Analysis of some representative cases is presented in Chapter 6.

## 2.2 Introduction to multiscale finite element methods

We start our discussion with the MsFEM for linear elliptic equations

$$Lp = f \text{ in } \Omega, \tag{2.1}$$

where $\Omega$ is a domain in $\mathbb{R}^d$ ($d = 2, 3$), $Lp := -\text{div}(k(x)\nabla p)$, and $k(x)$ is a heterogeneous field varying over multiple scales. We note that MsFEM can be easily extended to systems such as elasticity equations, as well as to nonlinear problems (see Section 2.4 and Chapter 3). The choice of the notations $k(x)$ and $p(x)$ in (2.1) is used because of the applications of the method to porous media flows later in the book. We note that the tensor $k(x) = (k_{ij}(x))$ is assumed to be symmetric and satisfies $\alpha|\xi|^2 \leq k_{ij}\xi_i\xi_j \leq \beta|\xi|^2$, for all $\xi \in \mathbb{R}^d$ and with $0 < \alpha < \beta$. We omit $x$ dependence when there is no ambiguity and assume the summation over repeated indices (Einstein summation convention) unless otherwise stated.

Y. Efendiev, T.Y. Hou, *Multiscale Finite Element Methods: Theory and Applications*, 13
Surveys and Tutorials in the Applied Mathematical Sciences 4,
DOI 10.1007/978-0-387-09496-0_2, © Springer Science+Business Media LLC 2009

MsFEMs consist of two major ingredients: multiscale basis functions and a global numerical formulation that couples these multiscale basis functions. Basis functions are designed to capture the multiscale features of the solution. Important multiscale features of the solution are incorporated into these localized basis functions which contain information about the scales that are smaller (as well as larger) than the local numerical scale defined by the basis functions. A global formulation couples these basis functions to provide an accurate approximation of the solution. Next, we discuss some basic choices for multiscale basis functions and global formulations.

*Basis functions.* First, we discuss the basis function construction. Let $T_h$ be a usual partition of $\Omega$ into finite elements (triangles, quadrilaterals, and so on). We call this partition the coarse grid and assume that the coarse grid can be resolved via a finer resolution called the fine grid. For clarity of this exposition, we plot rectangular coarse and fine grids in Figure 2.1 (left figure). Let $x_i$ be the interior nodes of the mesh $T_h$ and $\phi_i^0$ be the nodal basis of the standard finite element space $W_h = \text{span}\{\phi_i^0\}$. For simplicity, one can assume that $W_h$ consists of piecewise linear functions if $T_h$ is a triangular partition. Denote by $S_i = \text{supp}(\phi_i^0)$ (the support of $\phi_i^0$) and define $\phi_i$ with support in $S_i$ as follows

$$L\phi_i = 0 \text{ in } K, \quad \phi_i = \phi_i^0 \text{ on } \partial K, \quad \forall K \in T_h, \ K \subset S_i; \qquad (2.2)$$

that is multiscale basis functions coincide with standard finite element basis functions on the boundaries of a coarse-grid block $K$, and are oscillatory in the interior of each coarse-grid block. Throughout, $K$ denotes a coarse-grid block. Note that even though the choice of $\phi_i^0$ can be quite arbitrary, our main assumption is that the basis functions satisfy the leading-order homogeneous equations when the right-hand side $f$ is a smooth function (e.g., $L^2$ integrable). We would like to remark that MsFEM formulation allows one to take advantage of scale separation which is discussed later in the book. In particular, $K$ can be chosen to be a domain smaller than the coarse grid as illustrated in Figure 2.1 (right figure) if the small region can be used to represent the heterogeneities within the coarse-grid block. In this case, the basis function has the formulation (2.2), except $K$ is replaced by a smaller region, $K_{\text{loc}}$, $L(\phi_i) = 0$ in $K_{\text{loc}}$, $\phi_i = \phi_i^0$ on $\partial K_{\text{loc}}$, where the values of $\phi_i^0$ inside $K$ are used in imposing boundary conditions on $\partial K_{\text{loc}}$. In general, one solves (2.2) on the fine grid to compute basis functions. In some cases, the computations of basis functions can be performed analytically. To illustrate the basis functions, we depict them in Figure 2.2. On the left, the basis function is constructed when $K$ is a coarse partition element, and on the right, the basis function is constructed by taking $K$ to be an element smaller than the coarse-grid block size. Note that a bilinear function in Figure 2.2 (right figure) is used to demonstrate boundary conditions on a small computational domain and this bilinear function is not a part of the basis function. In this case, the assembly of the stiffness matrix uses only the information in small computational regions and the basis function can be "periodically" extended

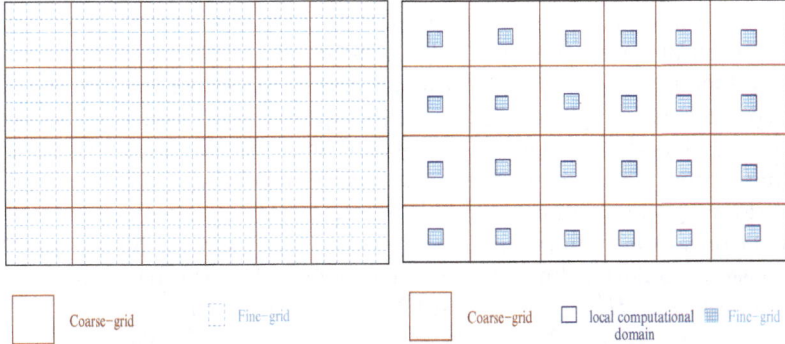

Coarse–grid      Fine–grid        Coarse–grid      local computational   Fine–grid
                                                   domain

**Fig. 2.1.** Schematic description of a coarse grid.

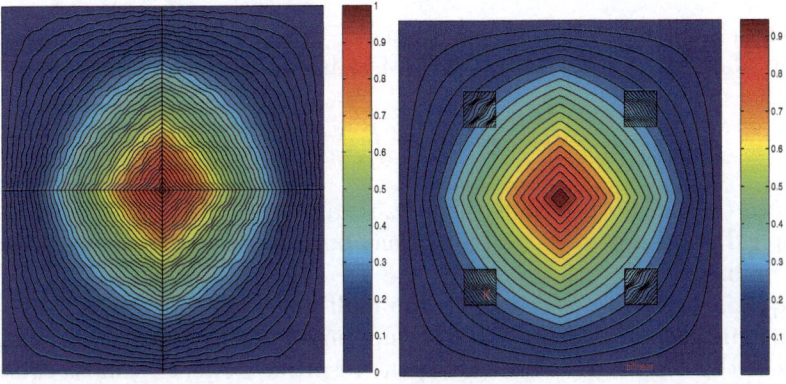

**Fig. 2.2.** Example of basis functions. Left: basis function with $K$ being a coarse element. Right: basis function with $K$ being RVE (bilinear function demonstrates only the boundary conditions on RVE and is not a part of the basis function).

to the coarse-grid block, if needed (see later discussions and Section 2.6). Computational regions smaller than the coarse-grid block are used if one can use smaller regions to characterize the local heterogeneities within the coarse grid block (e.g., periodic heterogeneities). We call such regions Representative Volume Element (RVE) following standard practice in engineering. More precisely, we assume that the size of the RVE is much larger than the characteristic length scale. In this case, one can use the solution in RVE with prescribed boundary conditions to represent the solution in the coarse block as is done in homogenization (e.g., [43, 164]). Later, we briefly discuss an extension of the method to problems with singular right-hand sides. In this case, it is necessary to include basis functions with singular right-hand sides. Once the basis functions are constructed, we denote by $\mathcal{P}_h$ the finite element space spanned by $\phi_i$

$$\mathcal{P}_h = \text{span}\{\phi_i\}.$$

*Global formulation.* Next, we discuss the global formulation of MsFEM. The representation of the fine-scale solution via multiscale basis functions allows reducing the dimension of the computation. When the approximation of the solution $p_h = \sum_i p_i \phi_i(x)$ ($p_i$ are the approximate values of the solution at coarse-grid nodal points) is substituted into the fine-scale equation, the resulting system is projected onto the coarse-dimensional space to find $p_i$. This can be done by multiplying the resulting fine-scale equation with coarse-scale test functions. Other approaches can be taken for general nonlinear problems. In the case of Galerkin finite element methods, when the basis functions are conforming ($\mathcal{P}_h \subset H_0^1(\Omega)$), the MsFEM is to find $p_h \in \mathcal{P}_h$ such that

$$\sum_K \int_K k \nabla p_h \cdot \nabla v_h dx = \int_\Omega f v_h dx, \quad \forall\, v_h \in \mathcal{P}_h. \tag{2.3}$$

One can choose the test functions from $W_h$ (instead of $\mathcal{P}_h$) and arrive at the Petrov–Galerkin version of the MsFEM as introduced in [143]. Find $p_h \in \mathcal{P}_h$ such that

$$\sum_K \int_K k \nabla p_h \cdot \nabla v_h dx = \int_\Omega f v_h dx, \quad \forall\, v_h \in W_h. \tag{2.4}$$

We note that in both formulations (2.3) and (2.4), the fine-scale system is multiplied by coarse-scale test functions and, thus, the resulting system is coarse-dimensional.

Equation (2.3) or (2.4) couples the multiscale basis functions. This gives rise to a linear system of equations for finding the values of the solution at the nodes of the coarse-grid block, thus, the resulting system of linear equations determines the solution on the coarse grid. To show this, for simplicity, we consider the Petrov–Galerkin formulation of the MsFEM (see (2.4)). Representing the solution in terms of multiscale basis functions, $p_h = \sum_i p_i \phi_i$, it is easy to show that (2.4) is equivalent to the following linear system,

$$A p_{\text{nodal}} = b, \tag{2.5}$$

where $A = (a_{ij})$, $a_{ij} = \sum_K \int_K k \nabla \phi_i \nabla \phi_j^0 dx$. $p_{\text{nodal}} = (p_1, ..., p_i, ...)$ are the nodal values of the coarse-scale solution, and $b = (b_i)$, $b_i = \int_\Omega f \phi_i^0 dx$. Here, we do not consider the discretization of boundary conditions. As in the case of standard finite element methods, the stiffness matrix $A$ has sparse structure. We note that the computation of the stiffness matrix requires the integral computation for $a_{ij}$ and $b_i$. The computation of $a_{ij}$ requires the evaluation of the integrals on the fine grid. One can use a simple quadrature rule, for example, one point per fine grid cell. In this case, $\int_K k \nabla \phi_i \nabla \phi_j^0 dx \approx \sum_{\tau \subset K} (k \nabla \phi_i)|_\tau \nabla \phi_j^0$, where $\tau$ denotes a fine grid block and $(k \nabla \phi_i)|_\tau$ is the value of the flux within a fine grid block $\tau$. Note that when source terms or mobilities change, one can pre-compute the stiffness matrix once and re-use it. For example, if the source terms change, the stiffness matrix will remain the same and one needs to re-compute $b_i$. If mobilities ($\lambda(x)$

in (1.2)) change and remain a smooth function, one can modify $a_{ij}$ using a piecewise constant approximation for $\lambda(x)$. In this case, the modified stiffness matrix elements $a_{ij}^{\lambda}$ have the form $a_{ij}^{\lambda} \approx \sum_K \lambda_K \int_K k \nabla \phi_i \nabla \phi_j^0 dx$, where $a_{ij}^{\lambda}$ are the elements of the stiffness matrix corresponding to (1.2), $\lambda_K$ are approximate values of $\lambda(x)$ in $K$, and the integrals $\int_K k \nabla \phi_i \nabla \phi_j^0 dx$ are pre-computed. Later on, we derive some explicit expressions for the elements of the stiffness matrix in the one-dimensional case.

If the local computational domain is chosen to be smaller than the coarse-grid block, then one can use an approximation of the basis functions in RVE (local domain) to represent the left-hand side of (2.4) (or (2.3)). We assume that the information within RVE can be used to characterize the local solution within the coarse-grid block such that

$$\frac{1}{|K|} \int_K k \nabla \phi_i dx \approx \frac{1}{|K_{\text{loc}}|} \int_{K_{\text{loc}}} k \nabla \tilde{\phi}_i dx, \qquad (2.6)$$

where $K_{\text{loc}}$ refers to local computational region (RVE) and $\tilde{\phi}_i$ is the basis function defined in $K_{\text{loc}}$ and given by the solution of $\text{div}(k \nabla \tilde{\phi}_i) = 0$ in $K_{\text{loc}}$ with boundary conditions $\tilde{\phi}_i = \phi_i^0$ on $\partial K_{\text{loc}}$. Equation (2.6) holds, for example, in the general $G$-convergence setting where homogenization by periodization (also called the principle of periodic localization) can be performed (see [164]) and the size of the RVE is assumed to be much larger than the characteristic length scale. One can approximate the left-hand side of (2.4) based on RVE computations. In particular, the elements of the stiffness matrix (see (2.5)) $a_{ij} = \sum_K \int_K k \nabla \phi_i \nabla \phi_j^0 dx$ can be approximated using

$$\frac{1}{|K|} \int_K k \nabla \phi_i \nabla \phi_j^0 dx \approx \frac{1}{|K_{\text{loc}}|} \int_{K_{\text{loc}}} k \nabla \tilde{\phi}_i \nabla \phi_j^0 dx.$$

A similar approximation can be done for (2.3). In the general $G$-convergence setting, this approximation holds in the limit $\lim_{h \to 0} \lim_{\epsilon \to 0}$ (see Section 2.6 for details), and for periodic problems, one can justify this approximation in the limit $\lim_{\epsilon/h \to 0}$. In periodic problems, one can also take advantage of two-scale homogenization expansion and this is discussed in Section 2.6 along with further discussions on the use of smaller regions. Similar approximation can be done for the right-hand side of (2.4) (or (2.3)).

As we discussed earlier, using multiscale basis functions, a fine-scale approximation of the solution can be easily computed. In particular, $p_h = \sum_i p_i \phi_i$ provides an approximation of the solution, where $p_i$ are the values of the solution at the coarse nodes obtained via (2.5). When regions smaller than the coarse-grid block are used for computing basis functions, $p_h = \sum_i p_i \phi_i$ provides approximate fine-scale details of the solution only in RVE regions. One can use the periodic homogenization concept to extend the fine-scale features in RVE to the entire domain. This is discussed in Section 2.6.

In the above discussion, we presented the simplest basis function construction and a global formulation. In general, the global formulation can be easily modified and various global formulations based on finite volume, mixed finite element, discontinuous Galerkin finite element, and other methods can be derived. Many of them are studied in the literature and some of them are discussed here.

As for basis functions, the choice of boundary conditions in defining the multiscale basis functions plays a crucial role in approximating the multiscale solution. Intuitively, the boundary condition for the multiscale basis function should reflect the multiscale oscillation of the solution $p$ across the boundary of the coarse grid element. By choosing a linear boundary condition for the basis function, we create a mismatch between the exact solution $p$ and the finite element approximation across the element boundary. In Section 2.3, we discuss this issue further and introduce a technique to alleviate this difficulty. We would like to note that in the one-dimensional case this issue is not present because the boundaries of the coarse element consist of isolated points.

The MsFEM can be naturally extended to solve nonlinear partial differential equations. As in the case of linear problems, the main idea of MsFEM remains the same with the exception of basis function construction. Because of nonlinearities, the multiscale basis functions are replaced by multiscale maps, which are in general nonlinear maps from $W_h$ to heterogeneous fields (see Chapter 3).

**Pseudo-code.** MsFEM can be implemented within an existing finite element code. Below, we present a simple pseudo-code that outlines the implementation of MsFEM. Here, we do not discuss coarse-grid generation. We note that the latter is important for the accuracy, robustness, and efficient parallelization of MsFEM and is briefly discussed in Section 2.9.3.

---

**Algorithm 2.2.1**

---

Set coarse mesh configuration from fine-scale mesh information.
For each coarse grid block $n$ do
- For each vertex $i$
- Solve for $\phi_n^i$ satisfying $- L(\phi_n^i) = 0$ and boundary conditions (see (2.2))
- End for.
End do
Assemble stiffness matrix on the coarse mesh (see (2.5), also (2.3) or (2.4)).
Assemble the external force on the coarse mesh (see (2.5), also (2.3) or (2.4)).
Solve the coarse formulation.

---

*Comments on the assembly of stiffness matrix.* One can use the representation of multiscale basis functions via fine-scale basis functions to assemble the stiffness matrix. This is particularly useful in code development. Assume that multiscale basis function (in discrete form) $\phi_i$ can be written as

$$\phi_i = d_{ij}\phi_j^{0,f},$$

where $D = (d_{ij})$ is a matrix and $\phi_j^{0,f}$ are fine-scale finite element basis functions (e.g., piecewise linear functions). The $i$th row of this matrix contains the fine-scale representation of the $i$th multiscale basis function. Substituting this expression into the formula for the stiffness matrix $a_{ij}$ in (2.3), we have

$$a_{ij} = \int_\Omega k\nabla\phi_i\nabla\phi_j dx = d_{il}\int_\Omega k\nabla\phi_l^{0,f}\nabla\phi_m^{0,f} dx\; d_{jm}.$$

Denoting the stiffness matrix for the fine-scale problem by $A^f = (a_{lm}^f)$, $a_{lm}^f = \int_\Omega k\nabla\phi_l^{0,f}\nabla\phi_m^{0,f} dx$ we have

$$A = DA^f D^T.$$

Similarly, for the right-hand side, we have $b = \int_\Omega \phi_i f dx = Db^f$, where $b^f = (b_i^f)$, $b_i^f = \int_\Omega f\phi_i^{0,f} dx$. This simplification can be used in the assembly of the stiffness matrix. The similar procedure can be done for the Petrov–Galerkin MsFEM (see (2.4)).

*One-dimensional example.* In one-dimensional case, the basis functions and the stiffness matrix (see (2.5)) can be computed almost explicitly. For simplicity, we consider

$$-(k(x)p')' = f,$$

$p(0) = p(1) = 0$, where $'$ refers to the spatial derivative. We assume that the interval $[0, 1]$ is divided into $N$ segments $0 = x_0 < x_1 < x_2 < \cdots < x_i < x_{i+1} < \cdots < x_N = 1$. The multiscale basis function for the node $i$ is given by

$$(k(x)\phi_i')' = 0 \tag{2.7}$$

with the support in $[x_{i-1}, x_{i+1}]$. In the interval $[x_{i-1}, x_i]$, the boundary conditions for the basis function $\phi_i$ are defined as $\phi_i(x_{i-1}) = 0$, $\phi_i(x_i) = 1$. In the interval $[x_i, x_{i+1}]$, the boundary conditions for the basis function $\phi_i$ are defined as $\phi_i(x_i) = 1$, $\phi_i(x_{i+1}) = 0$. Note that for the computation of the elements of the stiffness matrix, we do not need an explicit expression of $\phi_i$ and instead, we simply need to compute $k(x)\phi_i'$. From (2.7), it is easy to see that $k(x)\phi_i' = \text{const}$, where the constants are different in $[x_{i-1}, x_i]$ and $[x_i, x_{i+1}]$. This constant can be easily computed by writing $\phi_i' = \text{const}/k(x)$ and integrating it over $[x_{i-1}, x_i]$. This yields

$$k(x)\phi_i' = \frac{1}{\int_{x_{i-1}}^{x_i}\frac{dx}{k(x)}}$$

on $[x_{i-1}, x_i]$ and

$$k(x)\phi_i' = -\frac{1}{\int_{x_i}^{x_{i+1}}\frac{dx}{k(x)}}$$

on $[x_i, x_{i+1}]$. Then, the elements of the stiffness matrix $A$ (see (2.4)) are given by

$$
\begin{aligned}
a_{ij} &= \int_{x_{i-1}}^{x_i} k(x)\phi_i^{'}(\phi_j^0)^{'} dx + \int_{x_i}^{x_{i+1}} k(x)\phi_i^{'}(\phi_j^0)^{'} dx \\
&= \frac{1}{\int_{x_{i-1}}^{x_i} \frac{dx}{k(x)}} \int_{x_{i-1}}^{x_i} (\phi_j^0)^{'} dx - \frac{1}{\int_{x_i}^{x_{i+1}} \frac{dx}{k(x)}} \int_{x_{i-1}}^{x_i} (\phi_j^0)^{'} dx.
\end{aligned}
\tag{2.8}
$$

Taking into account that $\int_{x_{i-1}}^{x_i} (\phi_{i-1}^0)^{'} dx = -1$, $\int_{x_{i-1}}^{x_i} (\phi_i^0)^{'} dx = 1$, $\int_{x_i}^{x_{i+1}} (\phi_i^0)^{'} dx = -1$, $\int_{x_i}^{x_{i+1}} (\phi_{i+1}^0)^{'} dx = 1$, we have

$$
a_{i,i-1} = -\frac{1}{\int_{x_{i-1}}^{x_i} \frac{dx}{k(x)}}, \quad a_{ii} = \frac{1}{\int_{x_{i-1}}^{x_i} \frac{dx}{k(x)}} + \frac{1}{\int_{x_i}^{x_{i+1}} \frac{dx}{k(x)}}, \quad a_{i,i+1} = -\frac{1}{\int_{x_i}^{x_{i+1}} \frac{dx}{k(x)}}.
$$

Consequently, the stiffness matrix has a tridiagonal form and the linear system is (2.5), where $b_i = \int_0^1 f\phi_i^0 dx$.

In Figure 2.3, we illustrate the solution and a few multiscale basis functions. We refer to [147] for the analysis in the one-dimensional case.

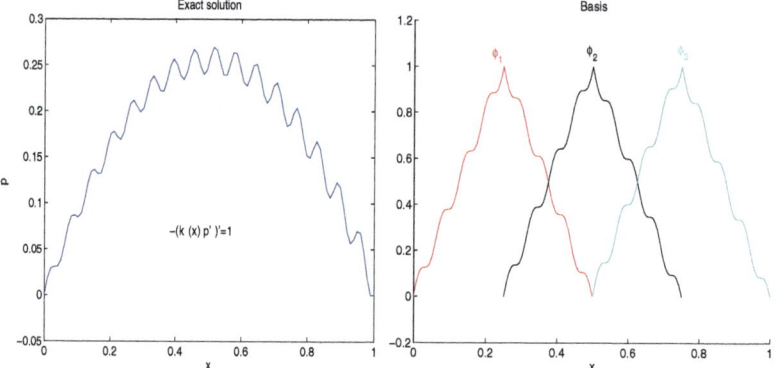

**Fig. 2.3.** An illustration of one-dimensional basis functions and the solution.

## 2.3 Reducing boundary effects

### 2.3.1 Motivation

The boundary conditions for the basis functions play a crucial role in capturing small-scale information. If the local boundary conditions for the basis functions do not reflect the nature of the underlying heterogeneities, MsFEMs can have large errors. These errors result from the resonance between the coarse-grid size and the characteristic length scale of the problem. When the coefficient $k(x)$ is a periodic function varying over the $\epsilon$ scale $(k(x) = k(x/\epsilon))$, the convergence rate of MsFEM contains a term $\epsilon/h$ (see [147]), which is large when $h \approx \epsilon$. Recall that $h$ is the coarse mesh size. As illustrated by the error analysis of [147], the error due to the resonance manifests as a ratio between the wavelength of the small-scale oscillation and the grid size; the error becomes large when the two scales are close. A deeper analysis based on the homogenization theory shows the main source of the resonance effect. By a judicious choice of boundary conditions for basis functions, we can reduce the resonance errors significantly. Some approaches including the use of reduced problems based on the solutions of one-dimensional problems along the boundaries (e.g., [145, 159, 160]) and oversampling methods (e.g., [145, 107, 73]) are studied in the literature with the goal of reducing resonance errors. In general, one can construct multiscale basis functions in various different ways (see, e.g., [266, 273] for energy minimizing basis functions). Here, we focus on oversampling methods.

Next, we present an outline of the analysis that motivates the oversampling method. We consider a simple case with two distinct scales (i.e., $k(x) = k(x, x/\epsilon)$) and assume that $k$ is a periodic function with respect to $x/\epsilon$. In this case, the solution has a well-known multiscale expansion (see, e.g., [43, 164] or Appendix B)

$$p = p_0 + \epsilon \chi^j \frac{\partial}{\partial x_j} p_0 + \epsilon \theta_\epsilon^p,$$

where $p_0$ satisfies the homogenized equation $-\text{div}(k^*(x)\nabla p_0) = f$. The homogenized coefficients are defined via an auxiliary (cell) problem over a period of size $\epsilon$. To illustrate this, we denote the fast variable by $y = x/\epsilon$ and, thus, the coefficients have the form $k(x, y)$. Then, $k^*(x) = (1/|Y|) \int_Y k(x, y)(I + \nabla_y \chi(x, y))dy$, where $\chi = (\chi^1, ..., \chi^d)$ is a solution of

$$\text{div}_y(k(x, y)(I + \nabla_y \chi(x, y))) = 0 \qquad (2.9)$$

in the period $Y$ for a fixed $x$ (see [43, 164] for more details). For simplicity, one can assume that $x$ represents a coarse grid block. If there is no slow dependence with respect to $x$ in the coefficients, $k = k(x/\epsilon) = k(y)$, then there is only one cell problem (2.9) for the entire domain $\Omega$. It can be shown that $p_0 + \epsilon \chi^j \partial p_0/\partial x_j$ approximates the solution $p$ in $H^1$ norm for small $\epsilon$ (see [43, 164] or Appendix B for the details).

Following multiple scale expansion, as discussed above (see also Appendix B), we can write a similar expansion for the basis function

$$\phi_i = \phi_i^1 + \epsilon \theta_\epsilon, \qquad (2.10)$$

where $\phi_i^1 = \phi_i^0 + \epsilon \chi^j \partial \phi_i^0/\partial x_j$ is the part of the basis function that has the same nature of oscillations near boundaries as the approximation of the fine-scale solution $p_0 + \epsilon \chi^j \partial p_0/\partial x_j$. Assuming $\phi_i^0$ is a linear function, it can be

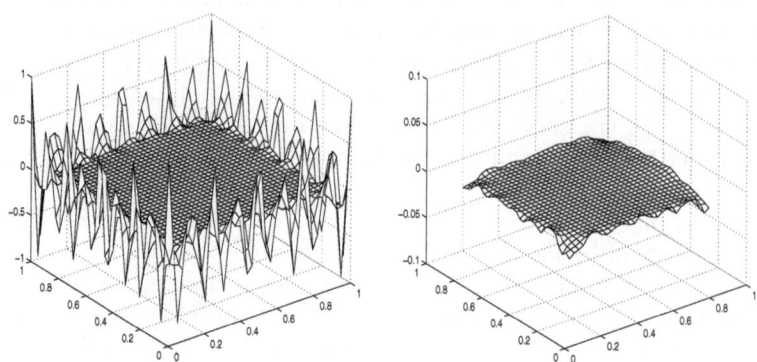

**Fig. 2.4.** An illustration of boundary layer function $\theta_\epsilon$. Left: $\theta_\epsilon$ in the coarse element with oscillatory boundary conditions. Right: $\theta_\epsilon$ in $\epsilon$ distance away from the boundaries.

easily shown that $\theta_\epsilon$ satisfies $\mathrm{div}(k\nabla\theta_\epsilon) = 0$ in $K$ and $\theta_\epsilon = -\chi^j \partial\phi_i^0/\partial x_j$ on $\partial K$. If one can ignore $\epsilon\theta_\epsilon$ in (2.10), then MsFEM will converge independently of the resonance error. The term $\epsilon\theta_\epsilon$ is due to the mismatch between the fine-scale solution and multiscale finite element solution along the boundaries of the coarse-grid block where the multiscale finite element solution is linear. This mismatch error propagates into the interior of the coarse-grid block. The analysis shows that the MsFEM error is dominated by $\theta_\epsilon$. In Figure 2.4, we depict $\theta_\epsilon(x)$ and the same $\theta_\epsilon(x)$ which is $\epsilon$ distance away from the boundaries. It is clear from this figure that the oscillations decay quickly as we move away from the boundaries. To avoid these oscillations, one needs to sample a larger domain and use only interior information to construct the basis functions. The decay of these oscillations basically dictates how large the sampling region should be chosen.

### 2.3.2 Oversampling technique

Motivated by the above discussion and the convergence analysis of [147], Hou and Wu proposed an oversampling method in [145] to overcome the difficulty due to scale resonance. Because the boundary layer in the first-order corrector is thin, we can sample in a domain with the size larger than $h$ and use only the interior sampled information to construct the basis functions. By doing this, we can reduce the influence of the boundary layer in the larger sample domain on the basis functions significantly. It is intuitively clear from Figure 2.4 that the effects of artificial boundary conditions are significantly reduced for this special two-scale example.

Specifically, let $\phi_j^E$ be the basis functions satisfying the homogeneous elliptic equation in the larger domain $K_E \supset K$ (see Figure 2.5). We then form the actual basis $\phi_i$ by linear combination of $\phi_j^E$,

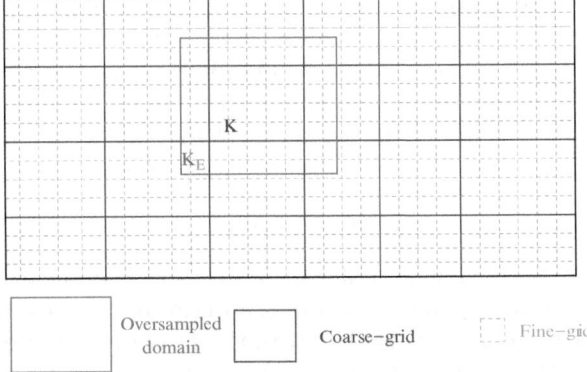

**Fig. 2.5.** Schematic description of oversampled region.

$$\phi_i = \sum_{j=1}^{d} c_{ij} \phi_j^E \ .$$

The coefficients $c_{ij}$ are determined by condition $\phi_i(x_j) = \delta_{ij}$, where $x_j$ are nodal points. Extensive numerical experiments have demonstrated that the oversampling technique does improve the numerical error substantially in many applications. Some numerical examples are presented in Section 2.9. On the other hand, the oversampling technique results in a nonconforming MsFEM method, where the basis functions are discontinuous along the edges of coarse-grid blocks. In [107] we perform a careful estimate of the nonconforming errors. The analysis shows that the nonconforming error is indeed small, and consistent with our numerical results [145, 146]. Our analysis also reveals another source of resonance, which is the mismatch between the mesh size and the "perfect" sample size. In the case of a periodic structure, the "perfect" sample size is the length of an integer multiple of the period. We call the new resonance the "cell resonance". In the error expansion, this resonance effect appears as a higher-order correction. In numerical computations, we found that the cell resonance error is generally small, and is rarely observed in practice. Nonetheless, it is possible to completely eliminate this cell resonance error by using the oversampling technique to construct the basis functions, but using piecewise linear functions as test functions. This reduces the resonance error further (see [143]).

## 2.4 Generalization of MsFEM: A look forward

Next, we present a general framework of MsFEMs (following Efendiev, Hou, and Ginting [104]) which is further discussed in Chapter 3. Consider

$$Lp = f, \tag{2.11}$$

where $L : X \rightarrow Y$ is an operator. The objective of the MsFEM is to approximate $p$ on the coarse grid. Denote by $W_h$ a family of finite-dimensional space such that it possesses an approximation property (see [274], [229]), as before. Here $h$, as before, is a scale of computation (coarse grid). In general, multiscale basis functions are replaced by multiscale maps defined as $E^{MsFEM} : W_h \rightarrow \mathcal{P}_h$. For each element $v_h \in W_h$, $v_{r,h} = E^{MsFEM} v_h$ is defined as

$$L^{\mathrm{map}} v_{r,h} = 0 \text{ in } K, \qquad (2.12)$$

where $L^{\mathrm{map}}$ can be, in general, different from $L$ (e.g., can be a discrete operator). Note that $v_h$ (the quantity with the subscript $h$) denotes the coarse-scale approximation and $v_{r,h}$ (the quantity with the subscript $r, h$) denotes the fine-scale approximation. For linear problems, we simply used the subscript $h$ to denote fine-scale approximations.

Note that $L^{\mathrm{map}}$ allows us to capture the effects of the small scales. Moreover, the domains different from the target coarse block $K$ can be used in the computations of the local solutions. To solve (2.12) one needs to impose boundary and initial conditions. This issue needs to be resolved on a case-by-case basis, and the main idea is to interpolate $v_h$ onto the underlying fine grid.

To find a solution of (2.11) in $\mathcal{P}_h$, one can substitute $p_h$ (which denotes a coarse-scale solution defined in $W_h$) into (2.11) discretized on the fine grid. Because $p_h$ is defined on the coarse grid, the resulting system is projected onto the coarse-dimensional space. This can be done in various ways. A common approach is to multiply the resulting fine-scale system by coarse-scale test functions; that is find $p_h \in W_h$ (consequently $p_{r,h} \in \mathcal{P}_h$) such that

$$\langle L^{\mathrm{global}} p_{r,h}, v_h \rangle = \langle f, v_h \rangle, \quad \forall v_h \in W_h, \qquad (2.13)$$

where $\langle \cdot, \cdot \rangle$ denotes a duality between $X$ and $Y$ (defined for the discrete variational formulation), and $L^{\mathrm{global}}$ can be, in general, different from $L$. We note that the fine-scale system $L^{\mathrm{global}} p_{r,h} - f$ is multiplied by coarse-scale test functions. One can also minimize the residual $L^{\mathrm{global}} p_{r,h} - f$ at some nodes to obtain a coarse-dimensional problem. Other approaches based on upscaled equations can also be used (see Section 5.4). In general, $L^{\mathrm{map}}$ and $L^{\mathrm{global}}$ can be different for nonlinear problems. Moreover, $p_h$ can represent only some of the physical variables involved in the simulations (see Section 5.4).

The convergence of the MsFEM is to show that $p_h \approx p^*$ and $p_{r,h} \approx p$ in appropriate spaces for small $h$, where $p_{r,h} = E^{MsFEM} p_h$. Here $p^*$ is a coarse-scale solution of (2.11). The correct choices of $L^{\mathrm{map}}$ and $L^{\mathrm{global}}$ are the essential part of MsFEM and guarantee the convergence of the method. We note that for linear elliptic equations, $L^{\mathrm{map}}$ is a linear map, and consequently, $\mathcal{P}_h$ is a linear space spanned by $E^{MsFEM} \phi_j^0$, where $\phi_j^0 \in W_h$. This formulation is equivalent to linear MsFEMs introduced earlier.

MsFEMs can be easily extended to the system of linear equations, such as elasticity equations (e.g., [235]),

$$\text{div}(\mathcal{C} : E(u)) = f,$$

where $\mathcal{C}$ is the fourth-order stiffness tensor representing material properties, $u$ is the displacement field (vector), and $E(u) = \frac{1}{2}(\nabla u + (\nabla u)^T)$ is the small strain tensor. In this case, multiscale basis functions will satisfy the local homogeneous equations $L\phi_i = 0$ in $K$, $\phi_i = \phi_i^0$ on $\partial K$, where $L$ is the elasticity operator, $Lu = \text{div}(\mathcal{C} : (\frac{1}{2}(\nabla u + (\nabla u)^T)))$. Note that the basis functions are vector fields. Vector fields $\phi_i^0$ are standard finite element basis functions used for solving the system of equations. For example, for elasticity equations, $\phi_i^0$ are linear functions for each element of the vector field (see [235]). The variational formulation that couples these basis functions will remain similar to (2.3) (or (2.4)).

We note that a main feature of MsFEMs presented in this book is the use of a variational formulation at the coarse scale that allows us to couple multiscale basis functions. Multiscale basis functions or multiscale maps defined by (2.12) are not necessarily based on partial differential equations and can have a discrete structure and satisfy a discrete equation at the fine grid. It is evident from the above abstract formulation that $L^{\text{map}}$ is used only for the computation of $p_{r,h}$ (given $p_h$) and the variational formulation (2.13) can be chosen in different ways depending on the problem. One can consider general applications of MsFEMs involving discrete problems where the basis functions satisfy discrete systems. For example, one can consider an application where the coarse-scale equations have a continuum formulation and describe porous media flows, whereas the local problems are discrete and solved via the pore network model. MsFEMs can be used to deal with these problems.

## 2.5 Brief overview of various global couplings of multiscale basis functions

### 2.5.1 Multiscale finite volume (MsFV) and multiscale finite volume element method (MsFVEM)

Mass conservative schemes play a central role in subsurface applications. For this reason, it is important to consider methods that can provide a mass conservative approximation for the flux defined by $v = -k\nabla p$. One of these methods within a finite volume context was first proposed in [159]. The main idea of this approach is to use a finite volume global formulation with multiscale basis functions and obtain a mass conservative velocity field on a coarse grid. A similar approach was independently proposed later in [104, 133] where a finite volume element method was used. These approaches differ in their details as discussed later. In these approaches, the finite volume element method is taken as a global coupling mechanism for multiscale basis functions. The construction of basis functions remains the same as discussed earlier.

To demonstrate the concept of MsFV as well as MsFVEM, we assume $\mathcal{T}_h$ is the collection of coarse elements $K$. We introduce a dual grid and denote

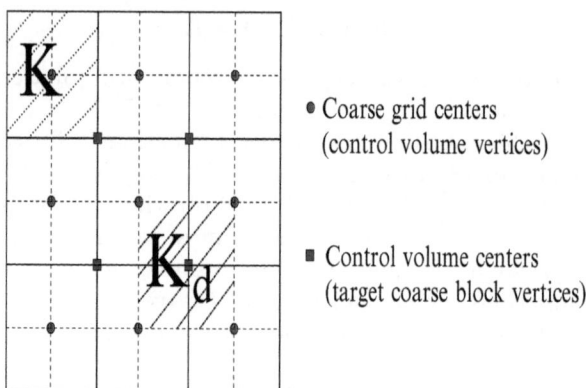

**Fig. 2.6.** Schematic of nodal points and coarse grids.

it by $K_d$ (see Figure 2.6 for illustration in the case of simple rectangular grids). Furthermore, we denote the vertices of dual coarse grids by $x_{K_d}$ (their collection by $Z_{K_d}$) and the vertices of target coarse-grid blocks by $x_K$ (their collection by $Z_K$).

As before, the key idea of the method is the construction of basis functions on the coarse grids, such that these basis functions capture the small-scale information. As in the case of MsFEM, the basis functions are constructed from the solution of the leading-order homogeneous elliptic equation on each coarse element with some specified boundary conditions. To demonstrate MsFV, we denote by $\mathcal{P}_h$ the space spanned by the basis functions $\{\phi_j\}_{x_j \in Z_{K_d}}$ as defined before (see (2.2)). In MsFV, the basis functions on the dual grid are used and a mass conservation equation is set up on the target coarse-grid blocks. In particular, we seek $p_h \in \mathcal{P}_h$ with $p_h = \sum_{x_j \in Z_{K_d}^0} p_j \phi_j$ (where $p_j$ are the approximate values of the solution at $x_{K_d}$ and $Z_{K_d}^0$ is the collection of interior vertices) such that

$$\int_{\partial K} k \nabla p_h \cdot n \, dl = \int_K f \, dx, \tag{2.14}$$

for every target coarse-grid block $K \in \mathcal{T}_h$. Here $n$ defines the normal vector on the boundary. The equation (2.14) results in a system of linear equations for the solution values at the nodal points of the coarse mesh. In particular, we have

$$A p_{\text{nodal}} = b,$$

where $A = (a_{ij})$, $a_{ij} = \sum_j \int_{\partial K_j} k \nabla \phi_i \cdot n \, dl$, $b_j = \int_{K_j} f \, dx$. Here $j$ refers to the index of the coarse-grid block $K_j$.

In MsFVEM, the basis functions on the target coarse-grid blocks are chosen and the mass conservation equation is set up on the dual grid. We do not repeat the formulation here. The resulting multiscale method differs from the MsFEM, because it employs the finite volume or finite volume element method as a global solver. We would like to note that the coarse-scale velocity field obtained using MsFVEM is conservative in control volume elements, whereas the velocity field obtained using MsFV is conservative in coarse elements. Further treatment is needed to obtain a conservative velocity field on the fine grid (see [159]).

*Pseudo-code.* We present a pseudo-code for the implementation of MsFV.

---

**Algorithm 2.5.1**

---

Set coarse mesh configuration from fine-scale mesh information.
For each coarse grid block $n$ do
- For each control volume element $i$ associated with the coarse block $n$
- Solve for $\phi_n^i$ satisfying - $L(\phi_n^i) = 0$ and boundary conditions (see (2.2))
- End for.
End do.
Assemble the mass balance equation on the coarse grid according to (2.14).
Assemble the external force on the coarse mesh according to (2.14).
Solve the coarse grid formulation.

---

### 2.5.2 Mixed multiscale finite element method

MsFV and MsFVEM introduced earlier provide a mass conservative velocity field (defined as $v = -k\nabla p$) on the coarse grid. However, the reconstructed fine-scale velocity field (using multiscale basis functions) is not conservative for the fine grid elements adjacent to coarse grid boundaries. For multiphase flow and transport simulations, the conservative fine-scale velocity is often needed. A treatment within MsFV is proposed in [159]. In this section, we present a mixed MsFEM where multiscale basis functions for the velocity field, which is highly heterogeneous, are constructed. This method allows us to achieve a mass conservative fine-scale velocity field and is used in Chapter 5 for multiphase flow simulations in heterogeneous porous media.

Our presentation of mixed MsFEM follows [71] (see also [25], [1], and [26]). First, we re-write the elliptic equation in the form

$$k^{-1}v + \nabla p = 0 \quad \text{in} \quad \Omega$$
$$\text{div}(v) = f \quad \text{in} \quad \Omega \tag{2.15}$$

with non-homogeneous Neumann boundary conditions $v \cdot n = g$ on $\partial\Omega$. In mixed multiscale finite element methods, the basis functions for the velocity field, $v = -k\nabla p$, are needed. As in the case of MsFEM, one can use known mixed finite element spaces to construct these basis functions. For simplicity, we consider multiscale basis functions corresponding to lowest-order Raviart–Thomas elements (following [71], [25]). The basis functions for the velocity in each coarse block $K$ is given by

$$
\begin{aligned}
\operatorname{div}(k\nabla\phi_i^K) &= \frac{1}{|K|} \quad \text{in} \quad K \\
k\nabla\phi_i^K \cdot n &= \begin{cases} g_i^K & \text{on } e_i^K \\ 0 & \text{else,} \end{cases}
\end{aligned}
\tag{2.16}
$$

where $g_i^K = 1/|e_i^K|$ and $e_i^K$ are the edges of $K$ (see Figure 2.7 for the illustration). Note that these basis functions are defined for each edge by imposing constant flux along an edge (constant Neumann boundary condition) and zero flux over all other edges of the coarse-grid block. In order to preserve the total mass and have a well-posed system, some source term is needed. The source term is taken to be constant.

We define the finite-dimensional space for the velocity by

$$
\mathcal{V}_h = \operatorname{span}\{\psi_i^K\},
$$

where $\psi_i^K = k\nabla\phi_i^K$. For each edge $e_i$, one can combine the basis functions in adjacent coarse-grid blocks and obtain the basis function for the edge $e_i$ denoted by $\psi_i$ (or $\psi_{e_i}$). More precisely, if we denote by $K_1$ and $K_2$ adjacent coarse-grid blocks, then $\psi_i$ solves (2.16) in $K_1$ and solves $\operatorname{div}(\psi_i) = -1/|K_2|$ in $K_2$, and $g_i^{K_2} = -1/|e_i|$ on $e_i^{K_2}$ and 0 otherwise. In other words, $\psi_i = \psi_i^{K_1}$ in $K_1$ and $\psi_i = -\psi_i^{K_2}$ in $K_2$, where $\psi_i^K$ is defined via the solution of (2.16). This is illustrated in Figure 2.8. In [1], the author proposes a different construction for mixed multiscale basis functions by solving the local problem in two adjacent coarse grid blocks with zero Neumann boundary conditions and imposing positive and negative source terms. For example, $\operatorname{div}(\psi_i) = 1/|K_1|$ in $K_1$, $\operatorname{div}(\psi_i) = -1/|K_2|$ in $K_2$, and $\psi_i \cdot n = 0$ on outer boundaries of $K_1 \bigcup K_2$.

The basis functions for the pressure are piecewise constant functions over each $K$. We denote the span of these basis functions by $Q_h$. The multiscale basis functions, as in MsFEM, attempt to capture the small-scale information of the media. The functions $\psi_i^K$ are the basis functions for the velocity field and conservative both on the fine and coarse grids provided the local problems are solved using a conservative scheme. An approximation of the fine-scale velocity field can be obtained if average fluxes along the coarse edges are known, that is if $v_e$ is the average normal flux along the edge $e$ and $\psi_e$ is the corresponding basis function, then $v \approx \sum_e v_e \psi_e$ is an approximation of the fine-scale velocity field. These average fluxes, for example, can be also obtained from MsFV or MsFVEM or by using upscaling methods as in, for example, [91]. A similar idea is presented in [159] and [1]. The mixed finite element framework, presented next, couples the velocity and pressure basis functions and provides an approximation of the global solution (both $p$ and $v$).

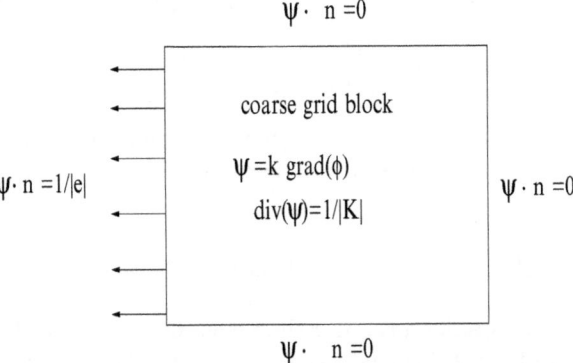

**Fig. 2.7.** Schematic description of a velocity basis function construction in a coarse grid block.

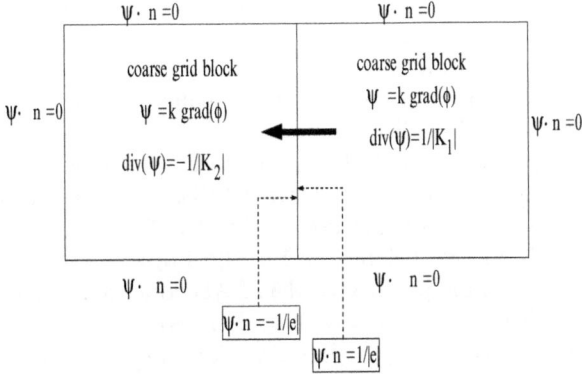

**Fig. 2.8.** Schematic description of a velocity basis function for an edge combining adjacent basis functions.

To formulate the mixed MsFEM, we use the numerical approximation associated with the lowest-order Raviart–Thomas mixed finite element to find $\{v_h, p_h\} \in \mathcal{V}_h \times Q_h$ such that $v_h \cdot n = g_h$ on $\partial\Omega$, where $g_h = g_{0,h} \cdot n$ on $\partial\Omega$ and $g_{0,h} = \sum_{e \in \{\partial K \cap \partial\Omega, K \in T_h\}} (\int_e g ds) \psi_e$, $\psi_e \in \mathcal{V}_h$ is the corresponding basis function to edge $e$,

$$\int_\Omega k^{-1} v_h \cdot w_h dx - \int_\Omega \operatorname{div}(w_h) p_h dx = 0, \quad \forall w_h \in \mathcal{V}_h^0$$

$$\int_\Omega \operatorname{div}(v_h) q_h dx = \int_\Omega f q_h dx, \quad \forall q_h \in Q_h, \tag{2.17}$$

where $\mathcal{V}_h^0$ is a subspace of $\mathcal{V}_h$ with elements that satisfy homogeneous Neumann boundary conditions. The above formulation was the mixed MsFEM introduced in [71].

The discrete formulation of (2.17) can be easily written down as

$$
\begin{bmatrix} A & C \\ C^T & 0 \end{bmatrix} \begin{bmatrix} v^D \\ -p^D \end{bmatrix} = \begin{bmatrix} 0 \\ b \end{bmatrix}, \tag{2.18}
$$

where $v = \sum_i v_i^D \psi_i$ and $p = \sum_i p_i^D q_i$ with $v_i^D$ being normal interface fluxes and $p_i^D$ being the cell average solution. Here $A = (a_{ij})$, $C = (c_{ik})$, and $b = (b_k)$ are defined by

$$
a_{ij} = \int_\Omega \psi_i \cdot (k)^{-1} \psi_j dx, \quad c_{ik} = \int_\Omega q_k \, \mathrm{div}(\psi_i) dx \text{ and } b_k = \int_\Omega q_k f dx.
$$

This linear system is indefinite, and thus it is in general harder to solve than the positive definite systems that arise (e.g., from Galerkin finite element discretizations). However, it is common to solve the mixed linear system (2.18) by using a so-called hybrid formulation. In the hybrid formulation the system (2.18) is localized by introducing an extra set of unknowns representing $p$ at the grid cell interfaces. By performing some simple algebraic manipulations, we then obtain a positive definite system that is solved for the interface pressures. Finally, the solution to (2.18) is computed from the solution to the hybrid system by performing local algebraic calculations.

We note that in [10], a discontinuous Galerkin method has been used as a global coupling for multiscale basis functions and discontinuous Galerkin MsFEM is proposed. We refer to [10] for details. We also refer to [16], for the use of discontinuous Galerkin method within the framework of HMM.

*Pseudo-code.* Next, we briefly outline the implementation of mixed MsFEM. We have put simple prototype MATLAB codes for solving elliptic equations with mixed MsFEM (courtesy of J. Aarnes) at
$http : //www.math.tamu.edu/ \sim yalchin.efendiev/codes/.$

---

**Algorithm 2.5.2**

---

Set coarse mesh configuration from fine-scale mesh information.
For each coarse-grid block $n$ do
  – For each edge of a coarse-grid block
  – Solve for $\psi_n^i$ according to (2.16)
  – End for
End do.
Assemble the coarse-scale system according to (2.17).
Assemble the external force on the coarse mesh according to (2.17).
Solve the coarse grid formulation.

---

*Comments on the assembly of stiffness matrix.* Similar to the Galerkin MsFEM, one can use the representation of multiscale basis functions via fine-scale basis functions to assemble the matrices in (2.18). It can be shown that

$A = D A^f D^T$, where $D = (d_{ij})$ is a matrix defined by

$$\psi_i = d_{ij} \psi_j^{0,f}$$

with $\psi_j^{0,f}$ being fine-scale basis functions, $A^f = (a_{lm}^f)$, $a_{lm}^f = \int_\Omega \psi_l^{0,f} \cdot (k)^{-1} \psi_m^{0,f} dx$. This simplification can be used in the assembly of the stiffness matrix.

## 2.6 MsFEM for problems with scale separation

In some applications, regions smaller than the coarse-grid block are sufficient to represent the small-scale effects. In these applications, one can use the basis functions constructed in a smaller region, instead of the coarse-grid block, to capture the small-scale effects. The basic idea behind this localization is that $(1/|K|) \int_K k \nabla \phi_i dx$ in the computation of the stiffness matrix (2.4) can be approximated by $(1/|K_{\text{loc}}|) \int_{K_{\text{loc}}} k \nabla \tilde{\phi}_i dx$, where $\phi_i$ is the solution of $\text{div}(k \nabla \phi_i) = 0$ in $K$, $\phi_i = \phi_i^0$ on $\partial K$, and $\tilde{\phi}_i$ is the solution of $\text{div}(k \nabla \tilde{\phi}_i) = 0$ in $K_{\text{loc}}$, $\tilde{\phi}_i = \phi_i^0$ on $\partial K_{\text{loc}}$. Here, $K_{\text{loc}}$ refers to a smaller region (RVE) as before. Next, we briefly discuss this approximation. Within the general $G$-convergence theory (e.g., [164]), it can be shown that (e.g., [164])

$$\lim_{\epsilon \to 0} \frac{1}{|K|} \int_K k \nabla \phi_i dx = \frac{1}{|K|} \int_K k^* \nabla \phi_i^0 dx, \qquad (2.19)$$

where $\epsilon$ is the characteristic length scale and $\phi_i^0$ is the homogenized part of the basis function. Note that the $G$-convergence theory does not assume periodicity and $k^*(x)$ are homogenized coefficients independent of $\epsilon$ such that the solution and the fluxes of the homogenized equation $-\text{div}(k^*(x) \nabla p^*) = f$ approximate the solution of fine-scale equation $-\text{div}(k(x) \nabla p) = f$ in appropriate norms (we refer to, e.g., [164] for details). The same result holds for $(1/|K_{\text{loc}}|) \int_{K_{\text{loc}}} k \nabla \tilde{\phi}_i dx$; that is

$$\lim_{\epsilon \to 0} \frac{1}{|K_{\text{loc}}|} \int_{K_{\text{loc}}} k \nabla \tilde{\phi}_i dx = \frac{1}{|K_{\text{loc}}|} \int_{K_{\text{loc}}} k^* \nabla \tilde{\phi}_i^0 dx. \qquad (2.20)$$

Assuming that $k^*(x)$ is sufficiently smooth, one can approximate $k^*(x)$ within each coarse block by a constant and show that the right-hand sides of (2.19) and (2.20) are close for small $h$ (note that $K_{\text{loc}} \subset K$). Consequently, for small $\epsilon$, the left-hand sides of (2.19) and (2.20) will be close; that is

$$\lim_{|K| \to 0} \lim_{\epsilon \to 0} \frac{1}{|K|} \int_K k \nabla \phi_i dx = \lim_{|K_{\text{loc}}| \to 0} \lim_{\epsilon \to 0} \frac{1}{|K_{\text{loc}}|} \int_{K_{\text{loc}}} k \nabla \tilde{\phi}_i dx. \qquad (2.21)$$

The relation (2.21) shows that $\int_{K_{\text{loc}}} k \nabla \tilde{\phi}_i dx$ can be used to approximate $\int_K k \nabla \phi_i dx$ in the limit $\lim_{h \to 0} \lim_{\epsilon \to 0}$ (limit of scale separation). From here, one can show that the multiscale finite element solution approximates the fine-scale solution in $\lim_{h\to 0} \lim_{\epsilon\to 0}$. We refer to [108] for the details where the more general problem is studied. For periodic problems, this approximation can be shown in the limit $\epsilon/h_{\mathrm{loc}} \to 0$ (where $h_{\mathrm{loc}}$ is the size of $K_{\mathrm{loc}}$) if $K_{\mathrm{loc}}$ is much larger than the period size. If $k^*(x)$ is sufficiently smooth and $\phi_i^0$ is piecewise linear, then (2.21) is equal to $k^*(x_0)\nabla\phi_i^0$, where $x_0$ is the point to which the region $K$ or $K_{\mathrm{loc}}$ contracts (see [164] for a more precise definition of $x_0$). In this case, the location of RVE within the coarse grid block is not very important and one can choose RVE, for example, at the center of the mass of the coarse element.

In the case of periodic heterogeneities, where the period is known, the basis functions can be approximated using homogenization theory by

$$\phi_j \approx \phi_j^0 + \epsilon\chi^i\frac{\partial}{\partial x_i}\phi_j^0, \qquad (2.22)$$

where the summation over repeated indices occurs. This approach derives from homogenization and $\chi$ is a periodic solution (with average zero) of (2.9). Consequently, the approximation of the basis functions can be carried out in a domain of the size of the period that characterizes the small-scale oscillation of $k(x)$. This reduces the computational cost if the period is much smaller than the coarse-grid block. With this approximation, the stiffness matrix (see (2.5)) can be assembled in the periods instead of the coarse grid blocks:

$$a_{ij} = \sum_K \int_K k\nabla\phi_i \cdot \nabla\phi_j^0 dx \approx \sum_K \frac{|K|}{|Y|} \int_Y k\nabla(\phi_j^0 + \epsilon\chi^i\frac{\partial}{\partial x_i}\phi_j^0) \cdot \nabla\phi_j^0 dx, \qquad (2.23)$$

where $Y$ is the period within $K$. One can further approximate this expression and show that $a_{ij} \approx \sum_K (|K|/|Y|) \int_Y k(I + \nabla\chi)\nabla\phi_j^0 \cdot \nabla\phi_j^0 dx$. In [269], the author uses the approximation of the basis functions based on (2.22) for periodic coefficients and shows that MsFEM converges as the period size and the mesh size go to zero.

As we discussed earlier, using multiscale basis functions, a fine-scale approximation of the solution can be easily computed, by $p_h = \sum_i p_i\phi_i$. When regions smaller than the coarse-grid block are used for computing basis functions, then the basis functions can be extended to a coarse-grid block based on homogenization expansion. In particular, from the problem in $K_{\mathrm{loc}}$ (RVE) one can easily extract $\chi$ and use it to construct an extension of the basis function to the coarse-grid block. These basis functions can be further used to obtain an approximate fine-scale solution in the entire domain.

We would like to note that this approximation procedure is not limited to periodic problems and can be applied to problems where homogenization by periodization (see the principle of periodic localization [164]) is true. The random homogeneous case with ergodicity is one of these cases. The techniques discussed in this section can be also used in MsFV, MsFEM, mixed MsFEM, and other multiscale methods when the problem has some special features such as periodicity or strong scale separation.

## 2.7 Extension of MsFEM to parabolic problems

MsFEM can be naturally extended to parabolic equations. In this section, we briefly describe the extension of MsFEM to the parabolic equation

$$\frac{\partial}{\partial t} p(x,t) - \text{div}(k(x,t)\nabla p(x,t)) = f \qquad (2.24)$$

with appropriate boundary conditions on the finite time interval $[0,T]$ and smooth initial conditions. In general, when there are space and time heterogeneities, basis functions are the solutions of the leading-order homogeneous parabolic equations. In the absence of time heterogeneities (i.e., $k(x,t) = k(x)$), one can use spatial basis functions developed for elliptic equations. To introduce MsFEM, we assume for simplicity that the interval $[0,T]$ is divided into $M$ equal parts $0 = t_0 < t_1 < \cdots < t_M = T$. These intervals are coarse-scale intervals; that is $\Delta t = t_{i+1} - t_i$ is larger than the characteristic time scale. The basis functions are constructed in $[t_n, t_{n+1}]$ as the solution of

$$\frac{\partial}{\partial t} \phi_i(x,t) - \text{div}(k(x,t)\nabla \phi_i(x,t)) = 0 \qquad (2.25)$$

in each $K$ such that $\phi_i = \phi_i^0$ on $\partial K$ and $\phi_i(x, t = t_n) = \phi_i^0$. Here, $\phi_i^0 \in W_h$ are standard finite element basis functions (e.g., piecewise linear functions). We seek the finite-dimensional approximation of the solution in $[t_n, t_{n+1}]$ as

$$p_h^{n+1}(x,t) = \sum_i p_i^{n+1} \phi_i(x,t), \qquad (2.26)$$

where $p_i^{n+1}$ (approximate nodal values of the solution) will be determined. Then, substituting (2.26) into the original equation, multiplying it by $\phi_i^0$ (as in the Petrov–Galerkin formulation) and integrating over the space and $[t_n, t_{n+1}]$, we have

$$p_i^{n+1} \int_\Omega \phi_i(x, t_{n+1}) \phi_j^0(x)dx - p_i^n \int_\Omega \phi_i(x, t_n) \phi_j^0(x)dx$$
$$+ \int_{t_n}^{t_{n+1}} \sum_K \int_K k(x,t)\nabla p_h(x,t) \cdot \nabla \phi_j^0(x)dxdt = \int_{t_n}^{t_{n+1}} \int_\Omega f \phi_j^0(x)dxdt, \qquad (2.27)$$

where $p_h(x,t) = \sum_i p_i(t)\phi_i(x,t)$. The third term on the right-hand side can be treated implicitly or explicitly. In particular, the implicit method is given by

$$p_i^{n+1} \int_\Omega \phi_i(x, t_{n+1}) \phi_j^0(x) dx - p_i^n \int_\Omega \phi_i(x, t_n) \phi_j^0(x) dx$$

$$+ p_i^{n+1} \int_{t_n}^{t_{n+1}} \sum_K \int_K k(x, t) \nabla \phi_i(x, t) \cdot \nabla \phi_j^0(x) dx dt = \int_{t_n}^{t_{n+1}} \int_\Omega f \phi_j^0(x) dx dt.$$

$$\tag{2.28}$$

If the third term is evaluated explicitly, that is it is replaced by

$$p_i^n \int_{t_n}^{t_{n+1}} \sum_K \int_K k(x, t) \nabla \phi_i(x, t) \cdot \nabla \phi_j^0(x) dx dt,$$

then the resulting method is explicit. One can easily write down the corresponding discrete formulation which we omit here.

We note that if there are no temporal heterogeneities (i.e., $k(x, t) = k(x)$), the basis functions can be the solutions of elliptic equations as in the case of elliptic equations. The equations for the basis functions can be simplified depending on the relation between spatial and temporal heterogeneities (see Section 3.5). Equation (2.25) defines the basis functions independent of the relation between spatial and temporal heterogeneities. Furthermore, in the case of scale separation, (2.25) can be solved in a smaller volume, RVE, and this solution can be used in Equation (2.27) in a manner similar to the elliptic case.

Finally, we would like to note that one can couple the basis functions using different methods, such as finite volume element methods and so on. For example, the mixed MsFEM for parabolic equations (with time-independent coefficients, $k(x, t) = k(x)$) has the following formulation. We seek $\{v_h, p_h\} \in \mathcal{V}_h \times Q_h$, such that

$$\int_\Omega \frac{\partial p_h}{\partial t} q_h dx + \int_\Omega \text{div}(v_h) \, q_h dx = \int_\Omega f q_h dx, \quad \forall q_h \in Q_h$$

$$\int_\Omega k^{-1} v_h \cdot w_h dx - \int_\Omega \text{div}(w_h) \, p_h dx = 0, \quad \forall w_h \in \mathcal{V}_h^0,$$

$$\tag{2.29}$$

where $\mathcal{V}_h$, $\mathcal{V}_h^0$, and $Q_h$ are defined as before (cf. (2.17)) for elliptic equations.

## 2.8 Comparison to other multiscale methods

MsFEMs share similarities with many other multiscale methods. One of the early approaches is the upscaling technique (e.g., [91, 260, 47]) which is based on the homogenization method. The main idea of upscaling techniques is to form a coarse-scale equation and pre-compute the effective coefficients. In the case of linear elliptic equations, the coarse-scale equation has the same form as the fine-scale equation except that the coefficients are replaced by effective homogenized coefficients. The effective coefficients in upscaling methods are computed using the solution of the local problem in a representative volume. Various boundary conditions can be used for solving the local problems and, for simplicity, we consider

$$\text{div}(k\nabla\phi_e) = 0 \text{ in } K \tag{2.30}$$

with $\phi_e(x) = x \cdot e$ on $\partial K$, where $e$ is a unit vector. It is sufficient to solve (2.30) for $d$ linearly independent vectors $e_1, ..., e_d$ in $\mathbb{R}^d$ because $\phi_e = \sum_i \beta_i \phi_{e_i}$ if $e = \sum_i \beta_i e_i$. Here $K$ denotes a coarse-grid block, although one can use a smaller region as discussed in Section 2.6. The effective coefficients are computed in each $K$ as

$$\tilde{k}^* e = \frac{1}{|K|} \int_K k\nabla\phi_e dx. \tag{2.31}$$

We note that $\tilde{k}^*$ (which is not the same as the homogenized coefficients) is a symmetric matrix provided $k$ is symmetric and (2.31) can be computed for any point in the domain by placing the point at the center of $K$, i.e.,

$$\tilde{k}^*(x_0)e = \frac{1}{|K_{x_0}|} \int_{K_{x_0}} k\nabla\phi_e dx,$$

where $K_{x_0}$ is the RVE with the center at $x_0$ and $\phi_e$ is the local solution defined by (2.30) in $K_{x_0}$. One can use various boundary conditions, including periodic boundary conditions as well as oversampling methods. We refer to [91, 260] for the discussion on the use of various boundary conditions. Once the effective coefficients are calculated, the coarse-scale equation

$$-\text{div}(\tilde{k}^*\nabla p^*) = f \tag{2.32}$$

is solved over the entire region.

To show the similarity to MsFEMs, we write down the discretization of (2.32) using the Galerkin finite element method. Find $p_h^* \in W_h$, such that

$$\sum_K \int_K \tilde{k}^*\nabla p_h^* \cdot \nabla v_h dx = \int_\Omega f v_h dx, \quad \forall v_h \in W_h. \tag{2.33}$$

Next, we write down the Petrov–Galerkin MsFEM discretization (see (2.4))

$$a_{ij}p_i = b_j, \tag{2.34}$$

where $a_{ij} = \sum_K \int_K k\nabla\phi_i dx \cdot \nabla\phi_j^0$ (assuming $\phi_j^0$ is piecewise linear) and $b_j = \int_\Omega f\phi_j^0 dx$. One can show that

$$a_{ij} = \sum_K \int_K \tilde{k}^*\nabla\phi_i^0 \cdot \nabla\phi_j^0 dx$$

because $(1/|K|)\int_K k\nabla\phi_i dx = \tilde{k}^*\nabla\phi_i^0$. We assumed that $\phi_i^0$ are piecewise linear functions. Thus, (2.34) and (2.33) are equivalent. This shows that the

MsFEM can be derived from traditional upscaling methods. However, the concept of MsFEMs differs from traditional upscaling methods, because the local information is directly coupled via a variational formulation and we do not assume a specific form for coarse-scale equations. Moreover, MsFEMs allow us to recover the local information adaptively which makes it a powerful tool (e.g., for porous media flow simulations). More advantages of MsFEM are discussed in later chapters.

Next, we briefly discuss the relation between variational multiscale approaches and MsFEM. These similarities are also shown in [26] in the context of mixed finite element methods. Here, we discuss Galerkin finite element methods. We assume that the fine-scale solution space $X_F$ is partitioned into the coarse-dimensional space $X_C$ (e.g., $W_h$), and the space containing the unresolved scales $X_U$,

$$X_F = X_C \oplus X_U.$$

We assume also that these spaces are the subspaces of $H_0^1(\Omega)$, for simplicity. The fine-scale solution can be written accordingly as

$$p = p_C + p_U.$$

Substituting this into the original equation and multiplying by the test functions from $X_C$, we obtain the equation for the coarse-scale solution

$$\int_\Omega k\nabla(p_C + p_U) \cdot \nabla v_h dx = \int_\Omega f v_h dx, \quad \forall v_h \in X_C. \qquad (2.35)$$

Similarly, multiplying the original equation by the test functions from $X_U$, we obtain the equation for the unresolved part of the solution

$$\int_\Omega k\nabla p_U \cdot \nabla v_h dx = \int_\Omega f v_h dx - \int_\Omega k\nabla p_C \cdot \nabla v_h dx, \quad \forall v_h \in X_U. \qquad (2.36)$$

To find the coarse-scale solution, $p_C$, one first solves $p_U$ from (2.36) and substitutes it into (2.35) to compute $p_C$. We note that (2.35) is exact and the solution of (2.36) is nonlocal. In general, the solution of (2.36) can be localized by imposing local boundary conditions. One can use various choices for the boundary conditions. Noting the solution of the local problem can be written via generic basis functions, one can derive a formulation similar to MsFEM.

To show the similarity between MsFEMs and variational multiscale methods, as an example, we consider the localization based on $p_U = 0$ on the boundaries of the coarse-grid block $K$. In this case, it is evident that the solution $p_C + p_U$ satisfies the local problem $\mathrm{div}(k\nabla(p_C + p_U)) = f$ in $K$ and $p_C + p_U$ is a piecewise linear function on $\partial K$. This solution can be approximated by multiscale finite element basis functions defined by (2.2). Thus, we can seek $p_C + p_U = \sum_i p_i \phi_i$. Substituting this expression into (2.35), we obtain a Petrov–Galerkin formulation of MsFEM if $\phi_i$ are chosen with zero right-hand side. We note that one of the differences between the variational multiscale method and MsFEM is that the former uses source terms in the formulation of the local problems. The representation of source terms with MsFEMs in the context of subsurface flows has been extensively studied in the literature (e.g., see [13, 175], Sections 5.5, and 5.6) within the context of subsurface flows. Typically, only singular source terms require special treatment with multiscale basis functions.

As we mentioned earlier, one can take advantage of scale separation in MsFEM. There are various ways to do so and these approaches will share similarities, for example, with the application of heterogeneous multiscale methods (HMM) ([97]), and multiscale enrichment methods ([121]). HMM has been extensively studied in the literature (e.g., see [17, 15, 98, 203] for the applications to elliptic equations). The main idea of this approach is to use small regions at quadrature points for the computation of effective coefficients. This is performed on-the-fly when the stiffness matrix corresponding to the coarse-scale problem is assembled. As mentioned above, multiscale basis functions can be approximated when there is scale separation. The basic idea behind this localization is that $(1/|K|) \int_K k\nabla\phi_i dx$ (in the stiffness matrix, see (2.5)) can be approximated by $(1/|K_{\mathrm{loc}}|) \int_{K_{\mathrm{loc}}} k\nabla\tilde{\phi}_i dx$, where $\phi_i$ is the solution of $\mathrm{div}(k\nabla\phi_i) = 0$ in $K$, $\phi_i = \phi_i^0$ on $\partial K$, and $\tilde{\phi}_i$ is the solution of $\mathrm{div}(k\nabla\tilde{\phi}_i) = 0$ in $K_{\mathrm{loc}}$, $\tilde{\phi}_i = \phi_i^0$ on $\partial K_{\mathrm{loc}}$. Using the general $G$-convergence theory (e.g., [164]), one can show (2.21). This result holds when $k^*(x)$ is a smooth function and (2.21) is equal to $k^*(x_0)\nabla\phi_i^0$ (assuming $\phi_i^0$ is piecewise linear) at almost every point $x_0$ to which the region $K$ and $K_{\mathrm{loc}}$ contract. For periodic problems, $k = k(x/\epsilon)$, it is not difficult to show that

$$|\frac{1}{|K|} \int_K k\nabla\phi_i dx - k^*\nabla\phi_i^0| \leq C(\frac{\epsilon}{h} + h),$$

where $k^*$ is computed for the coarse-grid block according to (2.31). Similarly,

$$|\frac{1}{|K_{\mathrm{loc}}|} \int_{K_{\mathrm{loc}}} k\nabla\tilde{\phi}_i dx - k^*\nabla\phi_i^0| \leq C(\frac{\epsilon}{h_{\mathrm{loc}}} + h_{\mathrm{loc}}).$$

Based on these results, one can show the convergence of MsFEMs using the local information in $K_{\mathrm{loc}}$. We refer to [108] for the details where a more general problem is studied. This approximation of the basis functions and the corresponding approximation of the stiffness matrix elements can save CPU time if there is a strong scale separation. The method obtained in this way is very similar to the application of HMM to elliptic equations, although it differs in some details (e.g., the computations are not performed at quadrature points). We would like to note that one can also use first-order corrector approximation for the basis functions as discussed earlier. In this case, the local solution in RVE can be used as a cell problem solution $\epsilon\chi$ in (2.22). We would like to mention that there are other approaches (e.g., [121, 122, 153]) which use the solution of the cell problem to construct multiscale basis functions based on the partition of unity method.

As we mentioned in Section 1.2, multiscale methods considered in this book differ from domain decomposition methods (e.g., [257]) where the local problems are solved many times iteratively to obtain an accurate approximation of a fine-scale solution. Multiscale methods studied in this book share similarities with upscaling/homogenization methods, where the basis functions are computed based on coarse-grid information. Figure 2.9 illustrates the main concept of the MsFEM and its advantages (see also Section 2.9). The multiscale methods attempt to find the coarse-scale solution and can also compute an approximation of the fine-scale solution via downscaling. One can use iterations (e.g., [93]) similar to domain decomposition methods or some type of global information to improve the accuracy of multiscale methods when there is no scale separation (see Chapter 4).

Finally, we remark that we restricted ourselves only to a few multiscale methods due to the page limitation. One can find similarities between multiscale finite element methods and other multiscale methods known in the literature. Some of these similarities may not be so apparent. Some of these algorithms are designed for periodic problems and have advantages when the underlying heterogeneities are periodic. For example, the approach proposed in [198] is based on two-scale convergence concept ([21]). This approach is generalized to problems with multiple separable scales ([139]) using hierarchical basis functions. In this book, we do not want to discuss the similarities between different multiscale methods to a great extent and instead focus on our work on extensions and applications of various MsFEMs. We again stress that the main idea of MsFEM stems from the earlier work of Babuška and Osborn [33]. In Chapters 3 and 4, we show that the MsFEM can take advantage of global information and can be naturally extended to nonlinear problems.

## 2.9 Performance and implementation issues

We outline the implementation of a Galerkin MsFEM for a simple test problem (following [145]) and define some notations that are used in the discussion below. We consider solving problems in a unit square domain. Let $N$ be the number of elements in each coordinate direction. The mesh size is thus $h = 1/N$. To compute the basis functions, each element is discretized into $M \times M$ subcell elements with mesh size $h_f = h/M$. To implement the oversampling method, we partition the domain into sampling domains where each of them contains many elements. Analysis and numerical tests indicate that the size of the sampling domains can be chosen freely as long as the boundary layer is avoided. In practice, though, one wants to maximize the efficiency of oversampling by choosing the largest possible sample size that reduces the redundant computation of overlapping domains to a minimum.

In general, the multiscale basis functions are constructed numerically, except for certain special cases. They are solved in each $K$ or $K_E$ using a standard FEM. The global linear system on $\Omega$ is solved using the same method.

Numerical tests show that the accuracy of the final solution is weakly insensitive to the accuracy of basis functions.

Because the basis functions are independent of each other, their construction can be carried out in parallel perfectly. In a parallel implementation of oversampling, the sample domains are chosen such that they can be handled within each processor without communication. More implementation details can be found in [145].

### 2.9.1 Cost and performance

In computations, a large amount of overhead time comes from constructing the basis functions. This is also true for classical upscaling methods discussed in Section 2.8. On a sequential machine, the operation count of the MsFEM is about twice that of a conventional FEM for a 2D problem. However, due to good parallel efficiency, this difference is reduced significantly on a massively parallel computer (see [145] for a detailed study of the MsFEM's parallel efficiency). This overhead can be reduced if there is scale separation.

In practice, multiple solves are often required for different source terms, boundary conditions, mobilities and so on. MsFEMs have advantages in such situations and the overhead of basis construction can be negligible because the basis functions can be re-used. This is illustrated in Figure 2.9, where pre-computed multiscale basis functions can be re-used for different external parameters such as source terms, boundary conditions and the like. Moreover, multiscale basis functions can be used to re-construct the fine-scale features of the solution in the regions of interest. This adaptivity is often used in subsurface applications where the fine-scale features of the velocity $(-k\nabla p)$ are re-constructed in some regions where the detailed velocity information is needed, for example, for updating sharp interfaces. In summary, MsFEMs provide the following advantages in simulations: (1) parallel multiscale basis function construction (which can be very cheap if there is scale separation); (2) re-use of basis functions for different external parameters and inexpensive coarse-scale solve; and (3) adaptive downscaling of the fine-scale features of the solution in the regions of interest.

Significant computational savings are obtained for time-dependent problems such as those that occur in subsurface applications. In these problems, the heterogeneities representing porous media properties do not change and the basis functions are pre-computed at the initial time. These basis functions are used throughout the simulations and the elliptic equations are solved on the coarse grid each time. In this sense, our approaches are similar to classical upscaling methods where the upscaled quantities are pre-computed before solving the equations on the coarse grid. In some situations, local basis function update is required, for example, if there is a sharp interface dividing two propagating fluids. The interface modifies the permeability and this requires local updates of the basis functions.

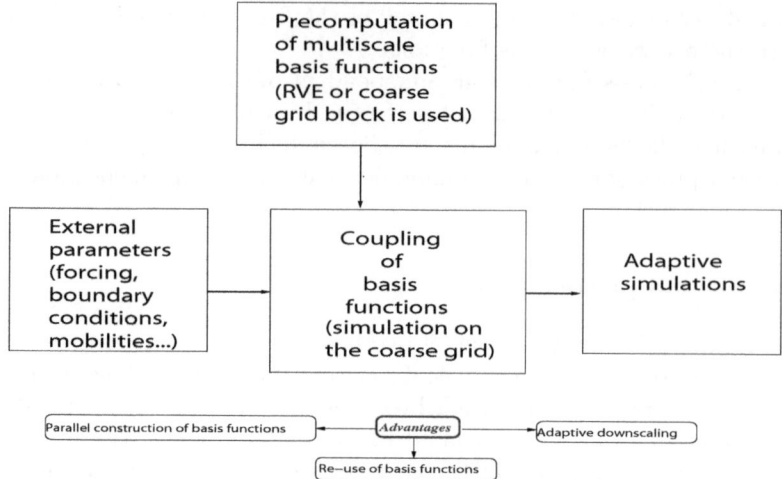

**Fig. 2.9.** A schematic illustration of multiscale simulations and advantages.

### 2.9.2 Convergence and accuracy

Because we need to use an additional grid to compute the basis function numerically, it makes sense to compare our MsFEM with a traditional FEM at the subcell (fine) grid, $h_f = h/M$. Note that the MsFEM captures the solution at the coarse grid $h$, whereas FEM tries to resolve the solution at the fine grid $h_f$. Our extensive numerical experiments demonstrate that the accuracy of the MsFEM on the coarse grid $h$ is comparable to that of FEM on the fine grid.

As an example, in Table 2.9.2 we present the results from [145] for

$$k(x/\epsilon) = \frac{2 + A\sin(2\pi x_1/\epsilon)}{2 + A\cos(2\pi x_2/\epsilon)} + \frac{2 + \sin(2\pi x_2/\epsilon)}{2 + A\sin(2\pi x_1/\epsilon)} \quad (A = 1.8), \quad (2.37)$$

$$f(x) = -1 \quad \text{and} \quad p|_{\partial\Omega} = 0. \quad (2.38)$$

The convergence of three different methods is compared for fixed $\epsilon/h = 0.64$, where "-L" indicates that a linear boundary condition is imposed on the multiscale basis functions, "os" indicates the use of oversampling, and LFEM stands for the standard FEM with bilinear basis functions.

We see clearly the scale resonance in the results of MsFEM-L and the (almost) first-order convergence (i.e., no resonance) in MsFEM-os-L. The error of MsFEM-os-L is smaller than that of LFEM obtained on the fine grid. In [147, 145], more extensive convergence tests have been presented.

There have been many numerical studies of MsFEM, in particular, in the context of multiphase flow simulations. Some of these results are presented and discussed in the book.

**Table 2.1.** Convergence for periodic case

| N | $\epsilon$ | MsFEM-L | | MsFEM-os-L | | LFEM | |
|---|---|---|---|---|---|---|---|
| | | $\|E\|_{l^2}$ | Rate | $\|E\|_{l^2}$ | Rate | $MN$ | $\|E\|_{l^2}$ |
| 16 | 0.04 | 3.54e-4 | | 7.78e-5 | | 256 | 1.34e-4 |
| 32 | 0.02 | 3.90e-4 | -0.14 | 3.83e-5 | 1.02 | 512 | 1.34e-4 |
| 64 | 0.01 | 4.04e-4 | -0.05 | 1.97e-5 | 0.96 | 1024 | 1.34e-4 |
| 128 | 0.005 | 4.10e-4 | -0.02 | 1.03e-5 | 0.94 | 2048 | 1.34e-4 |

### 2.9.3 Coarse-grid choice

We would like to remark that in MsFEM simulations, one is not restricted to rectangular or box-shaped coarse and fine grids. In fact, as demonstrated in a number of papers [11, 5], one can use an unstructured fine grid. Moreover, the coarse grid can have an arbitrary shape and the only requirement on the coarse grid is that every coarse grid consists of a connected union of fine-grid blocks. In Figure 2.10, we present an example from [11]. As one can observe from this figure the coarse-grid blocks have quite irregular shapes. In [9], the authors develop gridding techniques that use single-phase flow information (surrogate global information) to construct a coarse grid. The coarse grid is chosen such that it minimizes the magnitude of the single-phase velocity field variation within each coarse-grid block. This automatic coarse-grid generator allows one to use an optimal coarse grid for accurate simulation purposes in two-phase flow simulations. In Chapter 4, we discuss an extension of mixed MsFEM to unstructured coarse grids and include a few numerical examples to demonstrate its effectiveness. We would like to note that the fine grid blocks in neighboring coarse-grid blocks do not need to match along the interface.

In general, an appropriate choice of the coarse-grid will improve the efficiency and accuracy of multiscale methods. It is often possible that the solution may have smooth variation along coarse grid boundaries if the coarse grid is judiciously selected. This can lead to improved numerical results. Some of these issues are discussed in Chapter 4. For computational purposes, it is important that the coarse grid is more regular (for accuracy purposes) and the number of fine-grid blocks within a coarse grid is approximately the same (for load-balancing purposes).

## 2.10 An application to two-phase flow

MsFEMs and their modifications have been used with success in two-phase flow simulations through heterogeneous porous media. First, we briefly describe the underlying fine-scale equations. We present two-phase flow equations neglecting the effects of gravity, compressibility, capillary pressure, and dispersion on the fine scale. Porosity, defined as the volume fraction of the void

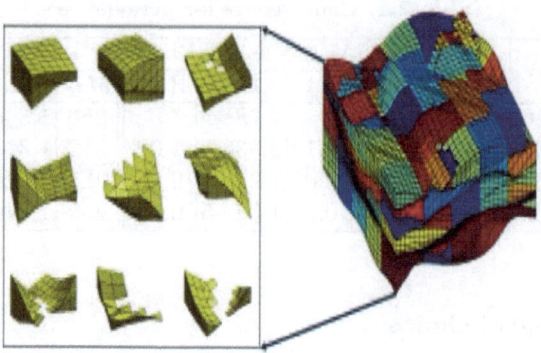

**Fig. 2.10.** Unstructured fine and coarse grids (from [11]).

space, is taken to be constant and therefore serves only to rescale time. The two phases are referred to as water and oil and designated by the subscripts $w$ and $o$, respectively. We can then write Darcy's law, with all quantities dimensionless, for each phase $j$ $(j = w, o)$ as follows;

$$v_j = -\lambda_j(S)k\nabla p, \tag{2.39}$$

where $v_j$ is phase velocity, $S$ is water saturation (volume fraction), $p$ is pressure, $\lambda_j = k_{rj}(S)/\mu_j$ is phase mobility, where $k_{rj}$ and $\mu_j$ are the relative permeability and viscosity of phase $j$, respectively, and $k$ is the permeability tensor.

Combining Darcy's law with conservation of mass, $\mathrm{div}(v_w + v_o) = 0$, allows us to write the flow equation in the following form

$$\mathrm{div}(\lambda(S)k\nabla p) = q_t, \tag{2.40}$$

where the total mobility $\lambda(S)$ is given by $\lambda(S) = \lambda_w(S) + \lambda_o(S)$ and $q_t$ is a source term representing wells/sources. The term $q_t = q_w + q_o$ represents the total volumetric source term. The saturation dynamics affects the flow equations. One can derive the equation describing the dynamics of the saturation

$$\frac{\partial S}{\partial t} + \mathrm{div}(v f_w(S)) = -q_w, \tag{2.41}$$

where $f_w(S)$ is the fractional flow of water, given by $f_w = \lambda_w/(\lambda_w + \lambda_o)$. The signs of the source terms that appear in (2.40) and (2.41) can be inter-changed. The total velocity $v$ is given by

$$v = v_w + v_o = -\lambda(S)k\nabla p. \tag{2.42}$$

In the presence of capillary effects, an additional degenerate diffusion term is present in (2.41).

If $k_{rw} = S$, $k_{ro} = 1 - S$, and $\mu_w = \mu_o$, then the flow equation reduces to

$$\text{div}(k\nabla p^{sp}) = q_t.$$

This equation is called single-phase flow equation.

For two-phase flow simulations, we first solve the coarse-scale pressure equation using MsFVEM. More precisely, assuming that the solution $S(x, t)$ is known at time $t = t_n$, we solve $\text{div}(\lambda(S(x, t_n))k\nabla p) = q_t$ using MsFVEM to compute $p(x, t_{n+1})$ on the coarse grid. The fine-scale velocity $v(x, t_{n+1})$ is then re-constructed by solving a local fine-scale problem over each dual cell with flux boundary conditions, as determined from the pressure solution. This velocity is then used in the explicit solution of the saturation equation using a first-order upwind method to compute $S(x, t_{n+1})$. The overall procedure is thus an IMPES (implicit pressure, explicit saturation) approach. We also consider an approach where the coarse-scale velocity is used to update the saturation field.

As we see from (2.40) and (2.41), the pressure equation is solved many times for different saturation profiles. Thus, computing the basis functions once at time zero is very beneficial and the basis functions are only updated near sharp interfaces. In fact, our numerical results show that only slight improvement can be achieved by updating the basis functions near sharp fronts.

We present a representative numerical example for a permeability field generated using two-point geostatistics. To generate this permeability field, we have used the GSLIB algorithm [85]. The permeability is log-normally distributed with prescribed variance $\sigma^2 = 1.5$ ($\sigma^2$ here refers to the variance of $\log k$) and some correlation structure. The correlation structure is specified in terms of dimensionless correlation lengths in the $x_1$- and $x_2$-directions, $l_1 = 0.4$, and $l_2 = 0.04$, nondimensionalized by the system length. Linear boundary conditions are used for constructing multiscale basis functions. A spherical variogram is used. In this numerical example, the fine-scale field is $120 \times 120$, and the coarse-scale field is $12 \times 12$ defined in the rectangle with the length 5 and the width 1. For the two-phase flow simulations, the system is considered to initially contain only oil ($S = 0$) and water is injected at inflow boundaries ($S = 1$ is prescribed); that is we specify $p = 1$, $S = 1$ along the $x = 0$ edge and $p = 0$ along the $x = 5$ edge, and no flow boundary conditions on the lateral boundaries. Relative permeability functions are specified as $k_{rw} = S^2$, $k_{ro} = (1 - S)^2$; water and oil viscosities are set to $\mu_w = 1$ and $\mu_o = 5$. Source terms $q_w$ and $q_t$ are zero. Results are presented in terms of the fraction of oil in the produced fluid, called fractional flow or oil-cut (designated $F$), against pore volume injected (PVI). Fractional flow is given by

$$F(t) = 1 - \frac{\int_{\partial\Omega^{\text{out}}}(v \cdot n)f(S)ds}{\int_{\partial\Omega^{\text{out}}} v \cdot n ds}, \tag{2.43}$$

where $\Omega^{\text{out}}$ refers to the part of the boundary with outer flow; that is $v \cdot n > 0$. PVI represents dimensionless time and is computed via

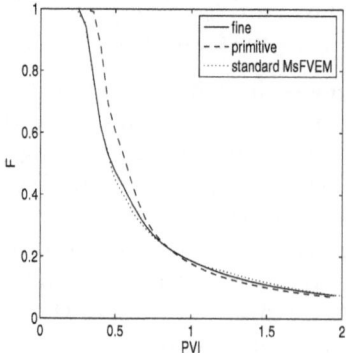

**Fig. 2.11.** Fractional flow comparison for a permeability field generated using two-point geostatistics.

$$\text{PVI} = \int Q \, dt / V_p, \tag{2.44}$$

where $V_p$ is the total pore volume of the system and $Q = \int_{\partial \Omega^{out}} v \cdot n \, ds$ is the total flow rate.

In Figure 2.11, we compare the fractional flows (oil-cut). The dashed line corresponds to the calculations performed using a simple saturation upscaling (no subgrid treatment) where (2.41) is solved with $v$ replaced by the coarse-scale $v$ obtained from MsFVEM. Note that the coarse-scale $v$ is defined as a normal flux, $\int_{\partial K} v \cdot n \, dl$ along the edge for each coarse-grid block. We call this the primitive model because it ignores the oscillations of $v$ within the coarse-grid block in the computation of $S$. The dotted line corresponds to the calculations performed by solving the saturation equation on the fine grid using the reconstructed fine-scale velocity field. The fine-scale details of the velocity are reconstructed using the multiscale basis functions. In these simulations, the errors are due to MsFVEM. In the primitive model, the errors are due to both MsFVEM for flow equations (2.40) and the upscaling in the saturation equation (2.41). We observe from this figure that the second approach, where the saturation equation is solved on the fine grid, is very accurate, but the first approach overpredicts the breakthrough time. We note that the second approach contains the errors only due to MsFVEM because the saturation equation is solved on the fine grid. The first approach contains in addition to MsFVEM's errors, the errors due to saturation upscaling which can be large if no subgrid treatment is performed. The saturation snapshots are compared in Figure 2.12. One can observe that there is a very good agreement between the fine-scale saturation and the saturation field obtained using MsFVEM.

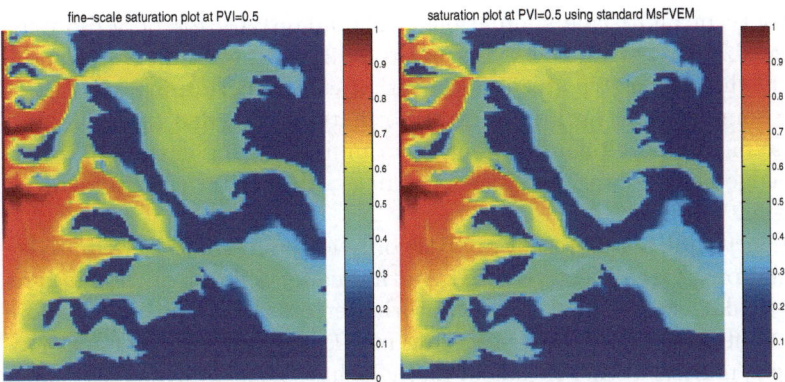

**Fig. 2.12.** Saturation maps at PVI = 0.5 for fine-scale solution (left figure) and standard MsFVEM (right figure).

## 2.11 Discussions

In this chapter, we presented an introduction to MsFEMs. We attempted to keep the presentation accessible to a broader audience and avoided some technical details in the presentation. One of the main ingredients of MsFEM is the construction of basis functions. Various approaches can be used to couple multiscale basis functions. This leads to multiscale methods, such as mixed MsFEM, MsFVEM, DG-MsFEM, and so on. Most of the discussions here focus on linear problems and local multiscale basis functions. We have discussed the effects of localized boundary conditions and the approaches to improve them. The relation to some other multiscale methods is discussed.

We would like to note that various multiscale methods are compared numerically in [167]. In particular, the authors in [167] perform comparisons of MsFVM, mixed MsFEM, and variational multiscale methods. Numerical results are performed for various uniform coarse grids and the sensitivities of these approaches with respect to coarse grids are discussed. As we mentioned earlier, one can use general, nonuniform, coarse grids to improve the accuracy of local multiscale methods.

We discussed approximations of basis functions in the presence of strong scale separation. In this case, the basis functions and the elements of the stiffness matrix can be approximated using the solutions in smaller regions (RVE). One can also approximate basis functions by solving the local problems approximately, for example, using approximate analytical solutions [250] for some types of heterogeneities. In [192], the authors propose an approach where the multiscale basis functions are computed inexpensively via multigrid iterations. They show that the obtained method gives nearly the same accuracy on the coarse grid as MsFEM with accurately resolved basis functions.

In this chapter, we did not discuss adaptivity issues that are important for multiscale simulations (see, e.g., [213, 38] for discussions on error estimates and adaptivity in multiscale simulations). The adaptivity for periodic numerical homogenization within HMM is studied in [214]. In general, one would like to identify the regions where the localization can be performed and the regions where some type of limited global information is needed (see Chapter 4 for the use of limited global information in multiscale simulations). To our best knowledge, such adaptivity issues are not addressed in the literature with a mathematical rigor.

In Section 6.1, we present analysis only for a few multiscale finite element formulations. Our objective is to give the reader a flavor of the analysis, and in particular, stress the subgrid capturing errors. We would like to note that a lot of effort has gone into analyzing multiscale finite element methods. For example, multiscale finite element methods have been analyzed for random homogeneous coefficients [99, 72], for highly oscillatory coefficients with multiple scales [99], for problems with discontinuous coefficients [99], and for various settings of MsFEMs. Our main objective in this book is to give an overview of multiscale finite element methods and present representative cases for the analysis. We believe the results presented in Section 6.1 will help the reader who is interested in the analysis of multiscale finite element methods and, in particular, in estimating subgrid capturing errors.

# 3

# Multiscale finite element methods for nonlinear equations

## 3.1 MsFEM for nonlinear problems. Introduction

The objective of this chapter is to present a generalization of the MsFEM to nonlinear problems ([110, 112, 113, 104]) which was first presented in [110]. This generalization, as the MsFEM for linear problems, has two main ingredients: a global formulation and multiscale localized "basis" functions. We discuss numerical implementation issues and applications.

Let $T_h$ be a coarse-scale partition of $\Omega$, as before. We denote by $W_h$ a usual finite-dimensional space, which possesses approximation properties, for example, piecewise linear functions over triangular elements, as defined before. In further presentation, $K$ is a coarse element that belongs to $T_h$. To formulate MsFEMs for general nonlinear problems, we need (1) a multiscale mapping that gives us the desired approximation containing the small-scale information and (2) a multiscale numerical formulation of the equation.

We consider the formulation and analysis of MsFEMs for general nonlinear elliptic equations

$$- \operatorname{div} k(x, p, \nabla p) + k_0(x, p, \nabla p) = f \text{ in } \Omega, \quad p = 0 \text{ on } \partial\Omega, \tag{3.1}$$

where $k(x, \eta, \xi)$ and $k_0(x, \eta, \xi)$, $\eta \in \mathbb{R}$, $\xi \in \mathbb{R}^d$ satisfy the general assumptions (6.42)–(6.46), which are formulated later. Note that here $k$ and $k_0$ are nonlinear functions of $p$ as well as $\nabla p$. Moreover, both $k$ and $k_0$ are heterogeneous spatial fields. Later, we extend the MsFEM to nonlinear parabolic equations where $k$ and $k_0$ are also heterogeneous functions with respect to the time variable.

*Multiscale mapping.* Unlike MsFEMs for linear problems, "basis" functions for nonlinear problems need to be defined via nonlinear maps that map coarse-scale functions into fine-scale functions. We introduce the mapping $E^{MsFEM} : W_h \to P_h$ in the following way. For each coarse-scale element $v_h \in W_h$, we denote by $v_{r,h}$ the corresponding fine-scale element ($r$ stands for resolved), $v_{r,h} = E^{MsFEM} v_h$. Note that for linear problems in Chapter

Y. Efendiev, T.Y. Hou, *Multiscale Finite Element Methods: Theory and Applications*, 47
Surveys and Tutorials in the Applied Mathematical Sciences 4,
DOI 10.1007/978-0-387-09496-0_3, © Springer Science+Business Media LLC 2009

2 (also in Chapter 4), we have used the subscript $h$ (e.g., $p_h$) to denote the approximation of the fine-scale solution, whereas for nonlinear problems $p_h$ stands for the approximation of the homogenized solution and $p_{r,h}$ is the approximation of the fine-scale solution. The spatial field $v_{r,h}$ is defined via the solution of the local problem

$$- \operatorname{div} k(x, \eta^{v_h}, \nabla v_{r,h}) = 0 \text{ in } K, \tag{3.2}$$

where $v_{r,h} = v_h$ on $\partial K$ and $\eta^{v_h} = (1/|K|) \int_K v_h dx$ for each $K$ (coarse element). The equation (3.2) is solved in each $K$ for given $v_h \in W_h$. Note that the choice of $\eta^{v_h}$ guarantees that (3.2) has a unique solution. In nonlinear problems, $\mathcal{P}_h$ is no longer a linear space (although we keep the same notation). We would like to point out that different boundary conditions can be chosen as in the case of linear problems to obtain more accurate solutions and this is discussed later. For linear problems, $E^{MsFEM}$ is a linear operator, where for each $v_h \in W_h$, $v_{r,h}$ is the solution of the linear problem. Consequently, $\mathcal{P}_h$ is a linear space that can be obtained by mapping a basis of $W_h$. This is precisely the construction presented in [143] for linear elliptic equations (see Section 3.3).

*An illustrating example.* To illustrate the multiscale mapping concept, we consider the equation

$$\operatorname{div}(k(x, p)\nabla p) = f. \tag{3.3}$$

In this case, the multiscale map is defined in the following way. For each $v_h \in W_h$, $v_{r,h}$ is the solution of

$$div(k(x, \eta^{v_h})\nabla v_{r,h}) = 0 \text{ in } K \tag{3.4}$$

with the boundary condition $v_{r,h} = v_h$ on $\partial K$. For example, if $K$ is a triangular element and $v_h$ are piecewise linear functions, then the nodal values of $v_h$ will determine $v_{r,h}$. Equation (3.4) is solved on the fine grid, in general. In the one-dimensional case, one can obtain an explicit expression for $E^{MsFEM}$ (see (3.12)). The map $E^{MsFEM}$ is nonlinear; however, for a fixed $v_h$, this map is linear. In fact, one can represent $v_{r,h}$ using multiscale basis functions as $v_{r,h} = \sum_i \alpha_i \phi_i^{v_h}$, where $\alpha_i = v_h(x_i)$ ($x_i$ being nodal points) and $\phi_i^{v_h}$ are multiscale basis functions defined by

$$\operatorname{div}(k(x, \eta^{v_h})\nabla \phi_i^{v_h}) = 0 \text{ in } K, \quad \phi_i^{v_h} = \phi_i^0 \text{ on } \partial K.$$

Consequently, linear multiscale basis functions can be used to represent $v_{r,h}$. We can further assume that the basis functions can be interpolated via a simple linear interpolation

$$\phi_i^{\eta_0} \approx \beta_1 \phi_i^{\eta_1} + \beta_2 \phi_i^{\eta_2}, \tag{3.5}$$

where $\beta_1$, $\beta_2$ are interpolation constants that depend on $\eta_0$, $\eta_1$, and $\eta_2$. For example, if $\eta_1 < \eta_0 < \eta_2$, then $\beta_1 = (\eta_0 - \eta_1)/(\eta_2 - \eta_1)$ and $\beta_2 = (\eta_2 - \eta_0)/(\eta_2 -$

$\eta_1$). In this case, one can compute the basis functions for some pre-defined values of $\eta$s and interpolate for any other $\eta$. We can also use the combined set of basis functions $\{\phi_i^{\eta_1}, \phi_i^{\eta_2}\}$ for representing the solution for a set of values of $\eta$.

*Multiscale numerical formulation.* As discussed earlier, one can use various global formulations for MsFEM. Our goal is to find $p_h \in W_h$ ( consequently, $p_{r,h}(= E^{MsFEM}p_h) \in \mathcal{P}_h$) such that $p_{r,h}$ "approximately" satisfies the fine-scale equations. When substituting $p_{r,h}$ into the fine-scale system, the resulting equations need to be projected onto coarse-dimensional space because $p_{r,h}$ is defined via $p_h$. This projection is done by multiplying the fine-scale equation with coarse-scale test functions. First, we present a Petrov–Galerkin formulation of MsFEM for nonlinear problems. The multiscale finite element formulation of the problem is the following. Find $p_h \in W_h$ (consequently, $p_{r,h}(= E^{MsFEM}p_h) \in \mathcal{P}_h$) such that

$$\langle \kappa_{r,h} p_h, v_h \rangle = \int_\Omega f v_h dx, \quad \forall v_h \in W_h, \tag{3.6}$$

where

$$\langle \kappa_{r,h} p_h, v_h \rangle = \sum_{K \in \mathcal{T}_h} \int_K (k(x, \eta^{p_h}, \nabla p_{r,h}) \cdot \nabla v_h + k_0(x, \eta^{p_h}, \nabla p_{r,h}) v_h) dx. \tag{3.7}$$

As we notice that the fine-scale equation is multiplied by coarse-scale test functions from $W_h$. Note that the above formulation of MsFEM is a generalization of the Petrov–Galerkin MsFEM introduced earlier for linear problems.

We note that the method presented above can be extended to systems of nonlinear equations.

*Pseudo-code.* In the computations, we seek $p_h = \sum_i p_i \phi_i^0 \in W_h$ which satisfies (3.6). This equation can be written as a nonlinear system of equations for $p_i$,

$$A(p_1, ..., p_i, ...) = b, \tag{3.8}$$

where $A$ is given by (3.7). Here, $p_i$ can be thought as nodal values of $p_h$ on the coarse grid. To find the form of $A$, we take $v_h = \phi_i^0$ in (3.7). This yields the $i$th equation of the system (3.8) denoted by $A_i(p_1, ..., p_i, ...) = b_i$, where $b_i = \int_\Omega f \phi_i^0 dx$. Denote by $K_i$ triangles with the common vertex $x_i$. Then,

$$A_i(p_1, ..., p_i, ...) = \sum_{K_i} \int_{K_i} (k(x, \eta^{p_h}, \nabla p_{r,h}) \cdot \nabla \phi_i^0 + k_0(x, \eta^{p_h}, \nabla p_{r,h}) \phi_i^0) dx.$$

In each $K_i$, $\eta^{p_h} = \sum_j p_j \int_{K_i} \phi_j^0 dx$, where $j$ are the nodes of the triangles with common vertex $i$. Later, we present a one-dimensional example, where an explicit expression for $A_i$ is presented. It is clear that $A_i$ will depend only on the nodal values of $p_j$ which are defined at the nodes of the triangles with common vertex $i$. This system is usually solved by an iterative method on a

---

**Algorithm 3.1.1**

---

Construct a coarse grid.
Until convergence, do
- For each coarse element, compute the multiscale map $E : W_h \rightarrow \mathcal{P}_h$ according to (3.2).
- Solve the coarse variational formulation using (3.6) and (3.7).

---

coarse grid and the local solutions can be re-used and treated independently in each coarse-grid block. In Section 3.7, we discuss some of the iterative methods.

*One dimensional example.* We consider a simple one-dimensional case

$$-(k(x,p)p')' = f,$$

$p(0) = p(1) = 0$, where $'$ refers to the spatial derivative. We assume that the interval $[0, 1]$ is divided into $N$ segments

$$0 = x_0 < x_1 < x_2 < \cdots < x_i < x_{i+1} < \cdots < x_N = 1.$$

For a given $p_h \in W_h$, $p_{r,h}$ is the solution of

$$(k(x, \eta^{p_h})p'_{r,h})' = 0, \tag{3.9}$$

where $p_{r,h}(x_i) = p_h(x_i)$ for every interior node $x_i$. In the interval $[x_{i-1}, x_i]$, (3.9) can be solved. To compute (3.7), we only need to evaluate $k(x, \eta^{p_h})p'_{r,h}$. Noting that this quantity is constant, $k(x, \eta^{p_h})p'_{r,h} = c(x_{i-1}, x_i)$ (directly follows from (3.9)), we can easily find that

$$p'_{r,h} = c(x_{i-1}, x_i)/k(x, \eta^{p_h}), \tag{3.10}$$

where $\eta^{p_h} = \frac{1}{2}(p_h(x_{i-1}) + p_h(x_i))$. Taking the integral of (3.10) over $[x_{i-1}, x_i]$, we have

$$p_h(x_i) - p_h(x_{i-1}) = c(x_{i-1}, x_i) \int_{x_{i-1}}^{x_i} \frac{1}{k(x, \eta^{p_h})} dx.$$

Consequently,

$$c(x_{i-1}, x_i) = k(x, \eta^{p_h})p'_{r,h} = \frac{p_h(x_i) - p_h(x_{i-1})}{\int_{x_{i-1}}^{x_i} \frac{1}{k(x, \eta^{p_h})} dx}.$$

To evaluate (3.7) (note that $k_0 = 0$) with $v_h = \phi_i^0$, we have

$$A_i(p_h) = \int_{x_{i-1}}^{x_i} c(x_{i-1}, x_i)(\phi_i^0)' dx + \int_{x_i}^{x_{i+1}} c(x_i, x_{i+1})(\phi_i^0)' dx$$

$$= \frac{p_h(x_i) - p_h(x_{i-1})}{\int_{x_{i-1}}^{x_i} \frac{1}{k(x, \eta^{p_h})} dx} \int_{x_{i-1}}^{x_i} (\phi_i^0)' dx + \frac{p_h(x_{i+1}) - p_h(x_i)}{\int_{x_i}^{x_{i+1}} \frac{1}{k(x, \eta^{p_h})} dx} \int_{x_i}^{x_{i+1}} (\phi_i^0)' dx.$$

Denoting $p_i = p_h(x_i)$ and taking into account that $\int_{x_{i-1}}^{x_i} (\phi_i^0)' dx = 1$, $\int_{x_i}^{x_{i+1}} (\phi_i^0)' dx = -1$, we have

$$A_i(p_{i-1}, p_i, p_{i+1}) = \frac{p_i - p_{i-1}}{\int_{x_{i-1}}^{x_i} \frac{1}{k(x, \frac{1}{2}(p_{i-1}+p_i))} dx} - \frac{p_{i+1} - p_i}{\int_{x_i}^{x_{i+1}} \frac{1}{k(x, \frac{1}{2}(p_i+p_{i+1}))} dx}. \quad (3.11)$$

Using the above calculations, one can easily write down an explicit expression for the multiscale map, $E^{MsFEM} : p_h \to p_{r,h}$. In particular, from (3.10), it can be shown that $p_{r,h}$ in $[x_{i-1}, x_i]$ is given by

$$p_{r,h}(x) = p_h(x_{i-1}) + \frac{p_h(x_i) - p_h(x_{i-1})}{\int_{x_{i-1}}^{x_i} \frac{1}{k(x, \eta^{p_h})} dx} \int_{x_{i-1}}^{x} \frac{dx}{k(x, \eta^{p_h})}. \quad (3.12)$$

One can use explicit solutions (see page 117, [220]) in a general case

$$-(k(x, p, p')') = f$$

to write down the variational formulation of (3.7) via the nodal values $p_i$. In particular, denote $\xi = \xi(x, \eta, c)$ to be the solution of

$$k(x, \eta, \xi) = c.$$

Then, from (3.2), we obtain $p'_{r,h} = \xi(x, \eta^{p_h}, c(x_{i-1}, x_i))$. Taking the integral of this equation over $[x_{i-1}, x_i]$, we obtain

$$p_h(x_i) - p_h(x_{i-1}) = \int_{x_{i-1}}^{x_i} \xi(x, \eta^{p_h}, c(x_{i-1}, x_i)) dx.$$

Because $\eta^{p_h} = \frac{1}{2}(p_{i-1} + p_i)$, we have the following implicit equation for $c(x_{i-1}, x_i)$

$$p_i - p_{i-1} = \int_{x_{i-1}}^{x_i} \xi(x, \frac{1}{2}(p_{i-1} + p_i), c(x_{i-1}, x_i)) dx.$$

With this implicit expression for $c(x_{i-1}, x_i)$, we have

$$A_i(p_h) = \int_{x_{i-1}}^{x_i} c(x_{i-1}, x_i)(\phi_i^0)' dx + \int_{x_i}^{x_{i+1}} c(x_i, x_{i+1})(\phi_i^0)' dx$$
$$= c(x_{i-1}, x_i) - c(x_i, x_{i+1}).$$

This expression shows that $A_i(p_h)$ nonlinearly depends on $p_{i-1}$, $p_i$, and $p_{i+1}$ and provides an explicit expression for the system of nonlinear equations that result from (3.7).

## 3.2 Multiscale finite volume element method (MsFVEM)

Next, we present a different formulation that provides a mass conservative method. By its construction, the finite volume method has local conservative properties [118] and it is derived from a local relation, namely the balance equation/conservation expression on a number of subdomains which are called control volumes. The finite volume element method can be considered as a Petrov–Galerkin finite element method, where the test functions are constants defined in a dual grid. For simplicity, we consider a triangular coarse grid. Consider a triangle $K$, and let $z_K$ be its barycenter. The triangle $K$ is divided into three quadrilaterals of equal area by connecting $z_K$ to the midpoints of its three edges. We denote these quadrilaterals by $K_z$, where $z \in Z_h(K)$ are the vertices of $K$. Also we denote $Z_h = \bigcup_K Z_h(K)$, and $Z_h^0$ are all vertices that do not lie on $\partial\Omega_D$, where $\partial\Omega_D$ are Dirichlet boundaries. In this case, the control volume $V_z$ is defined as the union of the quadrilaterals $K_z$ sharing the vertex $z$ (see Figure 3.1). The MsFVEM is to find $p_h \in W_h$ (consequently,

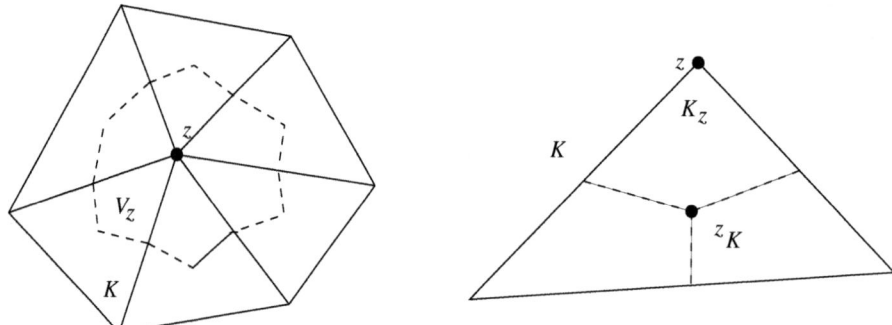

**Fig. 3.1.** *Left*: Portion of triangulation sharing a common vertex $z$ and its control volume. *Right*: Partition of a triangle K into three quadrilaterals.

$p_{r,h} = E^{MsFEM} p_h$) such that

$$-\int_{\partial V_z} k\left(x, \eta^{p_h}, \nabla p_{r,h}\right) \cdot n \, ds + \int_{V_z} k_0\left(x, \eta^{p_h}, \nabla p_{r,h}\right) dx = \int_{V_z} f \, dx \quad \forall z \in Z_h^0,$$

(3.13)

where $n$ is the unit normal vector pointing outward on $\partial V_z$. Note that the number of control volumes which satisfies (3.13) is the same as the dimension of $W_h$. The equation (3.13) gives rise to the finite-dimensional system of equations that provide the solution at the coarse nodes.

## 3.3 Examples of $\mathcal{P}_h$

*Linear case.* For linear operators, $\mathcal{P}_h$ can be obtained by mapping a basis of $W_h$ because $E^{MsFEM}$ is a linear operator. Define a basis of $W_h$, $W_h = \text{span}(\phi_i^0)$, where $\phi_i^0$ are standard linear basis functions (assuming $K$ is a triangular or tetrahedral element). Denote by $\phi_i$ the map of each basis function $\phi_i^0$ (i.e., $\phi_i = E^{MsFEM}\phi_i^0$). From the definition of $E^{MsFEM}$ it follows that $\phi_i$ satisfies

$$- \text{div}(k(x)\nabla\phi_i) = 0 \text{ in } K \in \mathcal{T}_h \tag{3.14}$$

and $\phi_i = \phi_i^0$ on $\partial K$. These are the basis functions defined for MsFEM in Chapter 2.

*Special nonlinear case.* For the special case, $k(x, p, \nabla p) = k(x)b(p)\nabla p$, $\mathcal{P}_h$ can be related to the linear case. Indeed, for this case, the local problems associated with the multiscale mapping $E^{MsFEM}$ (see (3.2)) have the form

$$-\text{div}(k(x)b(\eta^{v_h})\nabla v_{r,h}) = 0 \text{ in } K.$$

Because $\eta^{v_h}$ are constants over $K$, the local problems satisfy the linear equations,

$$- \text{div}(k(x)\nabla\phi_i) = 0 \text{ in } K,$$

and $\mathcal{P}_h$ can be obtained by mapping a basis of $\mathcal{T}_h$ as it is done in the linear case. Thus, for this case, the basis functions are the same as those for the linear problem.

*$\mathcal{P}_h$ using subdomain problems.* One can use the solutions of smaller subdomain (smaller than $K \in \mathcal{T}_h$), RVE, problems to approximate the solutions of the local problems (3.2). This can be done if the small region can be used to represent the heterogeneities within the coarse-grid block, for example, periodic heterogeneities when the size of the period is much smaller than the coarse-grid block size. As in the linear case, we would like to use the solution in smaller regions to approximate the integrals on the right-hand side of (3.7). In these cases, we can solve (3.2) in a subdomain RVE with boundary conditions $v_h$ restricted onto the subdomain boundaries as done in Section 2.6. More precisely, instead of (3.2), the following local problem is solved,

$$- \text{div}\, k(x, \eta^{v_h}, \nabla\tilde{v}_{r,h}) = 0 \text{ in } K_{\text{loc}}, \tag{3.15}$$

where $\tilde{v}_{r,h} = v_h$ on $\partial K_{\text{loc}}$ and $\eta^{v_h} = \frac{1}{|K|}\int_K v_h dx$ for each $K$ (coarse element). The integrals in (3.7) can be computed using $K_{\text{loc}}$,

$$\langle\kappa_{r,h}p_h, v_h\rangle \approx \sum_{K \in \mathcal{T}_h} \frac{|K|}{|K_{\text{loc}}|} \int_{K_{\text{loc}}} (k(x, \eta^{p_h}, \nabla\tilde{p}_{r,h})\cdot\nabla v_h + k_0(x, \eta^{p_h}, \nabla\tilde{p}_{r,h})v_h)dx,$$

$$\tag{3.16}$$

where $\tilde{p}_{r,h}$ are only computed in $K_{\text{loc}}$ using (3.15). The equations (3.15) and (3.16) provide the formulation of MsFEM when using regions smaller than the coarse-grid block.

As in the case of linear problems, it was shown that ([220])

$$\lim_{\epsilon \to 0} \frac{1}{|K|} \int_K k(x, \eta^{v_h}, \nabla v_{r,h}) dx = \frac{1}{|K|} \int_K k^*(x, \eta^{v_h}, \nabla v_{r,h}^0) dx, \qquad (3.17)$$

where $\epsilon$ is the characteristic length scale and $v_{r,h}^0$ is the homogenized part of $v_{r,h}$ defined in a $G$-convergence setting (e.g., [220]). In particular, $v_{r,h}^0$ satisfies $\operatorname{div} k^*(x, \eta^{v_h}, v_{r,h}^0) = 0$ in $K$, $v_{r,h}^0 = v_h$ on $\partial K$; a similar result holds in $K_{\mathrm{loc}}$ (cf. (2.21)). As in the linear case, it is easy to show that if $k^*(x, \eta, \xi)$ is smooth spatial function, then $v_{r,h}^0$ is approximately equal to $v_h$ for small $h$. From here, one can show that $(1/|K_{\mathrm{loc}}|) \int_{K_{\mathrm{loc}}} k(x, \eta^{v_h}, \nabla \tilde{v}_{r,h}) dx$ approximates $(1/|K|) \int_K k(x, \eta^{v_h}, \nabla v_{r,h}) dx$ in the limit $\lim_{h \to 0} \lim_{\epsilon \to 0}$. Based on (3.17), one can evaluate the integrals on the right-hand side of (3.7). To find the fine-scale approximation, the local solutions defined by (3.15) can be extended to the whole domain. This is based on the homogenization concept and $\nabla \tilde{v}_{r,h}$ is extended periodically in each coarse-grid block.

One can also use periodic homogenization and first-order correctors to approximate the solution of the local problem if $k(x, x/\epsilon, \eta, \xi)$ and $k_0(x, x/\epsilon, \eta, \xi)$ are locally periodic with respect to $y = x/\epsilon$. In this case, for each coarse grid block and $v_h \in W_h$, the following cell problem is solved,

$$\operatorname{div}_y(k(x, y, \eta^{v_h}, \nabla_x v_h + \nabla_y N_{v_h})) = 0 \ \ \text{in} \ Y, \qquad (3.18)$$

where $Y$ is the period and $N_{v_h}$ is the periodic function with zero average (assume $v_h$ is piecewise linear; i.e., $\nabla v_h$ is constant within $K$). Then, it can be shown that

$$\frac{1}{|K|} \int_K k(x, x/\epsilon, \eta^{v_h}, \nabla v_{r,h}) dx \approx \frac{1}{|Y|} \int_Y k(x, y, \eta^{v_h}, \nabla v_h + \nabla N_{v_h}) dx$$

in the limit as $\epsilon/h \to 0$. Consequently, the local periodic solution (3.18) can be used to approximate the right-hand side of (3.7). This provides CPU savings when there is strong scale separation (cf. Section 2.6).

## 3.4 Relation to upscaling methods

One can draw a parallel between multiscale methods and upscaling/homogenization techniques. First, we briefly describe an upscaling technique for (3.1) which is derived from homogenization methods (e.g., [220]). The main idea of upscaling techniques is to form a coarse-scale equation and pre-compute the effective coefficients. In the case of nonlinear elliptic equations, the coarse-scale equation has the same form as the fine-scale equation except that the fluxes $k(\cdot, \cdot, \cdot)$ and $k_0(\cdot, \cdot, \cdot)$ are replaced by effective homogenized fluxes. The effective coefficients in upscaling methods are computed using the solution of the local problem in a representative volume. For each $\eta \in \mathbb{R}$ and $e \in \mathbb{R}^d$, the following local problem is solved

$$\operatorname{div} k(x, \eta, \nabla \phi_e) = 0 \text{ in } K \tag{3.19}$$

with $\phi_e(x) = x \cdot e$ on $\partial K$. The effective coefficients are computed in each $K$ as

$$\tilde{k}^*(\eta, e) = \frac{1}{|K|} \int_K k(x, \eta, \nabla \phi_e) dx, \quad \tilde{k}_0^*(\eta, e) = \frac{1}{|K|} \int_K k_0(x, \eta, \nabla \phi_e) dx. \tag{3.20}$$

We note that $\tilde{k}^*$ and $\tilde{k}_0^*$ are not the same as the homogenized coefficients and (3.20) can be computed for any point in the domain by placing the point at the center of $K$. If $k$ and $k_0$ are periodic with respect to spatial variables, one can solve the local problems (3.19) over the period (with periodic boundary conditions) and perform averaging (3.20) over the period. One can also use various boundary conditions, including oversampling methods, when solving (3.19). Once the effective coefficients are calculated, the coarse-scale equation

$$\operatorname{div} k^*(x, p^*, \nabla p^*) + k_0^*(x, p^*, \nabla p^*) = f$$

with $k^* = \tilde{k}^*$ and $k_0^* = \tilde{k}_0^*$ is solved. In practice, one can pre-compute $k^*$ and $k_0^*$ for different values of $\eta \in \mathbb{R}$ and $e \in \mathbb{R}^d$, and use interpolation for evaluating $k^*$ and $k_0^*$ for other values of $\eta \in \mathbb{R}$ and $e \in \mathbb{R}^d$. Note that for linear problems, it is sufficient to solve (3.19) for $d$ linearly independent vectors $e_1, ..., e_d$ in $\mathbb{R}^d$ because $\phi_e = \sum_i \beta_i \phi_{e_i}$ if $e = \sum_i \beta_i e_i$. This is not the case for nonlinear problems and one needs to consider all possible $\eta \in \mathbb{R}$ and $e \in \mathbb{R}^d$.

MsFEMs do not compute effective parameters explicitly. One can show that, as in the case of linear problems, MsFEM for nonlinear problems is similar to upscaling methods. However, in MsFEMs, "the effective parameters" (in the form of local solutions) are computed on-the-fly. Note that one can compute the effective parameters based on these local solutions. The computation of the local solutions on-the-fly is more efficient when one deals with a limited range of values of $\eta = (1/|K|) \int_K v_h dx$ and $e = \nabla v_h$. Indeed, many simulations in practice do not require a lookup table of $k^*$ and $k_0^*$ for all possible values of $\eta \in \mathbb{R}$ and $e \in \mathbb{R}^d$, and the computation on-the-fly can save CPU time. Moreover, one can still use pre-computed local solutions to compute the effective coefficients, and then store them. These effective coefficients can be used in the simulation to approximate $k^*$ and $k_0^*$ for those values of $\eta \in \mathbb{R}$ and $e \in \mathbb{R}^d$ that are not computed. Moreover, in MsFEMs, one can use a larger set of multiscale basis functions for more accurate approximation. For example, for the simple nonlinear elliptic equation (3.3), one can use multiscale basis functions corresponding to several values of $\eta$ and avoid the interpolation step (cf. (3.5)).

## 3.5 Multiscale finite element methods for nonlinear parabolic equations

In this section, we present an extension of MsFEM to nonlinear parabolic equations. We consider

$$\frac{\partial}{\partial t} p - \operatorname{div} k(x, t, p, \nabla p) + k_0(x, t, p, \nabla p) = f. \tag{3.21}$$

For the nonlinear parabolic equations, the space–time operator $E^{MsFEM}$ is constructed in the following way. For each $v_h \in W_h$ there is a corresponding element $v_{r,h} = E^{MsFEM} v_h$ that is defined by

$$\frac{\partial}{\partial t} v_{r,h} - \operatorname{div} k(x, t, \eta^{v_h}, \nabla v_{r,h}) = 0 \text{ in } K \times [t_n, t_{n+1}], \tag{3.22}$$

with boundary condition $v_{r,h} = v_h$ on $\partial K$, and $v_{r,h}(t = t_n) = v_h$. Here $\eta^{v_h} = (1/|K|) \int_K v_h dx$.

Next, we present a global formulation of MsFEM. Our goal is to find $p_h \in W_h$ ($p_{r,h} = E^{MsFEM} p_h$) at time $t = t_{n+1}$ such that

$$\int_{t_n}^{t_{n+1}} \int_\Omega \left( \frac{\partial}{\partial t} p_h \right) v_h dx dt + \kappa(p_h, v_h) = \int_{t_n}^{t_{n+1}} \int_\Omega f v_h dx dt, \quad \forall v_h \in W_h, \tag{3.23}$$

where

$$\kappa(p_h, w_h) = \sum_K \int_{t_n}^{t_{n+1}} \int_K (k(x, t, \eta^{p_h}, \nabla p_{r,h}) \cdot \nabla w_h + k_0(x, t, \eta^{p_h}, \nabla p_{r,h}) w_h) dx dt.$$

The expression (3.23) can be further simplified to

$$\int_\Omega p_h(x, t_{n+1}) v_h dx - \int_\Omega p_h(x, t_n) v_h dx + \kappa(p_h, v_h) = \int_{t_n}^{t_{n+1}} \int_\Omega f v_h dx dt, \forall v_h \in W_h.$$

Here $p_{r,h}$ is the solution of the local problem (3.22) for a given $p_h$, $\eta^{p_h} = (1/|K|) \int_K p_h dx$, and $p_h$ is known at $t = t_n$. If $p_h$ at time $t = t_{n+1}$ is used in $\kappa(p_h, v_h)$, then the resulting method is implicit; that is

$$\int_\Omega p_h(x, t_{n+1}) v_h dx - \int_\Omega p_h(x, t_n) v_h dx + \kappa(p_h(x, t_{n+1}), v_h)$$

$$= \int_{t_n}^{t_{n+1}} \int_\Omega f v_h dx dt, \quad \forall v_h \in W_h.$$

If $p_h$ at time $t = t_n$ is used in $\kappa(p_h, v_h)$, then the resulting method is explicit. The Petrov–Galerkin formulation of the MsFEM can be replaced by the finite volume formulation as is done for nonlinear elliptic equations.

We would like to note that the operator $E^{MsFEM}$ can be constructed using larger domains as is done in MsFEMs with oversampling [145]. This way one reduces the effects of the boundary conditions and initial conditions. In particular, for temporal oversampling it is sufficient to start the computations before $t_n$ and end them at $t_{n+1}$. Consequently, the oversampling domain for $K \times [t_n, t_{n+1}]$ consists of $[\tilde{t}_n, t_{n+1}] \times K_E$, where $\tilde{t}_n < t_n$ and $K \subset K_E$. We would like to note that oscillatory initial conditions can be imposed (without using oversampling techniques) based on the solution of the elliptic part of the local problems (3.22). These initial conditions at $t = t_n$ are the solutions of

$$- \operatorname{div}(k(x, t, \eta, \nabla p_{r,h})) = 0 \text{ in } K, \qquad (3.24)$$

or

$$- \operatorname{div}(\overline{k}(x, \eta, \nabla p_{r,h})) = 0 \text{ in } K, \qquad (3.25)$$

where $\overline{k}(x, \eta, \xi) = (1/(t_{n+1} - t_n)) \int_{t_n}^{t_{n+1}} k(x, \tau, \eta, \xi) d\tau$ and $p_{r,h} = p_h$ on $\partial K$. The latter can become efficient depending on the interplay between the temporal and spatial scales.

Note that in the case of periodic media the local problems can be solved in a single period in order to construct $\kappa(p_h, w_h)$. In general, one can solve the local problems in a domain different from $K$ (an element) to calculate $\kappa(p_h, w_h)$. Note that the numerical advantages of our approach over the fine scale simulation are similar to those of MsFEMs. In particular, for each Newton's iteration a linear system of equations on a coarse grid is solved. Moreover, the local solutions can be re-used and treated independently in each coarse grid block.

For some special cases the operator $E^{MsFEM}$ introduced in the previous section can be simplified (see [112]). In general one can avoid solving the local parabolic problems if the ratio between temporal and spatial scales is known, and solve instead a simplified equation. For example, let the spatial scale be $\epsilon^\beta$ and the temporal scale be $\epsilon^\alpha$; that is, $k(x, t, \eta, \xi) = k(x/\epsilon^\beta, t/\epsilon^\alpha, \eta, \xi)$. If $\alpha < 2\beta$ one can solve instead of (3.22) the local problem $-\operatorname{div}(k(x, t, \eta^{p_h}, \nabla p_{r,h})) = 0$, if $\alpha > 2\beta$ one can solve instead of (3.22) the local problem $-\operatorname{div}(\overline{k}(x, \eta^{p_h}, \nabla p_{r,h})) = 0$, where $\overline{k}(x, \eta, \xi)$ is an average over time of $k(x, t, \eta, \xi)$, and if $\alpha = 2\beta$ we need to solve the parabolic equation in $K \times [t_n, t_{n+1}]$, (3.22).

We would like to note that, in general, one can use (3.24) or (3.25) as oscillatory initial conditions and these initial conditions can be efficient for some cases. For example, for $\alpha > 2\beta$ with initial conditions given by (3.25) the solutions of the local problems (3.22) can be computed easily because they are approximated by (3.25). Moreover, one can expect better accuracy with (3.25) for the case $\alpha > 2\beta$ because this initial condition is more compatible with the local heterogeneities compared to the artificial linear initial conditions (cf. (3.22)).

*One-dimensional example.* We consider a simple one-dimensional case

$$\frac{\partial p}{\partial t} - (k(x, t, p)p')' = f,$$

$p(0) = p(1) = 0$, $p(t = 0) = p_0(x)$. As before, we assume that the interval $[0, 1]$ is divided into $N$ segments $0 = x_0 < x_1 < x_2 < \cdots < x_i < x_{i+1} < \cdots < x_N = 1$ and the time interval $[0, T]$ is divided into $M$ segments $0 = t_0 < t_1 < t_2 < \cdots < t_i < t_{i+1} < \cdots < t_M = T$. We present a discrete formulation for the fully implicit method where the local problem (3.22) is elliptic (see discussion above); that is $(k(x, t, \eta^{v_h}, v'_{r,h}))' = 0$.

Denote by $p_i^n = p(x = x_i, t = t_n)$. Taking $v_h = \phi_i^0$ in (3.23) and using (3.11), one can easily get

$$(p_j^{n+1} - p_j^n) \int_\Omega \phi_j^0 \phi_i^0 dx + A(p_{i-1}^{n+1}, p_i^{n+1})(p_i^{n+1} - p_{i-1}^{n+1})$$

$$-A(p_i^{n+1}, p_{i+1}^{n+1})(p_{i+1}^{n+1} - p_i^{n+1}) = \int_\Omega f \phi_i^0 dx,$$

where

$$A(p_{i-1}^{n+1}, p_i^{n+1}) = \frac{1}{\int_{x_{i-1}}^{x_i} \frac{1}{k(x, t_{n+1}, \frac{1}{2}(p_{i-1}^{n+1} + p_i^{n+1}))} dx},$$

$$A(p_i^{n+1}, p_{i+1}^{n+1}) = \frac{1}{\int_{x_i}^{x_{i+1}} \frac{1}{k(x, t_{n+1}, \frac{1}{2}(p_i^{n+1} + p_{i+1}^{n+1}))} dx}.$$

## 3.6 Summary of convergence of MsFEM for nonlinear partial differential equations

The convergence of MsFEM for nonlinear problems has been studied for problems with scale separation (not necessarily periodic). These convergence results use homogenization or $G$-convergence results for nonlinear partial differential equations (see, e.g., [220] and Appendix B). To discuss these results, we assume that the fine scale is $\epsilon$. It can be shown that the solution $p$ converges (up to a subsequence) to $p_0$ in an appropriate norm, where $p_0 \in W_0^{1,\gamma}(\Omega)$ is a solution of a homogenized equation

$$- \operatorname{div} k^*(x, p_0, \nabla p_0) + k_0^*(x, p_0, \nabla p_0) = f, \qquad (3.26)$$

where $k^*$ and $k_0^*$ are homogenized coefficients.

In [112] it was shown using $G$-convergence theory that

$$\lim_{h \to 0} \lim_{\epsilon \to 0} \|p_h - p_0\|_{W_0^{1,\gamma}(\Omega)} = 0, \qquad (3.27)$$

(up to a subsequence) where $p_0$ is a solution of (3.26) and $p_h$ is a MsFEM solution given by (3.6). Here $\gamma$ is a parameter related to the monotonicity (see (6.42)–(6.46)). This result can be obtained without any assumption on the nature of the heterogeneities and cannot be improved because there could be infinitely many scales $\alpha(\epsilon)$ present such that $\alpha(\epsilon) \to 0$ as $\epsilon \to 0$.

For the periodic case, it can be shown that MsFEM converges in the limit as $\epsilon/h \to 0$. To show the convergence for $\epsilon/h \to 0$, we consider $h = h(\epsilon)$, such that $h(\epsilon) \gg \epsilon$ and $h(\epsilon) \to 0$ as $\epsilon \to 0$. We would like to note that this limit as well as the proof of the periodic case is different from (3.27), where the double-limit is taken. In contrast to the proof of (3.27), the proof of the periodic case requires the correctors for the solutions of the local problems.

We present the convergence results for MsFEM solutions. For general non-linear elliptic equations under the assumptions (stated later) (6.42)–(6.46) the strong convergence of MsFEM solutions can be shown. In the proof of this fact we show the form of the truncation error (in a weak sense) in terms of the resonance errors between the mesh size and small scale $\epsilon$ and explicitly derive the resonance errors. Under the general conditions, such as (6.42)–(6.46), one can prove strong convergence of MsFEM solutions without an explicit convergence rate (cf. [245]). To convert the obtained convergence rates for the truncation errors into the convergence rate of MsFEM solutions, additional assumptions, such as monotonicity, are needed.

Next, we formulate convergence theorems (see Section 6.2 and [104] for details).

**Theorem 3.1.** *Assume $k(x, \eta, \xi)$ and $k_0(x, \eta, \xi)$ are $\epsilon$ periodic functions with respect to $x$ and let $p_0$ be a solution of (3.26) and $p_h$ is a MsFEM solution given by (3.6). Moreover, we assume that $\nabla p_h$ is uniformly bounded in $L^{\gamma+\alpha}(\Omega)$ for some $\alpha > 0$. Then*

$$\lim_{\epsilon \to 0} \|p_h - p_0\|_{W_0^{1,\gamma}(\Omega)} = 0, \tag{3.28}$$

*where $h = h(\epsilon) \gg \epsilon$ and $h \to 0$ as $\epsilon \to 0$ (up to a subsequence).*

**Theorem 3.2.** *Let $p_0$ and $p_h$ be the solutions of the homogenized problem (3.26) and MsFEM (3.6), respectively, with the coefficient $k(x, \eta, \xi) = k(x/\epsilon, \xi)$ and $k_0 = 0$. Then*

$$\|p_h - p_0\|_{W_0^{1,\gamma}(\Omega)} \le C\left(\left(\frac{\epsilon}{h}\right)^\beta + h^\delta\right), \tag{3.29}$$

*where $\gamma$ and $\delta$ depend on operator constants defined in (6.42)–(6.46).*

We note that in Theorem 3.2, we assume that the equation is monotone, whereas in Theorem 3.1, we assume that the equation is pseudo-monotone. As discussed in Section 6.2, the monotonicity allows us to obtain explicit convergence rates. For parabolic equations, one can prove similar results (see [112]). One can also prove the convergence of $p_{r,h}$ to the fine-scale solution $p$ in $W^{1,\gamma}$ (e.g., [104, 112, 113]). Finally, we refer to [136] for convergence results of oversampling methods for nonlinear problems and applications to material science.

## 3.7 Numerical results

In this section we present some numerical results for MsFEMs for nonlinear elliptic equations. More numerical examples relevant to subsurface applications can be found in [104]. We present numerical results for both the MsFEM and MsFVEM. We use an inexact-Newton algorithm as an iterative technique to tackle the nonlinearity. For the numerical examples below, we use $k(x, p, \nabla p) = k(x, p)\nabla p$. Let $\{\phi_i^0\}_{i=1}^{N_{dof}}$ be the standard piecewise linear basis functions of $W_h$. Then MsFEM solution may be written as

$$p_h = \sum_{i=1}^{N_{dof}} \alpha_i \, \phi_i^0 \tag{3.30}$$

for some $\alpha = (\alpha_1, \alpha_2, ..., \alpha_{N_{dof}})^T$. Recall that $p_h \in W_h$ is an approximation for a homogenized solution, and $p_{r,h}$ is an approximation for a fine-scale solution. We need to find $\alpha$ such that

$$F(\alpha) = 0, \tag{3.31}$$

where $F : \mathbb{R}^{N_{dof}} \to \mathbb{R}^{N_{dof}}$ is a nonlinear operator such that

$$F_i(\alpha) = \sum_{K \in K^h} \int_K k(x, \eta^{p_h})\nabla p_{r,h} \cdot \nabla \phi_i^0 \, dx - \int_\Omega f \, \phi_i^0 \, dx. \tag{3.32}$$

We note that in (3.32) $\alpha$ is implicitly buried in $\eta^{p_h}$ and $p_{r,h}$. An inexact-Newton algorithm is a variation of Newton's iteration for a nonlinear system of equations, where the Jacobian system is only approximately solved. To be specific, given an initial iterate $\alpha^0$, for $k = 0, 1, 2, \cdots$ until convergence do the following

- Solve $F'(\alpha^k)\delta^k = -F(\alpha^k)$ by some iterative technique until $\|F(\alpha^k) + F'(\alpha^k)\delta^k\| \leq \beta_k \|F(\alpha^k)\|$.
- Update $\alpha^{k+1} = \alpha^k + \delta^k$.

In this algorithm $F'(\alpha^k)$ is the Jacobian matrix evaluated at iteration $k$. We note that when $\beta_k = 0$ we have recovered the classical Newton iteration. Here we have used

$$\beta_k = 0.001 \left( \frac{\|F(\alpha^k)\|}{\|F(\alpha^{k-1})\|} \right)^2, \tag{3.33}$$

with $\beta_0 = 0.001$. Choosing $\beta_k$ this way, we avoid oversolving the Jacobian system when $\alpha^k$ is still considerably far from the exact solution.

Next we present the entries of the Jacobian matrix. For this purpose, we use the following notations. Let $K_i^h = \{K \in \mathcal{T}_h : z_i \text{ is a vertex of } K\}$, $I^i = \{j : z_j \text{ is a vertex of } K \in K_i^h\}$, and $K_{ij}^h = \{K \in K_i^h : K \text{ shares } \overline{z_i z_j}\}$. We note that we may write $F_i(\alpha)$ as follows

$$F_i(\alpha) = \sum_{K \in K_i^h} \left( \int_K k(x, \eta^{p_h})\nabla p_{r,h} \cdot \nabla \phi_i^0 \, dx - \int_K f \, \phi_i^0 \, dx \right), \tag{3.34}$$

with

$$- \text{div}(k(x, \eta^{p_h}) \nabla p_{r,h}) = 0 \text{ in } K \text{ and } p_{r,h} = \sum_{z_m \in Z_K} \alpha_m \phi_m^0 \text{ on } \partial K, \quad (3.35)$$

where $Z_K$ is all the vertices of element $K$. It is apparent that $F_i(\alpha)$ is not fully dependent on all $\alpha_1, \alpha_2, ..., \alpha_d$. Consequently, $\partial F_i(\alpha)/\partial \alpha_j = 0$ for $j \notin I^i$. To this end, we denote $\psi_j = \partial p_{r,h}/\partial \alpha_j$. By applying the chain rule of differentiation to (3.35) we have the following local problem for $\psi_j$

$$- \text{div}(k(x, \eta^{p_h}) \nabla \psi_j) = \frac{1}{3} \text{div}(\frac{\partial k(x, \eta^{p_h})}{\partial p} \nabla p_{r,h}) \text{ in } K \text{ and } \psi_j = \phi_j \text{ on } \partial K.$$
$$(3.36)$$

The fraction $1/3$ comes from taking the derivative in the chain rule of differentiation. In the formulation of the local problem, we have replaced the nonlinearity in the coefficient by $\eta^{p_h}$, where for each triangle $K$ $\eta^{p_h} = 1/3 \sum_{i=1}^{3} \alpha_i^K$, which gives $\partial \eta^{p_h}/\partial \alpha_i = 1/3$. Moreover, for a rectangular element the fraction $1/3$ should be replaced by $1/4$.

Thus, provided that $v_{r,h}$ has been computed, then we may compute $\psi_j$ using (3.36). Using the above descriptions we have the expressions for the entries of the Jacobian matrix:

$$\frac{\partial F_i}{\partial \alpha_i} = \sum_{K \in K_i^h} \left( \frac{1}{3} \int_K \frac{\partial k(x, \eta^{p_h})}{\partial p} \nabla p_{r,h} \cdot \nabla \phi_i^0 \, dx + \int_K k(x, \eta^{p_h}) \nabla \psi_i \cdot \nabla \phi_i^0 \, dx \right)$$
$$(3.37)$$

$$\frac{\partial F_i}{\partial \alpha_j} = \sum_{K \in K_{ij}^h} \left( \frac{1}{3} \int_K \frac{\partial k(x, \eta^{p_h})}{\partial p} \nabla p_{r,h} \cdot \nabla \phi_i \, dx + \int_K k(x, \eta^{p_h}) \nabla \psi_j \cdot \nabla \phi_i^0 \, dx \right)$$
$$(3.38)$$

for $j \neq i$, $j \in I^i$.

The implementation of the oversampling technique is similar to the procedure presented earlier, except the local problems in larger domains are used. As in the nonoversampling case, we denote $\psi_j = \partial v_{r,h}/\partial \alpha_j$, such that after applying the chain rule of differentiation to the local problem we have:

$$-\text{div}(k(x, \eta^{p_h}) \nabla \psi_j) = \frac{1}{3} \text{div}(\frac{\partial k(x, \eta^{p_h})}{\partial p} \nabla v_{r,h}) \text{ in } K_E$$
$$\psi_j = \phi_j^0 \text{ on } \partial K_E, \quad (3.39)$$

where $\eta^{p_h}$ is computed over the corresponding element $K$ and $\phi_j^0$ is understood as the nodal basis functions on oversampled domain $K_E$. Then all the rest of the inexact-Newton algorithms are the same as in the nonoversampling case. Specifically, we also use (3.37) and (3.38) to construct the Jacobian matrix of the system. We note that we only use $\psi_j$ from (3.39) pertaining to the element $K$.

From the derivation (both for oversampling and nonoversampling) it is obvious that the Jacobian matrix is not symmetric but sparse. Computation of this Jacobian matrix is similar to computing the stiffness matrix resulting from standard finite elements, where each entry is formed by accumulation of element-by-element contributions. Once we have the matrix stored in memory, then its action to a vector is straightforward. Because it is a sparse matrix, devoting some amount of memory for entry storage is inexpensive. The resulting linear system is solved using a preconditioned biconjugate gradient stabilized method.

An an example to illustrate the convergence of the nonlinear MsFEM, we consider the following problem

$$-\text{div}(k(x/\epsilon, p)\nabla p) = -1 \quad \text{in } \Omega,$$
$$p = 0 \quad \text{on } \partial\Omega, \tag{3.40}$$

where $\Omega = [0, 1] \times [0, 1]$, $k(x/\epsilon, p) = k(x/\epsilon)/(1+p)^{l(x/\epsilon)}$, with

$$k(x/\epsilon) = \frac{2 + 1.8\sin(2\pi x_1/\epsilon)}{2 + 1.8\cos(2\pi x_2/\epsilon)} + \frac{2 + \sin(2\pi x_2/\epsilon)}{2 + 1.8\cos(2\pi x_1/\epsilon)} \tag{3.41}$$

and $l(x/\epsilon)$ is generated from $k(x/\epsilon)$ such that the average of $l(x/\epsilon) = Ck(x/\epsilon)$ over $\Omega$ is 2 with an appropriate choice of $C$. Here we use $\epsilon = 0.01$. Because the exact solution for this problem is not available, we use a well-resolved numerical solution using the standard finite element method as a reference solution. The resulting nonlinear system is solved using the inexact-Newton algorithm. The reference solution is solved on a $512 \times 512$ mesh. Tables 3.1 and 3.3 present the relative errors of the solution with and without oversampling, respectively. Here $N$ is the number of coarse blocks in each direction. In Tables 3.2 and 3.4, the relative errors for the multiscale finite volume element method are presented. The relative errors are computed as the corresponding error divided by the norm of the solution. In each table, the second, third, and fourth columns list the relative error in the $L^2$, $H^1$, and $L^\infty$ norm, respectively. As we can see from these two tables, the oversampling significantly improves the accuracy of the multiscale method.

In our next example, we consider the problem with nonperiodic coefficients, where $k(x, \eta) = k(x)/(1 + \eta)^{\alpha(x)}$. The coefficient $k(x) = \exp(\beta(x))$ is chosen such that $\beta(x)$ is a realization of a random field with the spherical variogram [85], the correlation lengths $l_1 = 0.2$, $l_2 = 0.02$, and with the variance $\sigma = 1$. The function $\alpha(x)$ is chosen such that $\alpha(x) = k(x) + \text{const}$ with the spatial average of 2. As for the boundary conditions we use "left-to-right flow" in the $\Omega = [0, 5] \times [0, 1]$ domain, $p = 1$ at the inlet ($x_1 = 0$), $p = 0$ at the outlet ($x_1 = 5$), and no flow boundary conditions on the lateral sides $x_2 = 0$ and $x_2 = 1$. In Table 3.5 we present the relative error for a multiscale method with oversampling. Similarly, in Table 3.6 we present the relative error for a multiscale finite volume method with oversampling. Clearly, the oversampling method captures the effects induced by the large correlation features. Both $H^1$ and horizontal flux errors are under five percent. Similar results have been observed for various kinds of nonperiodic heterogeneities.

In the next set of numerical examples, we test the MsFEM for problems with fluxes $k(x, \eta)$ that are discontinuous in space. The discontinuity in the fluxes is introduced by multiplying the underlying permeability function $k(x)$ by a constant in certain regions, while leaving it unchanged in the rest of the domain. As an underlying permeability field, $k(x)$, we choose the random field used for the results in Table 3.5. In the numerical example, the discontinuities are introduced along the boundaries of the coarse elements. In particular, $k(x)$ on the left half of the domain is multiplied by a constant $J$, where $J = \exp(4)$. The results in Table 3.7 show that the MsFEM converges and the error falls below five percent for relatively large coarsening. For the second numerical example (Table 3.8), the discontinuities are not aligned with the boundaries of the coarse elements. In particular, the discontinuity boundary is given by $x_2 = x_1\sqrt{2}+0.5$; that is the discontinuity line intersects the coarse-grid blocks. Similar to the aligned case, $\exp(4)$ jump magnitude is considered. The results presented above demonstrate the robustness and accuracy of our approach for anisotropic fields, where $h$ and $\epsilon$ are nearly the same, and the fluxes that are discontinuous spatial functions.

As for CPU comparisons, we have observed more than 92% CPU savings when using MsFEMs without oversampling. With the oversampling approach, the CPU savings depend on the size of the oversampled domain. For example, if the oversampled domain size is two times larger than the target coarse block (half coarse block extension on each side) we have observed 70% CPU savings for a $64 \times 64$ and 80% CPU savings for a $128 \times 128$ coarse grid. In general, the computational cost will decrease if the oversampled domain size is close to the target coarse block size, and this cost will be close to the cost of the MsFEM without oversampling. Conversely, the error decreases if the size of the oversampled domains increases. In the numerical examples, we have observed the same errors for the oversampling methods using either one coarse block extension or half coarse block extensions. The latter indicates that the leading resonance error is eliminated for the problems under consideration by using a smaller oversampled domain. Oversampled domains with one coarse block extension are previously used in simulations of flow through heterogeneous porous media. As indicated in [145], one can use large oversampled domains for simultaneous computations of the several local solutions. Moreover, parallel computations will improve the speed of the method because the MsFEM is well suited for parallel computation [145]. For the problems where $k(x, \eta, \xi) = k(x)b(\eta)\xi$ (see Section 3.3 and Section 5.3 for applications) our multiscale computations are very fast because the basis functions are built in the beginning of the computations. In this case, we have observed more than 95% CPU savings. We again would like to remark that the local solutions can be re-used in our multiscale simulations. This is similar to homogenization where the homogenized fluxes are computed once.

**Table 3.1.** Relative MsFEM Errors Without Oversampling

| N | $L^2$-Norm | | $H^1$-Norm | | $L^\infty$-Norm | |
|---|---|---|---|---|---|---|
| | Error | Rate | Error | Rate | Error | Rate |
| 32 | 0.029 | | 0.115 | | 0.03 | |
| 64 | 0.053 | -0.85 | 0.156 | -0.44 | 0.0534 | -0.94 |
| 128 | 0.10 | -0.94 | 0.234 | -0.59 | 0.10 | -0.94 |

**Table 3.2.** Relative MsFVEM Errors Without Oversampling

| N | $L^2$-Norm | | $H^1$-Norm | | $L^\infty$-Norm | |
|---|---|---|---|---|---|---|
| | Error | Rate | Error | Rate | Error | Rate |
| 32 | 0.03 | | 0.13 | | 0.04 | |
| 64 | 0.05 | -0.65 | 0.19 | -0.60 | 0.05 | -0.24 |
| 128 | 0.058 | -0.19 | 0.25 | -0.35 | 0.057 | -0.19 |

**Table 3.3.** Relative MsFEM Errors with Oversampling

| N | $L^2$-Norm | | $H^1$-Norm | | $L^\infty$-Norm | |
|---|---|---|---|---|---|---|
| | Error | Rate | Error | Rate | Error | Rate |
| 32 | 0.0016 | | 0.036 | | 0.0029 | |
| 64 | 0.0012 | 0.38 | 0.019 | 0.93 | 0.0016 | 0.92 |
| 128 | 0.0024 | -0.96 | 0.0087 | 1.14 | 0.0026 | -0.71 |

**Table 3.4.** Relative MsFVEM Errors with Oversampling

| N | $L^2$-Norm | | $H^1$-Norm | | $L^\infty$-Norm | |
|---|---|---|---|---|---|---|
| | Error | Rate | Error | Rate | Error | Rate |
| 32 | 0.002 | | 0.038 | | 0.005 | |
| 64 | 0.003 | -0.43 | 0.021 | 0.87 | 0.003 | 0.72 |
| 128 | 0.001 | 1.10 | 0.009 | 1.09 | 0.001 | 1.08 |

**Table 3.5.** Relative MsFEM Errors for Random Heterogeneities, Spherical Variogram, $l_1 = 0.20$, $l_2 = 0.02$, $\sigma = 1.0$

| N | $L^2$-Norm | | $H^1$-Norm | | $L^\infty$-Norm | | hor. flux | |
|---|---|---|---|---|---|---|---|---|
| | Error | Rate | Error | Rate | Error | Rate | Error | Rate |
| 32 | 0.0006 | | 0.0505 | | 0.0025 | | 0.025 | |
| 64 | 0.0002 | 1.58 | 0.029 | 0.8 | 0.001 | 1.32 | 0.017 | 0.57 |
| 128 | 0.0001 | 1 | 0.016 | 0.85 | 0.0005 | 1 | 0.011 | 0.62 |

**Table 3.6.** Relative MsFVEM Errors for Random Heterogeneities, Spherical Variogram, $l_1 = 0.20$, $l_2 = 0.02$, $\sigma = 1.0$

| N | $L^2$-Norm | | $H^1$-Norm | | $L^\infty$-Norm | | hor. flux | |
|---|---|---|---|---|---|---|---|---|
| | Error | Rate | Error | Rate | Error | Rate | Error | Rate |
| 32 | 0.0006 | | 0.0515 | | 0.0025 | | 0.027 | |
| 64 | 0.0002 | 1.58 | 0.029 | 0.81 | 0.0013 | 0.94 | 0.018 | 0.58 |
| 128 | 0.0001 | 1 | 0.016 | 0.85 | 0.0005 | 1.38 | 0.012 | 0.58 |

**Table 3.7.** Relative MsFEM Errors for Random Heterogeneities, Spherical Variogram, $l_1 = 0.20$, $l_2 = 0.02$, $\sigma = 1.0$, Aligned Discontinuity, Jump $= \exp(4)$

| N | $L^2$-Norm | | $H^1$-Norm | | $L^\infty$-Norm | | hor. flux | |
|---|---|---|---|---|---|---|---|---|
| | Error | Rate | Error | Rate | Error | Rate | Error | Rate |
| 32 | 0.0011 | | 0.1010 | | 0.0068 | | 0.195 | |
| 64 | 0.0006 | 0.87 | 0.0638 | 0.66 | 0.0045 | 0.59 | 0.109 | 0.84 |
| 128 | 0.0003 | 1.00 | 0.0349 | 0.87 | 0.0024 | 0.91 | 0.063 | 0.79 |

**Table 3.8.** Relative MsFEM Errors for Random Heterogeneities, Spherical Variogram, $l_1 = 0.20$, $l_2 = 0.02$, $\sigma = 1.0$, Nonaligned Discontinuity, Jump $= \exp(4)$

| N | $L^2$-Norm | | $H^1$-Norm | | $L^\infty$-Norm | | hor. flux | |
|---|---|---|---|---|---|---|---|---|
| | Error | Rate | Error | Rate | Error | Rate | Error | Rate |
| 32 | 0.0067 | | 0.1775 | | 0.1000 | | 0.164 | |
| 64 | 0.0016 | 2.07 | 0.0758 | 1.23 | 0.0288 | 1.80 | 0.077 | 1.09 |
| 128 | 0.0009 | 0.83 | 0.0687 | 0.14 | 0.0423 | -0.55 | 0.039 | 0.98 |

## 3.8 Discussions

An alternative approach for solving nonlinear problems using MsFEM is to
linearize them. For example, the nonlinear elliptic or parabolic equations considered in this chapter can be linearized, for example, as

$$\frac{\partial p^n}{\partial t} - \text{div}(b(x, t, p^{n-1}, \nabla p^{n-1}) \nabla p^n) + k_0(x, t, p^{n-1}, \nabla p^{n-1}) = f, \quad (3.42)$$

where $b(x, t, \eta, \xi) \cdot \xi = k(x, t, \eta, \xi)$. At this point, we assume that the solutions
of this linearized equation converge to a solution of the original nonlinear
equation. Then, at every iteration, one can apply the MsFEM where the basis
functions are constructed based on heterogeneous coefficients $b$ that change
at every iteration. Assuming that we can approximate the solution accurately
with the MsFEM at every iteration and the error of approximation does not
propagate, one can show that this procedure converges under some conditions.

One can also perform upscaling based on a linearized equation (3.42). In
this case, the approximation of the upscaled solution of the limiting equation

(in the sense of linearized iterations) can be obtained. More precisely, if one denotes $b_{n-1}(x,t) = b(x,t,p^{n-1}, \nabla p^{n-1})$ and $g_{n-1}(x,t) = k_0(x,t,p^{n-1}, \nabla p^{n-1})$, then the upscaled equation corresponding to (3.42) will have the form

$$\frac{\partial p^{*,n}}{\partial t} - \mathrm{div}(b^*_{n-1}(x,t)\nabla p^{*,n}) + g^*_{n-1}(x,t) = f.$$

Assuming that the solution of this iterative procedure converges, the final upscaled equation will have the form

$$\frac{\partial p^*}{\partial t} - \mathrm{div}(b^*(x,t)\nabla p^*) + g^*(x,t) = f.$$

Note that the obtained effective parameters (e.g., $b^*$) implicitly contain the information about the upscaled solution $p^*$. Because the upscaled solution contains the information about the global boundary conditions and source terms in a nonlinear fashion, one cannot re-use the upscaled coefficients (e.g., $b^*$) if the source or boundary conditions are changed.

The approximation of the local problems in the presence of scale separation or periodicity is discussed in this chapter. In particular, the local solutions and the evaluation of the variational formulation (see (3.7)) can be carried out in smaller regions. These issues are discussed in greater detail in [136]. One can attempt to use approximate solutions in the variational formulation (3.7) to compute the resulting system of nonlinear equations (e.g., fewer Newton iterations). To our best knowledge, these approximations are not considered in the literature.

In this section, we considered nonlinear elliptic and parabolic equations; however, the proposed methods can be applied in more general situations (see Section 2.4). In Chapter 5, we discuss the application of nonlinear MsFEMs to Richards' equations and to fluid flows in deformable porous media. The latter involves coupled nonlinear equations involving the interface dynamics between the fluid and solid components of the media. The methods discussed in this chapter can also be applied in material sciences. For example, in [136], the author applies numerical homogenization methods similar to those discussed in this chapter to nonlinear elasticity. In particular, the paper [136] explores oversampling techniques in nonlinear heterogeneous equations both numerically and analytically. The author proves the convergence of the method with oversampling, for convex and quasi-convex energies, in the context of general heterogeneities. This analysis provides an interesting variational interpretation of the Petrov–Galerkin formulation of the nonconforming multiscale finite element method for periodic problems.

# 4

# Multiscale finite element methods using limited global information

## 4.1 Motivation

Previously, we discussed multiscale methods that employ local information in computing basis functions. The accuracy of these approaches depends on local boundary conditions. Although effective in many cases, multiscale methods that only use local information may not accurately capture the solution at all scales. In particular, in regions with no scale separation, the local multiscale methods cannot accurately approximate the scales that are comparable to the computational coarse-grid size. A rich hierarchy of scales can introduce an important connectivity at different scales that need to be captured at larger scales. The natural question is how to incorporate the information from different scales into localized multiscale basis functions such that the resulting numerical solution provides an accurate approximation of the global solution.

In this chapter, we discuss how to take into account the information that is not captured by local basis functions. We call this information global information although it can be information in some large regions where important connectivity of the media may occur. For example, subsurface properties often do not have scale separation and high/low conductivity regions can be connected at various scales (e.g., Figures 1.2–1.4). The connectivity regions are often very complicated due to conductivity variations within these regions and their complex geometrical structures. Similar situations can occur in composite materials where the material properties can vary at different scales. These complex features are often incorporated into global fields which are used to construct localized multiscale basis functions. In this chapter, we discuss the concept of global multiscale methods and their applications.

We demonstrate the main idea of global multiscale methods on the example of porous media flow, although this concept can be generalized to many other applications such as composite materials. Consider

$$- \operatorname{div}(\lambda(x)k(x)\nabla p) = f, \tag{4.1}$$

Y. Efendiev, T.Y. Hou, *Multiscale Finite Element Methods: Theory and Applications*, 67
Surveys and Tutorials in the Applied Mathematical Sciences 4,
DOI 10.1007/978-0-387-09496-0_4, © Springer Science+Business Media LLC 2009

where $k(x)$ is a heterogeneous field and $\lambda(x)$ is assumed to be a smooth field. This equation is derived from two-phase flow equations when gravity and capillary effects are neglected (see Section 2.10). Our goal is to construct multiscale basis functions on the coarse grid (with grid size larger than the characteristic length scale of the problem) such that these basis functions can be used for various source terms $f(x)$, boundary conditions, and mobilities $\lambda(x)$. Here, $k(x)$ does not have scale separation and has a multiscale structure that may not be captured accurately via local basis functions.

In order to capture the multiscale structure of the media at different scales, one needs to embed the multiscale information into the global fields. More precisely, we assume that the solution can be represented by a number of fields $p_1, ..., p_N$, such that

$$p \approx G(p_1, ..., p_N), \tag{4.2}$$

where $G$ is a sufficiently smooth function, and $p_1, .., p_N$ are global fields. These fields typically contain the essential information about the heterogeneities at different scales and can also be local fields (see discussion below). In the above assumption (4.2), $p_i$ are solutions of elliptic equations. For the mixed MsFEM, one can formulate an assumption similar to (4.2) for velocities. We denote by $v_i = -k\nabla p_i$. Then, the above assumption can be written in the following way. There exist sufficiently smooth scalar functions $A_1(x), ..., A_N(x)$, such that the velocity corresponding to (4.1) ($v = -\lambda(x)k(x)\nabla p$) can be written as

$$v \approx A_1(x)v_1 + \cdots + A_N(x)v_N. \tag{4.3}$$

Note that it is important that $G$ (or $A_1$, ..., $A_N$) are smooth functions so that the multiscale basis functions which span $p_1, ..., p_N$ (or $v_1, ..., v_N$) can accurately approximate the global solution. More details on the assumption on $A_1, ..., A_N$ or $G$ are formulated later.

For problems without scale separation, the functions $p_1, ..., p_N$ are often the solutions of global problems or their approximations. These methods are effective when (4.1) is solved multiple times. For problems with scale separation, one can use the solutions of the local problems in constructing multiscale basis functions. We note that when only local information is used, one still needs (4.2) (or (4.3) for fluxes) in each coarse-grid block to guarantee that the local solutions can approximate the global solution in each coarse patch. Once these global fields are determined, the multiscale basis functions are constructed such that they span these global fields. Thus, the multiscale methods with limited global information can be regarded as an extension of MsFEM discussed in Chapter 2. One of the main challenges is to determine the global fields. This is discussed next.

In a general setting, it was shown by Owhadi and Zhang [218] that for an arbitrary smooth $\lambda(x)$, the solution of (4.1) is a smooth function of $d$ linearly independent solutions of single-phase flow equations ($N = d$), where $d$ is the space dimension. These results are shown under some suitable assumptions for the case $d = 2$ and more restrictive assumptions for the case $d = 3$. In [103], it was shown that for channelized permeability fields, $p$ is a smooth function of single-phase flow pressure (i.e., $N = 1$), where the single-phase pressure equation is described by $\mathrm{div}(k\nabla p^{sp}) = 0$ with boundary conditions as those corresponding to two-phase flow. Multiple global fields can be used for the system of equations or for the random coefficients. For the system of equations, these global fields are the solutions of the homogeneous system subject to boundary conditions $(0, ..., x_i, ..., 0)$ $(i = 1, ..., d)$, where $x_i$ is chosen for each component of the vector field solution and zero otherwise (as in homogenization; see [28]). When considering random permeability fields, the permeability field is typically parameterized with a parameter that represents the uncertainties. In this case, we deal with a family of heterogeneous permeability fields such as $k = k(x, \theta)$, where $\theta$ is in a high-dimensional space. For example, log-Gaussian permeability fields can be characterized using Karhunen–Loéve expansion (e.g., [182]) as

$$k(x, \theta_1, ..., \theta_M) = \exp(\sum_i \theta_i \Phi_i(x)),$$

where $\Phi_i(x)$ are pre-computed spatial fields that depend on a covariance matrix associated with $k$. In many of these parameterized cases, $k(x, \theta)$ is a smooth function of $\theta = (\theta_1, ..., \theta_M)$, and thus one can use the solutions corresponding to a few realizations of $k$ to represent the heterogeneities across the ensemble (see Section 5.7.1).

### 4.1.1 A motivating numerical example

In this section, we present a numerical example where the use of local boundary conditions does not perform well and there is a need to use some type of global information. We consider the two-phase immiscible flow and transport setting presented in Section 2.10 with quadratic relative permeability functions and neglect the effects of gravity and capillarity. Multiscale methods generally perform well for permeability fields generated using two-point correlation functions (e.g., [85]). However, the local multiscale methods do not perform well in the presence of strong nonlocal effects as do those that appear in channelized permeability fields. In our numerical example, we consider strongly channelized permeability fields, and in particular, show that the local basis functions cannot accurately capture the global effects. These permeability fields have been proposed in some recent benchmark tests, such as the Tenth SPE Comparative Solution Project [78].

In Figure 4.1, one of the layers of this 3D permeability field is depicted. All the layers have $60 \times 220$ fine-scale resolution, and we take the coarse grid to be $6 \times 22$. As can be observed, the permeability field contains a high-permeability channel, where most flow occurs in our simulation. In Figure 4.2, the saturation fields at time PVI = 0.5 are compared (see (2.44) for the definition

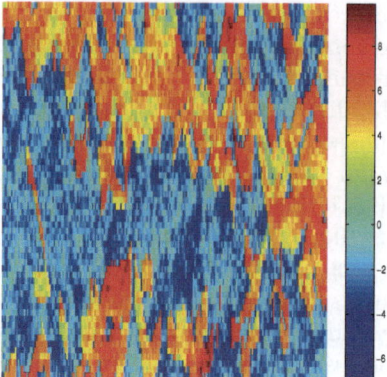

**Fig. 4.1.** Log-permeability for one of the layers of upper Ness.

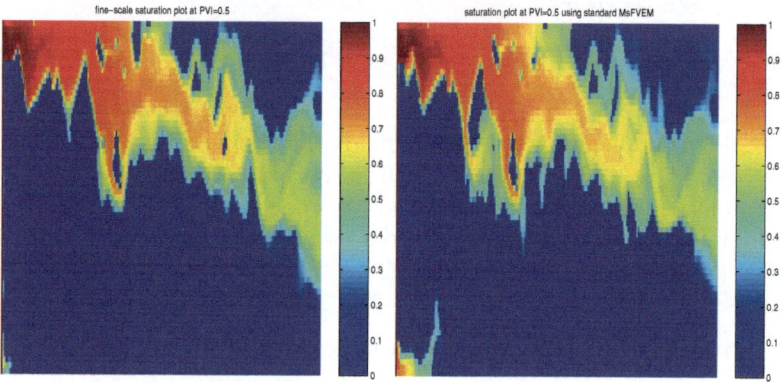

**Fig. 4.2.** Saturation maps at PVI $= 0.5$ for fine-scale solution (left figure) and standard MsFVEM (right figure).

of PVI). We use the MsFVEM as described in Section 2.10, where the elliptic equation is solved on the coarse grid, whereas the transport equation is solved on the fine grid with the fine-scale velocity field re-constructed using multiscale basis functions. Thus, the errors are due to the MsFVEM only. We see that MsFVEMs with local basis functions introduce some errors. In the bottom left corner, there is a saturation pocket that is not in the reference solution computed using the fine grid. This is because the local basis functions in the lower-left corner contain a high permeability region. However, this high permeability region does not have global connectivity, and the local basis functions cannot take this effect into account. Next, we discuss how global information can be incorporated into multiscale basis functions to improve the accuracy of the computations. Later in the book, we show that some more general multiphase flow and transport numerical results can be improved by using limited global information.

## 4.2 Mixed multiscale finite element methods using limited global information

### 4.2.1 Elliptic equations

In this section, we study mixed MsFEMs that employ global information. We consider elliptic equations with Neumann boundary conditions

$$-\text{div}(\lambda(x)k(x)\nabla p) = f(x) \quad \text{in} \quad \Omega$$
$$\lambda(x)k(x)\nabla p \cdot n = g \quad \text{on} \quad \partial\Omega, \tag{4.4}$$

where $k(x)$ is a heterogeneous field and $\lambda(x)$ is a smooth field, as before. We assume that $\int_\Omega p\,dx = 0$. Denote by $v = -\lambda(x)k(x)\nabla p$ the velocity. To construct basis functions for a global mixed MsFEM, we assume that the velocity field $v$ can be approximated by a priori defined global velocity fields, $v_1, ..., v_N$ in the following way. There exist functions $v_1, ..., v_N$ and $A_1(x), ..., A_N(x)$ such that

$$v(x) \approx \sum_{i=1}^N A_i(x)v_i(x), \tag{4.5}$$

where $A_i(x)$, $i = 1, ..., N$, are sufficiently smooth. The assumption (4.5) is made more precise in Section 6.3. We note that $v_i = -k\nabla p_i$ are, in general, solutions of $\text{div}(k\nabla p_i) = 0$, $v_i = -k\nabla p_i$, with some boundary conditions. In (4.5), we assume that the velocity field in each coarse-grid block can be approximated by a linear combination of a priori defined velocity fields.

Next, we construct the multiscale velocity basis functions using the information from $v_1, ..., v_N$. The main difference between this construction and the construction presented in Section 2.5.2 is the use of oscillatory boundary conditions that depend on $v_1, ..., v_N$. Specifically, we construct the basis functions for the velocity field as follows:

$$\text{div}(k(x)\nabla\phi_{ij}^K) = \frac{1}{|K|} \quad \text{in} \quad K$$
$$k\nabla\phi_{ij}^K \cdot n = \begin{cases} \frac{v_i \cdot n}{\int_{e_j} v_i \cdot n\,ds} & \text{on } e_j^K \\ 0 & \text{else,} \end{cases} \tag{4.6}$$

where $\int_K \phi_{ij}^K dx = 0$, $i = 1, ..., N$, $j = 1, ..., j_K$ ($j_K$ is the number of edges or faces of $K$), and $e_j^K$ are edges (or faces in $\mathbb{R}^3$) of $K$. In Figure 4.3, we schematically illustrate the basis function construction. Let $\psi_{ij}^K = k(x)\nabla\phi_{ij}^K$. We define the finite-dimensional space spanned by these basis functions by

$$\mathcal{V}_h = \text{span}\{\psi_{ij}^K\}.$$

We denote by $\mathcal{V}_h^0$ the span of $\psi_{ij}^K$ that satisfies homogeneous Neumann boundary conditions. We set $Q_h$ to be piecewise constant basis functions that are used to approximate the pressure $p$, as in Section 2.5.2.

As in Section 2.5.2, we can combine the basis functions in adjacent coarse-grid blocks with a common edge $e_j$ and obtain the multiscale basis function for the edge $e_j$ denoted by $\psi_{ij}$. Let $K_1$ and $K_2$ be adjacent coarse grid blocks. Then $\psi_{ij}$ solves (4.6) in $K_1$ and solves $\text{div}(\psi_{ij}) = -1/|K_2|$ in $K_2$, and $\psi_{ij} \cdot n = -v_i \cdot n / \int_{e_j^K} v_i \cdot n ds$ on $e_j^{K_2}$ and 0 otherwise. In other words, $\psi_{ij} = \psi_{ij}^{K_1}$ in $K_1$ and $\psi_{ij} = -\psi_{ij}^{K_2}$ in $K_2$, where $\psi_{ij}^K$ is defined via the solution of (2.16) (cf. Figure 2.8 for the illustration).

Let $\{v_h, p_h\}$ be the numerical approximation of $\{v, p\}$ with the basis functions defined previously. The numerical mixed formulation of (4.4) is to find $\{v_h, p_h\} \in \mathcal{V}_h \times Q_h$ such that

$$\int_\Omega (\lambda k)^{-1} v_h \cdot w_h dx - \int_\Omega \text{div}(w_h) p_h dx = 0 \quad \forall w_h \in \mathcal{V}_h^0$$
$$\int_\Omega \text{div}(v_h) q_h dx = \int_\Omega f q_h dx \quad \forall q_h \in Q_h. \tag{4.7}$$

The discrete formulation corresponding to the resulting system is similar to (2.18).

Note that for each edge, we have $N$ basis functions and we assume that $v_1, ..., v_N$ are linearly independent in order to guarantee that the basis functions are linearly independent. To ensure the boundary condition in (4.6) is well defined, we assume that $\int_{e_i^K} v_i \cdot n ds$ is not zero. To avoid the possibility that $\int_{e_i^K} |v_i \cdot n| ds$ is unbounded, we need to make certain assumptions that bound $\int_{e_i^K} |v_i \cdot n| ds$ from above. These assumptions are formulated in Section 6.3.

In Section 6.3, it is shown that the MsFEM using limited global information converges without any resonance error. We present numerical results in the next section as well as in Section 5.7.1 to demonstrate the importance of the use of global information.

*Remark 4.1.* We note that local mixed MsFEMs introduced in Section 2.5.2 can be obtained from mixed MsFEMs introduced in this section. To do this, one needs to use one global field $v_1$ which is a constant vector and $v_1 \cdot n \neq 0$ along each edge $e$. Taking into account that $v_1 \cdot n$ is constant along each edge, we have $v_1 \cdot n / \int_e v_1 \cdot n ds = 1/|e|$. This is the same as the boundary conditions introduced for local problems in Section 2.5.2.

*Remark 4.2.* The representative coarse grid $K$ can be nonconvex (c.f., Figure 4.3). The analysis presented in Section 6.3 implies that the global mixed multiscale finite element method works for nonconvex meshes. Strongly stretched meshes can have an impact on the convergence rate of the method following the analysis in Section 6.3.

*Remark 4.3.* The construction of velocity basis functions in (4.6) and the analysis in Section 6.3 imply that $K$ is not necessarily a polygon domain and the interface normal can be a spatial function.

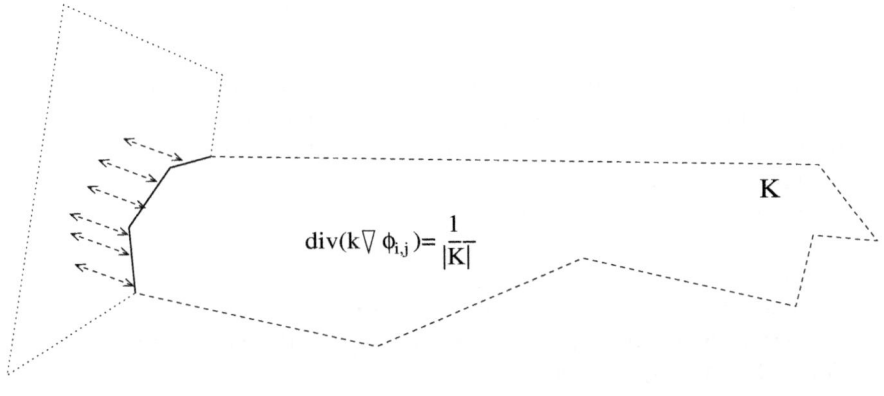

$$\operatorname{div}(k\nabla\,\phi_{i,j}) = \frac{1}{|K|}$$

K

_____ Nonzero flux boundary condition

- - - - - - - Zero flux boundary condition

.............. Neighboring coarse grid block

**Fig. 4.3.** Schematic description of velocity basis function construction.

*Remark 4.4.* We note that the global mixed MsFEMs presented in the book can be used when meshes have hanging nodes and when fine grids do not necessarily match across coarse grid interfaces.

*Pseudo-code.* Next, we briefly outline the implementation of mixed Ms-FEMs. We note that the implementation of mixed MsFEMs is similar to Algorithm 2.5.2, except one needs to compute or obtain global fields. Note that global fields can be defined iteratively (see Section 4.4). We have posted simple prototype MATLAB codes for solving elliptic equations with mixed MsFEMs (courtesy of J.E. Aarnes) at
$http://www.math.tamu.edu/ \sim yalchin.efendiev/codes/.$

### 4.2.2 Parabolic equations

Mixed MsFEMs using limited global information can be easily extended to parabolic equations. We consider the following model parabolic equation,

$$\frac{\partial}{\partial t}p - \operatorname{div}(\lambda(x)k(x)\nabla p) = f(x,t) \quad \text{in} \quad \Omega \times [0,T]$$

$$p = 0 \quad \text{on} \quad \partial\Omega \times [0,T] \tag{4.8}$$

$$p(t=0) = p(0) \quad \text{in} \quad \Omega,$$

where $k(x)$ is a bounded symmetric and positive definite matrix in $\Omega$, and $p(0)$ is a smooth spatial field. Denote by $v = -\lambda(x)k(x)\nabla p$.

---

**Algorithm 4.2.1**

---

Set coarse mesh configuration from fine-scale mesh information.
Define global fields $v_1, ..., v_i, ..., v_N$ used in the simulations.
For each coarse-grid block $n$ and for each global field $i$ do
- For each edge $j$ of a coarse-grid block
- Solve for $\psi_{ij}^n$ according to (4.6)
- End for
End do.
Assemble the coarse-scale system according to (4.7).
Assemble the external force on the coarse mesh according to (4.7).
Solve the coarse-grid formulation.

---

For parabolic equations, we assume that the velocity $v$ can be approximated by multiple global fields. In particular, we assume that there exist $v_1, ..., v_N$ and sufficiently smooth functions $A_1(t, x), ..., A_N(t, x)$ such that

$$v(t, x) \approx \sum_{i=1}^{N} A_i(t, x) v_i(x).$$

This assumption is made more precise in Section 6.3.1.

The mixed formulation associated with (4.8) is to find $\{v, p\}$ such that

$$\int_\Omega \frac{\partial}{\partial t} p \, q dx + \int_\Omega \mathrm{div}(v) \, q dx = \int_\Omega fq \quad \forall q \in L^2(\Omega)$$

$$\int_\Omega (\lambda k)^{-1} v \cdot w dx - \int_\Omega \mathrm{div}(w) \, p dx = 0 \quad \forall w \in H(\mathrm{div}, \Omega) \qquad (4.9)$$

$$p(t = 0) = p(0).$$

Let finite-dimensional space $V_h$ and $Q_h$ be defined as in the elliptic case. The space-discrete mixed formulation is to find $\{v_h, p_h\} : [0, T] \longrightarrow V_h \times Q_h$ such that

$$\int_\Omega \frac{\partial}{\partial t} p_h q_h dx + \int_\Omega \mathrm{div}(v_h) \, q_h dx = \int_\Omega fq_h dx \quad \forall q_h \in Q_h$$

$$\int_\Omega (\lambda k)^{-1} v_h \cdot w_h dx - \int_\Omega \mathrm{div}(w_h) \, p_h dx = 0 \quad \forall w_h \in V_h \qquad (4.10)$$

$$p_h(t = 0) = p_{0,h},$$

where $p_{0,h}$ is the $L^2$ projection of $p(0)$ onto $Q_h$. This problem can be also re-written in matrix form

$$A \frac{\partial}{\partial t} P + BU = F$$

$$B^T P - DU = 0 \qquad (4.11)$$

with $P(0)$ given, where $A$ and $D$ are symmetric positive and definite. After eliminating $U$, (4.11) is a linear system ODE for $P$,

$$A\frac{\partial}{\partial t}P + BD^{-1}B^T P = F.$$

The analysis of the method is presented in Section 6.3.

### 4.2.3 Numerical results

**The use of single global information**

In our numerical simulations, we perform two-phase flow and transport simulations with the same setting as before, except we assume that the source terms ($q_t$ in (2.40)) are given by a standard five-spot problem (see e.g., [1]), where the injection well is placed at the middle and the four production wells are placed at four corners of the rectangular global domain. We assume no flow along the boundaries. Initially, it is assumed that $S = 0$ in the whole domain. In the simulations, we solve the pressure equation on the coarse grid and reconstruct the fine-scale velocity field which is used to solve the saturation equation. The fine-scale velocity is reconstructed simply by using the multiscale basis function as

$$v = \sum_e v_e \psi_e,$$

where the sum is taken over all edges $e$ (or faces), $v_e$ is the coarse-scale normal velocity field for edge $e$ obtained via the solution of mixed MsFEMs, and $\psi_e$ is the velocity basis function for edge $e$. If there are several multiscale basis functions for each edge, then

$$v = \sum_{e,i} v_{e,i} \psi_{e,i},$$

where $i$ corresponds to the global field $v_i$ (see (4.3)). Because we use a reconstructed fine-scale velocity field, the errors will be due to mixed MsFEMs only. The basis functions are constructed at time zero and not changed throughout the simulations. As for permeability fields, we use heterogeneous permeability fields from the Tenth SPE Comparative Solution Project [78] (also referred to as SPE 10). Because of channelized structure of the permeability fields, the localized approaches do not perform well, as we observed earlier. On the other hand, the use of limited global information based on single-phase flow information improves the accuracy.

We first present numerical results where one global field (single-phase flow solution) is used ($N = 1$ in (4.3)). More precisely, $\operatorname{div}(k\nabla p_1) = q_t$, where $q_t$ represents the source terms corresponding to the five-spot problem and $v_1 = -k\nabla p_1$. We compare a mixed MsFEM with limited global information and a mixed MsFEM which uses only local information. In Tables 4.1–4.6 numerical results for different layers of SPE 10 using different viscosity ratios (see (2.39)) and different coarse grid sizes are shown. In these tables, the $L^1$ saturation errors over the time interval from 0 to 1 PVI as well as fractional flow errors are compared. It is evident from these tables that a mixed MsFEM using limited global information performs much better than a mixed MsFEM which only uses local information. Moreover, we observe that a mixed MsFEM converges as the mesh size decreases. We present saturation snapshots in Figure 4.4. These results indicate that for general complicated media such as SPE 10 with high contrast, one can expect the convergence of a mixed MsFEM as the coarse mesh size decreases when using limited global information.

**Table 4.1.** Relative Errors (Layer 40, $\mu_o/\mu_w = 3$)

| Coarse Grid | Frac. Flow Error (Global) | Saturation Error (Global) | Frac. Flow Error (Local) | Saturation Error (Local) |
|---|---|---|---|---|
| $6 \times 10$ | 0.0144 | 0.0512 | 0.1172 | 0.2755 |
| $12 \times 22$ | 0.0039 | 0.0370 | 0.1867 | 0.3158 |

**Table 4.2.** Relative Errors (Layer 50, $\mu_o/\mu_w = 3$)

| Coarse Grid | Frac. Flow Error (Global) | Saturation Error (Global) | Frac. Flow Error (Local) | Saturation Error (Local) |
|---|---|---|---|---|
| $6 \times 10$ | 0.0129 | 0.0871 | 0.1896 | 0.5061 |
| $12 \times 22$ | 0.0046 | 0.0568 | 0.1702 | 0.4578 |

**Table 4.3.** Relative Errors (Layer 70, $\mu_o/\mu_w = 3$)

| Coarse Grid | Frac. Flow Error (Global) | Saturation Error (Global) | Frac. Flow Error (Local) | Saturation Error (Local) |
|---|---|---|---|---|
| $6 \times 10$ | 0.0106 | 0.0562 | 0.0408 | 0.2291 |
| $12 \times 22$ | 0.0039 | 0.0421 | 0.0976 | 0.2530 |

**Mixed MsFEM on unstructured grids and the coupling to coarse-scale transport equation**

In [4], the mixed MsFEM is used for simulations on unstructured coarse grids. The use of unstructured coarse grids has advantages in subsurface simulations because they provide flexibility and can render more accurate upscaled

**Table 4.4.** Relative Errors (Layer 40, $\mu_o/\mu_w = 10$)

| Coarse Grid | Frac. Flow Error (Global) | Saturation Error (Global) | Frac. Flow Error (Local) | Saturation Error (Local) |
|---|---|---|---|---|
| 6 × 10 | 0.0080 | 0.0534 | 0.0902 | 0.2721 |
| 12 × 22 | 0.0026 | 0.0403 | 0.1414 | 0.3153 |

**Table 4.5.** Relative Errors (Layer 50, $\mu_o/\mu_w = 10$)

| Coarse Grid | Frac. Flow Error (Global) | Saturation Error (Global) | Frac. Flow Error (Local) | Saturation Error (Local) |
|---|---|---|---|---|
| 6 × 10 | 0.0049 | 0.0957 | 0.1577 | 0.5137 |
| 12 × 22 | 0.0041 | 0.0628 | 0.1404 | 0.4613 |

**Table 4.6.** Relative Errors (Layer 70, $\mu_o/\mu_w = 10$)

| Coarse Grid | Frac. Flow Error (Global) | Saturation Error (Global) | Frac. Flow Error (Local) | Saturation Error (Local) |
|---|---|---|---|---|
| 6 × 10 | 0.0044 | 0.0629 | 0.0280 | 0.2262 |
| 12 × 22 | 0.0025 | 0.0473 | 0.0678 | 0.2397 |

solutions for flow and transport equations. It is often necessary to use an un-structured coarse grid when highly heterogeneous reservoirs are discretized via irregular anisotropic fine grids. Our study is motivated by the development of coarse-scale models for coupled flow and transport equations in a multiphase system. An unstructured coarse grid is often used to upscale the transport equation with hyperbolic nature in a highly heterogeneous reservoir. Solving the flow equation on the same coarse grid provides a general robust coarse-scale model for the multiphase flow and transport at a low CPU cost. We note that most of the previous studies employ a two-grid approach where the flow equation is solved on a coarse grid and the transport equation is solved on a fine grid. We consider the nonuniform coarsening developed in [9] for the transport equation (also described in Section 5.5). The coarse grid we obtain is highly anisotropic and is not quasi-uniform. We present numerical results when both the flow and transport equations are solved on the coarse grid. In [4], numerical examples involving highly channelized permeability as well as a 3D reservoir model using an unstructured fine grid are presented. Next, we present a few numerical examples.

For our numerical example, we consider layer 65 of SPE 10. Using the algorithm for upscaling of the transport equation ([9]), we generate a coarse grid. In Figure 4.5, the fine-scale permeability and the coarse grid are plotted. In Figure 4.6, we present the results for the saturation fields at PVI= 1 when both the flow and the transport equations are solved on the coarse grid. One can see from this figure that the saturation profile looks realistic when an adaptive coarse grid is used and we preserve the geological realism reasonably

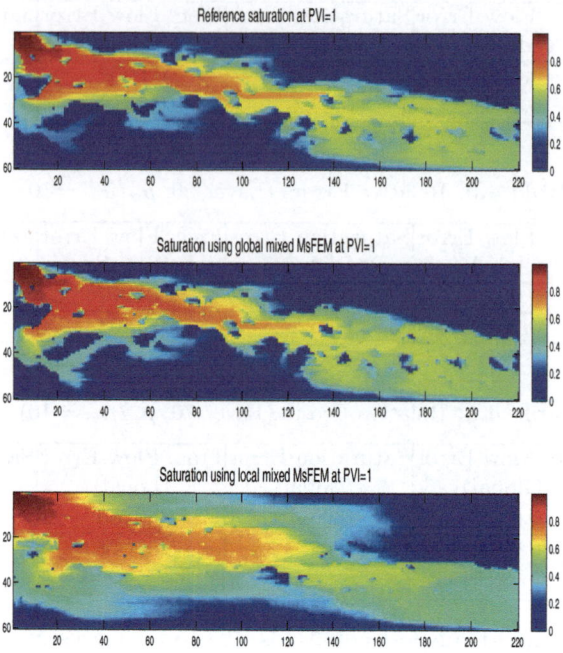

**Fig. 4.4.** Comparison of saturation fields between reference solution and MsFEM solution at PVI = 1, layer 50, 12 × 11 coarse grid and $\mu_o/\mu_w = 10$; top: reference saturation; middle: saturation using global mixed MsFEM; bottom: multiscale saturation using local mixed MsFEM.

well. In Table 4.7 we present $L^1$ relative errors for the saturation when different resolutions of the coarse grid are used. In the same table, we show the errors corresponding to the structured grids with a comparable number of coarse-grid blocks (shown in parentheses). We can make two important observations from this table. First, the errors are small (less than 1%). Second, the mixed MsFEM on an unstructured grid performs better. The latter is due to the fact that the unstructured grid is constructed using some relevant limited global information which usually increases the accuracy of the method.

In our next numerical example, we test the method on a synthetic reservoir with a corner-point grid geometry. The corner-point grid has vertical pillars, as shown in Figure 4.7, 100 layers, and 29,629 active cells (cells with positive volume). The permeability ranges from 0.1 mD to 1.7 D and the porosity is assumed to be constant. The corner-point grid (or pillar grid) format [231] is a very flexible grid format that is used in many commercial geomodeling softwares. Essentially a corner-point grid consists of a set of hexahedral cells

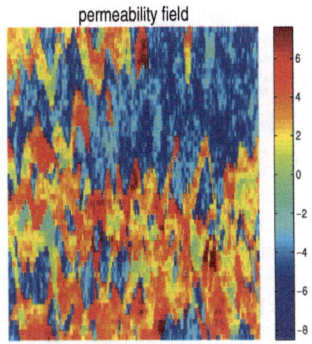

**Fig. 4.5.** $60 \times 220$ permeability field and the coarse grid with 180 blocks. A random color is assigned to each coarse-grid block.

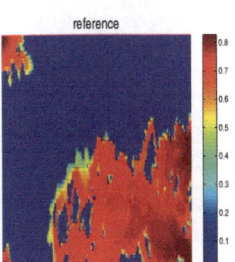
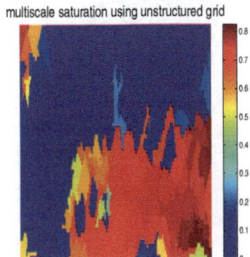

**Fig. 4.6.** Saturation comparisons.

**Table 4.7.** Relative $L^1$ Errors (Layer 65)

| Unstruct. Coarse (Number of Blocks) | Sat. Err. (Total) | Sat. Err. (Struct. Grid) (Total) |
|---|---|---|
| 180 | 0.0097 | 0.0130 (10×20) |
| 299 | 0.0080 | 0.0125 (15×22) |
| 913 | 0.0062 | 0.009 (20×44) |

that are aligned in a logical Cartesian fashion where one horizontal layer in the logical grid is assigned to each sedimentary bed to be modeled. In its simplest form, a corner-point grid is specified in terms of a set of vertical or inclined pillars defined over an areal Cartesian 2D mesh in the lateral direction. Each cell in the volumetric corner-point grid is restricted by four pillars and is defined by specifying the eight corner points of the cell, two on each pillar.

We consider only 60 vertical layers of the permeability field. The coarse grid is constructed by subdividing the fine-scale model on 30-by-30-by-60 corner-point cells into 202 coarse-grid blocks. In Figure 4.8, we plot: coarse-grid

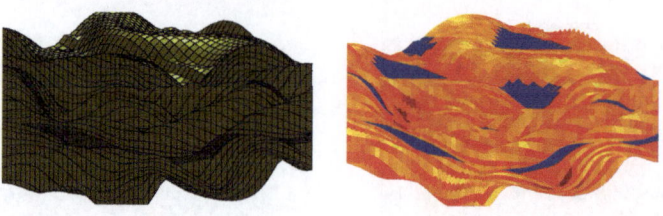

**Fig. 4.7.** A corner-point model with vertical pillars and 100 layers. To the right is a plot of the permeability field on a logarithmic scale. The model is generated with $SBED^{TM}$, and is courtesy of Alf B. Rustad at STATOIL.

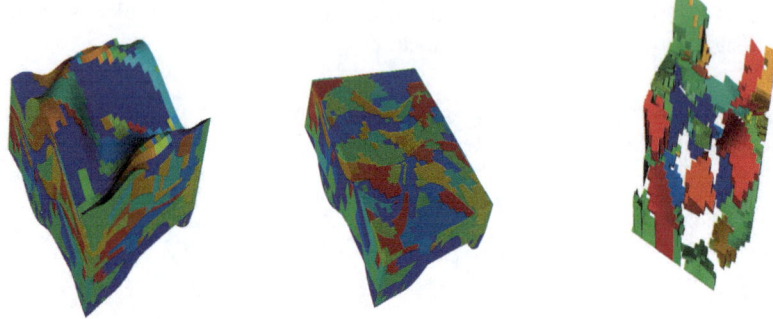

**Fig. 4.8.** Left: schematic description of unstructured coarsening (each coarse grid block is assigned a random color). Middle: a horizontal slice of unstructured coarsening presented on the left. Right: a coarse grid block (enlarged).

partitioning (left plot) where a random color is assigned to each coarse-grid block; a horizontal slice of coarse partitioning presented on the left plot; and several coarse-grid blocks. In Figure 4.9, we plot the water-cut curves. As we see from this figure, our method provides an accurate approximation of water-cut data. The error that is due to the mixed MsFEM is only 2% (here, we consider $L^1$ error in the saturation field at PVI = 0.5). We have observed 17% error in the saturation field when both flow and transport equations are solved on the coarse grid. This error is mainly due to the saturation upscaling. The detailed numerical studies when both flow and transport equations are coarsened can be found in [4]. In particular, we show that the errors due to mixed MsFEMs for solving the flow equation are much smaller than the errors due to upscaling of the transport equation. This suggests that more accurate upscaling methods for transport equations are needed. Multiscale methods for transport equations are discussed in Section 5.2.

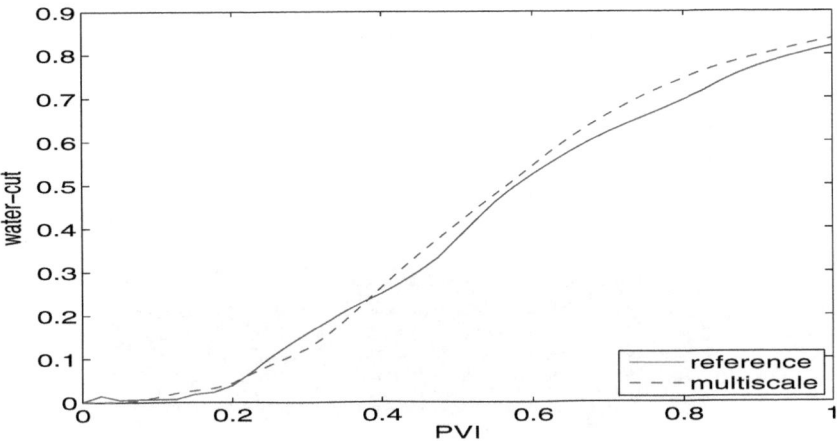

**Fig. 4.9.** Water-cut for reference and multiscale solutions.

## The use of multiple global information in parameter-dependent permeability

In our next set of numerical experiments, we consider a case where one needs to use multiple global fields to construct multiscale basis functions. One can consider this case as a simplified case for the more general stochastic case which is presented in Section 5.7. In our numerical experiments, we consider $k(x, \theta) = \exp(\theta Y(x))$. We investigate a range of $\theta$, $\theta_1 \leq \theta \leq \theta_2$, and use the global single-phase flow solutions corresponding to endpoints $\theta = \theta_1$ and $\theta = \theta_2$ to construct the multiscale basis functions. In particular, $v_1 = -k(x, \theta_1)\nabla p(x, \theta_1)$ and $v_2 = -k(x, \theta_2)\nabla p(x, \theta_2)$ (where $p(x, \theta_1)$ and $p(x, \theta_2)$ solve the global single-phase flow problem) are used to construct mixed multiscale basis functions as described earlier.

In Figure 4.10, the water-cut (which is equal to $1-F$, $F$ being the fractional flow) and the saturation profiles for a value of $\theta = 0.75$ are compared. The global fields corresponding to single-phase flow solutions are computed at $\theta_1 = 0.5$ and $\theta_2 = 1$. The simulations are run with $\mu_o/\mu_w = 5$. We note that the value of $\theta$ is different from the values used in generating basis functions. We observe from these figures that the mixed MsFEM provides an accurate representation of the solution. In particular, there is almost no difference in the water-cut curve and the error in the saturation profile at PVI = 1 is less than 5%. This observation is consistent for all other values between $\theta_1$ and $\theta_2$, and it is demonstrated next.

In our next set of numerical experiments, water-cut errors and saturation errors for values of $\theta$ between $\theta_1 = 0.5$ and $\theta_2 = 1.5$ are presented. We also compare these results with the results obtained using only one value of $\theta$,

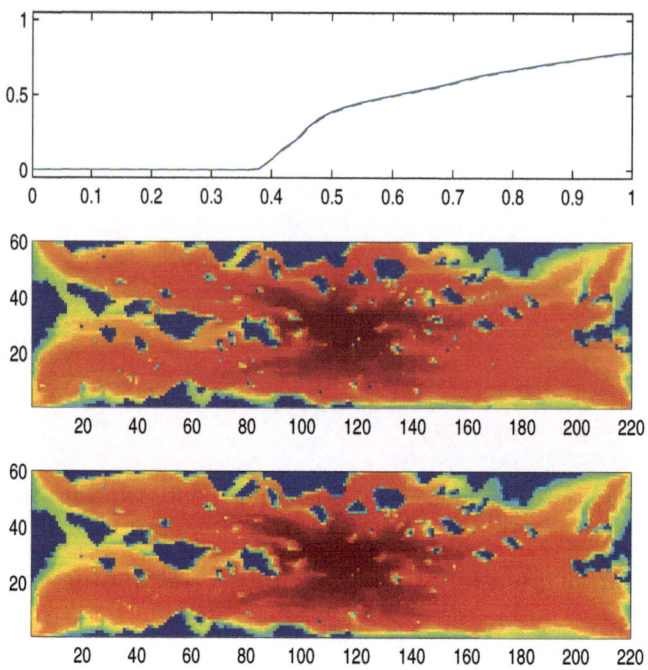

**Fig. 4.10.** Top: comparison of water-cut between reference solution and multiscale solution; middle: the reference saturation at PVI = 1; bottom: multiscale saturation at PVI = 1 (layer 85).

$\theta = 1$. More precisely, we only use the global solution corresponding to $\theta = 1$ to construct multiscale basis functions. Furthermore, these basis functions are used for solving the two-phase flow on the coarse grid for other values of $\theta$. We observe from Figures 4.11 and 4.12, that the results are substantially better if two global solutions are employed in characterizing the solutions for the entire range of $\theta$. In Figure 4.11, $\mu_o/\mu_w = 0.1$ is taken and in Figure 4.12, $\mu_o/\mu_w = 10$ is taken. It is clear from these figures that the use of two global solutions in mixed MsFEMs gives us an accurate approximation. The presented numerical results show that one can use a few realizations of the permeability field to construct basis functions that can be employed for solving two-phase flow and transport on the coarse grid accurately. Similar ideas have been used in applications of mixed MsFEMs to stochastic equations (see Section 5.7.1).

One of our goals with presented numerical results is to show that the solution can be approximated using multiple global fields. Next, we discuss the numerical convergence of global mixed MsFEMs with limited global information. For this reason, we consider different coarse grids, $6 \times 22$, $12 \times 44$, and

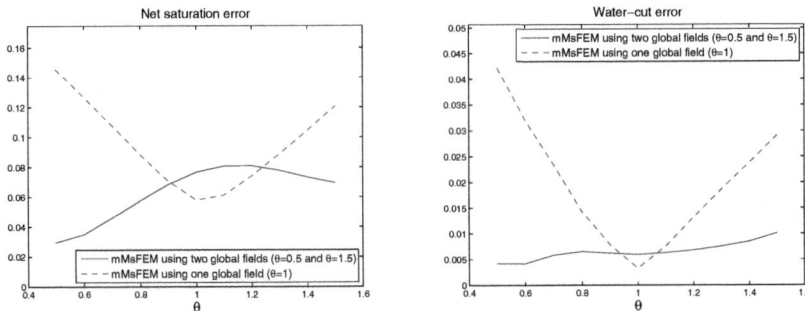

**Fig. 4.11.** $L^1$ saturation error and water-cut error using one single-phase flow solution and two single-phase flow solutions, $\mu_o/\mu_w = 0.1$ (layer 85).

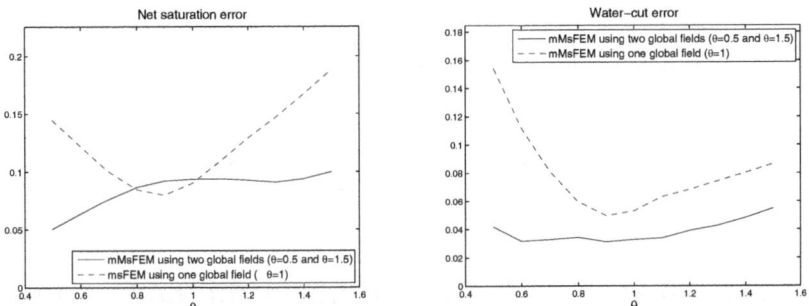

**Fig. 4.12.** $L^1$ saturation error and water-cut error using one single-phase flow solution and two single-phase flow solutions, $\mu_o/\mu_w = 10$ (layer 85).

$15 \times 55$ for the previous example with $\mu_o/\mu_w = 10$. Our convergence analysis (see Section 6.3) indicates that the proposed method converges up to a small parameter that represents how well the two-phase velocity field can be approximated by a single-phase velocity field in each coarse patch. Moreover, the convergence rate also depends on the smoothness of $A_i$ in (4.5). One can consider an ideal toy problem where the convergence rate can be verified by specifying the form of the solution up to smooth functions $A_i$ (see (4.5)). Instead, we would like to consider the SPE 10 example and show that as the coarse mesh size decreases the error decreases. We note that this is in contrast to standard MsFEMs where one can observe the resonance error. As a result, the mixed MsFEM does not converge as $h$ approaches zero. As we see from Figure 4.13, the mixed MsFEM using limited global information converges as the coarse mesh size decreases. This is again an indication that for general complicated media such as SPE 10 with high contrast, one can expect the convergence of mixed MsFEMs using limited global information.

We note that the method can be used for stochastic flow equations. This is presented in Section 5.7.1. In this case, one can take $v_i$ to be the realizations

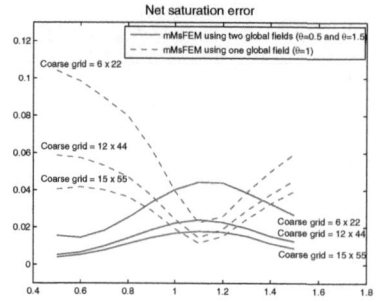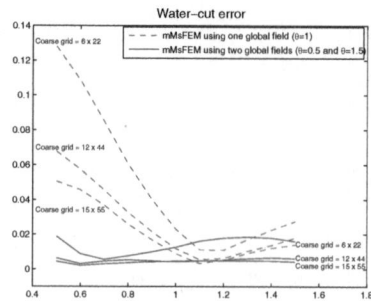

**Fig. 4.13.** $L^1$ saturation error and water-cut error using one single-phase flow solution and two single-phase flow solutions, $\mu_o/\mu_w = 10$ for different degrees of coarsening (layer 85).

of the random fields. This way multiscale basis functions capture the small-scale information across the realizations of stochastic porous media equations. Because these approaches do not necessarily require global information and can be considered as an application of MsFEMs, we present them in the last chapter of the book.

## 4.3 Galerkin multiscale finite element methods using limited global information

### 4.3.1 A special case

First, we consider a special case where only one global field is used for generating multiscale basis functions. We denote the solution of the pressure equation at time zero by $p^{sp}(x)$, where the superscript $sp$ refers to single-phase flow ($\lambda = 1$ in (4.1)). In defining $p^{sp}(x)$, we use the actual boundary conditions of the global problem. The boundary conditions for modified basis functions are defined in the following way. For simplicity of the presentation, we consider a rectangular partition in 2D. For each rectangular element $K$ with vertices $x_i$ ($i = 1, 2, 3, 4$), denote by $\phi_i(x)$ a restriction of the nodal basis on $K$, such that $\phi_i(x_j) = \delta_{ij}$. At the edges where $\phi_i(x) = 0$ at both vertices, we take the boundary condition for $\phi_i(x)$ to be zero. Consequently, the basis functions are localized. We only need to determine the boundary condition at two edges that have the common vertex $x_i$ ($\phi_i(x_i) = 1$). Denote these two edges by $[x_{i-1}, x_i]$ and $[x_i, x_{i+1}]$ (see Figure 4.14). We only need to describe the boundary condition $g_i(x)$ for the basis function $\phi_i(x)$ along the edges $[x_i, x_{i+1}]$ and $[x_i, x_{i-1}]$. If $p^{sp}(x_i) \neq p^{sp}(x_{i+1})$, then

$$g_i(x)|_{[x_i, x_{i+1}]} = \frac{p^{sp}(x) - p^{sp}(x_{i+1})}{p^{sp}(x_i) - p^{sp}(x_{i+1})}, \quad g_i(x)|_{[x_i, x_{i-1}]} = \frac{p^{sp}(x) - p^{sp}(x_{i-1})}{p^{sp}(x_i) - p^{sp}(x_{i-1})}.$$

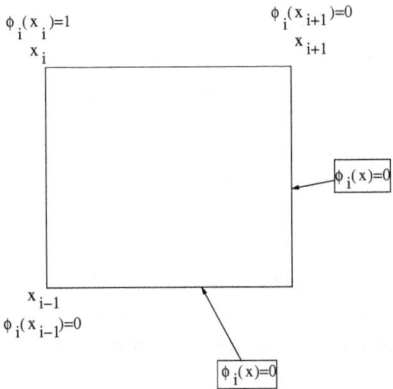

**Fig. 4.14.** Schematic description of nodal points.

If $p^{sp}(x_i) = p^{sp}(x_{i+1}) \neq 0$, then

$$g_i(x)|_{[x_i, x_{i+1}]} = \phi_i^0(x) + \frac{1}{2p^{sp}(x_i)} (p^{sp}(x) - p^{sp}(x_{i+1})),$$

where $\phi_i^0(x)$ is a linear function on $[x_i, x_{i+1}]$ such that $\phi_i^0(x_i) = 1$ and $\phi_i^0(x_{i+1}) = 0$. Similarly,

$$g_{i+1}(x)|_{[x_i, x_{i+1}]} = \phi_{i+1}^0(x) + \frac{1}{2p^{sp}(x_{i+1})} (p^{sp}(x) - p^{sp}(x_{i+1})),$$

where $\phi_{i+1}^0(x)$ is a linear function on $[x_i, x_{i+1}]$ such that $\phi_{i+1}^0(x_{i+1}) = 1$ and $\phi_{i+1}^0(x_i) = 0$. If $p^{sp}(x_i) = p^{sp}(x_{i+1}) \neq 0$, then one can also use simply linear boundary conditions. If $p^{sp}(x_i) = p^{sp}(x_{i+1}) = 0$ then linear boundary conditions are used. Finally, the basis function $\phi_i(x)$ is constructed by solving the leading-order homogeneous equation $\text{div}(k\nabla\phi_i) = 0$. The choice of the boundary conditions for the basis functions is motivated by the analysis. In particular, we would like our basis functions to span the fine-scale solution $p^{sp}(x)$. Using this property and Cea's lemma one can show that the pressure obtained from the numerical solution is equal to the underlying fine-scale pressure. The latter combined with the fact that the two-phase flow solution $p$ is a smooth function of $p^{sp}$ (see [103]) allows us to show that the proposed multiscale finite element method converges independent of resonance error. This approach is effective when the solution of (4.1) is a smooth function of $p^{sp}$.

### 4.3.2 General case

The MsFEMs considered above employ information from only one single-phase flow solution. In general, it might be necessary to use information from multiple global solutions for the computation of an accurate two-phase flow solution. The previous MsFEMs can be extended to take into account additional global information. Next, we present an extension of the Galerkin MsFEM that is based on the partition of unity method [32] (also see e.g., [248], [121], [153]).

As we mentioned before, we assume (4.2); that is $\|p - G(p_1, ..., p_N)\|_{L^2(\Omega)}$ is sufficiently small for a priori selected global fields $p_1, ..., p_N$. Here $G$ is a smooth function. Here, $p_1, ..., p_N$ are global (or local) fields that can approximate the solution.

Let $\omega_i$ be a coarse-grid patch (see Figure 4.15), and define $\phi_i^0$ to be partition of unity functions (e.g., piecewise linear basis functions) such that $\phi_i^0(x_j) = \delta_{ij}$. For simplicity of notation, denote $p_1 = 1$. Then, the MsFEM for each patch $\omega_i$ is constructed by

$$\Psi_{ij} = \phi_i^0 p_j,$$

where $j = 1, .., N$ and $i$ is the index of nodes (see Figure 4.15). We note that in each coarse patch $\sum_{i=1}^{n} \Psi_{ij} = p_j$ is the desired global field. Because the solution can be approximated by $p_j$, one can show that the MsFEM converges independent of resonance errors ([162]). Note that the form of the function $G$ is not important for the computations; however, it is crucial that the basis functions span $p_1, ..., p_N$ in each coarse block. The convergence results are presented in Section 6.3.

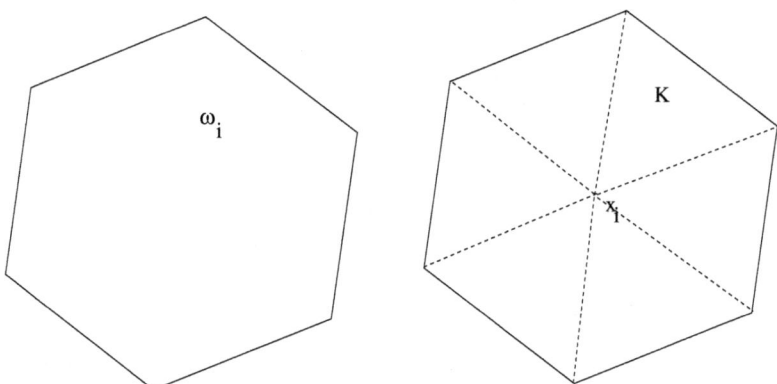

**Fig. 4.15.** Schematic description of patch.

### 4.3.3 Numerical results

Next, we show numerical results obtained for MsFEMs using limited global information presented in Section 4.3.1. We consider two-phase flow and transport and use only one global field, single-phase flow information ($N = 1$), as

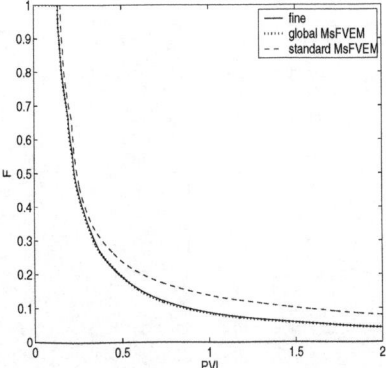

**Fig. 4.16.** Fractional flow comparison for standard MsFVEM and global MsFVEM.

described in Section 4.3.1 in the construction of basis functions. The basis functions are coupled via the finite volume formulation of the problem (see Section 2.5.1). We refer to this method as the global MsFVEM. Our first numerical example is for the permeability layer depicted in Figure 4.1 and two-phase flow parameters presented earlier in Section 2.10. As before, we specify $p = 1$, $S = 1$ along the $x = 0$ edge and $p = 0$ along the $x = 5$ edge. On the rest of the boundaries, we assume a no-flow boundary condition. Results are also presented in terms of the fraction of oil in the produced fluid (i.e., oil-cut, designated by $F$) against pore volume injected (PVI). Recall that PVI represents dimensionless time and is computed via $\int Q dt / V_p$, where $V_p$ is the total pore volume of the system and $Q$ is the total flow rate (see (2.44) for the definition of PVI).

In Figure 4.16, the fractional flows are plotted for standard and global MsFVEMs. We observe from this figure that the global MsFVEM is more accurate and provides nearly the same fractional flow response as the direct fine-scale calculations. In Figure 4.17, we compare the saturation fields at PVI = 0.5. As we see, the saturation field obtained using the global MsFVEM is very accurate and there is no longer the saturation pocket at the left bottom corner (cf. Section 4.1.1). Thus, the global MsFVEM captures the connectivity of the media accurately.

In the next set of numerical results, we test global MsFVEMs for a different layer (layer 40) of the SPE comparative solution project. In Figures 4.18 and 4.19, the fractional flows and total flow rates ($Q$) are compared for two different boundary conditions. One can see clearly that the global MsFVEM gives nearly exact results for these integrated responses. The standard MsFVEM tends to overpredict the total flow rate at time zero. This initial error persists at later times. This phenomenon is often observed in the upscaling of two-phase flows. More numerical results and discussions can be found in [103]. These numerical results demonstrate that global MsFEMs which use limited

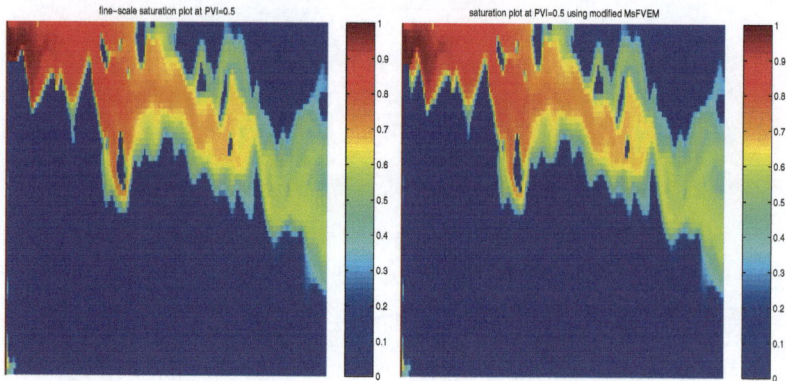

**Fig. 4.17.** Saturation maps at PVI $= 0.5$ for fine-scale solution (left figure) and global MsFVEM (right figure).

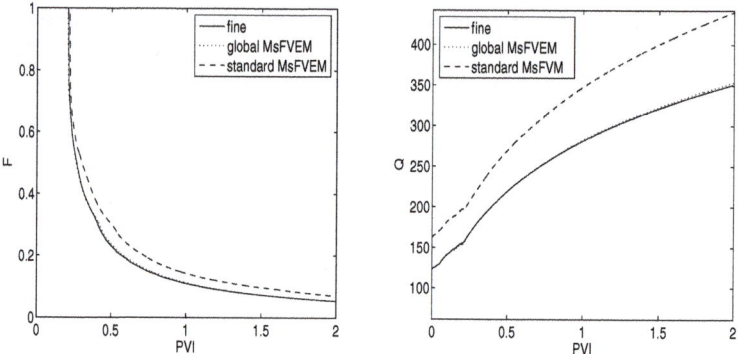

**Fig. 4.18.** Fractional flow (left figure) and total production (right figure) comparison for standard MsFVEM and global MsFVEM (layer 40).

global information are more accurate. Moreover, global MsFEMs are capable of capturing long-range flow features accurately for channelized permeability fields.

In the next set of numerical results, we consider another layer of the upper Ness (layer 59). In Figure 4.20, both fractional flow (left figure) and total flow (right figure) are plotted. We observe that the global MsFVEM gives almost the exact results for these quantities, whereas the standard MsFVEM overpredicts the total flow rate, and there are deviations in the fractional flow curve around PVI $\approx 0.6$. Note that unlike the previous case, fractional flow for standard MsFVEM is nearly exact at later times (PVI $\approx 2$). In Figure 4.21, the saturation maps are plotted at PVI $= 0.5$. The left figure represents the fine scale, the middle figure represents the results obtained using a standard MsFVEM, and the right figure represents the results obtained using a global

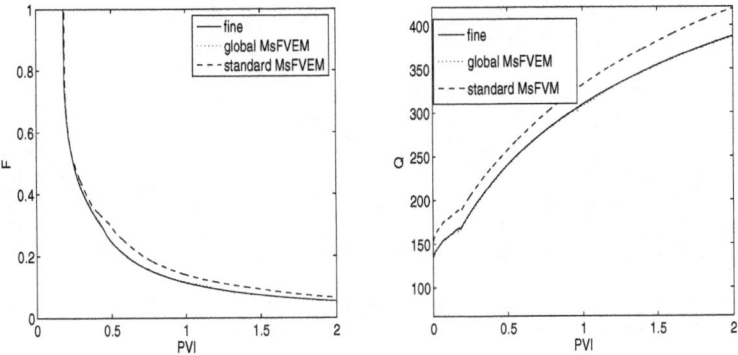

**Fig. 4.19.** Fractional flow (left figure) and total production (right figure) comparison for the standard MsFVEM and global MsFVEM (layer 40).

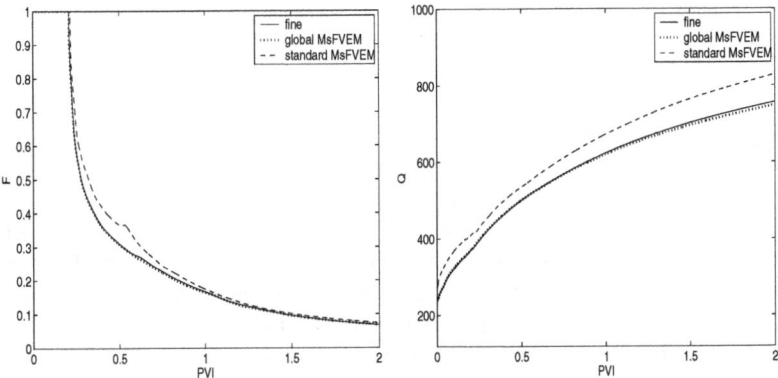

**Fig. 4.20.** Fractional flow (left figure) and total production (right figure) comparison for the standard MsFVEM and global MsFVEM.

MsFVEM. We observe from this figure that the saturation map obtained using a standard MsFVEM has some errors. These errors are more evident near the lower left corner. The results of the saturation map obtained using the global MsFVEM are almost the same as the fine-scale saturation field. It is evident from these figures that the global MsFVEM performs better than the standard MsFVEM.

## 4.4 The use of approximate global information

In the above discussions, the global fields are computed by solving simplified fine-scale equations. One can also use approximate global solutions instead of solving fine-scale elliptic equations. There are various ways one can attempt to

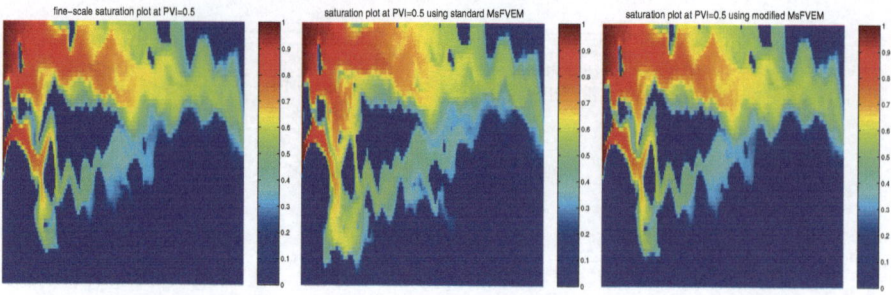

**Fig. 4.21.** Saturation maps at PVI = 0.5 for fine-scale solution (left figure), standard MsFVEM (middle figure), and global MsFVEM (right figure).

approximate the global fields and we briefly discuss two types of approximate global solutions. In the first approach, approximate global fields that capture nonlocal effects are computed iteratively. In the second approach, we attempt to compute global fields with fewer fine-scale details by homogenizing some small-scale features that can be localized.

### 4.4.1 Iterative MsFEM

One can attempt to capture nonlocal effects iteratively by using the approximate solutions obtained from MsFEMs. This procedure is schematically presented in Figure 4.22. At each iteration, approximations of the global solutions obtained via MsFEMs are used in the computation of multiscale basis functions. The computations of multiscale basis functions are the same as discussed above (e.g., (4.6)). Once the basis functions are computed, the global problem is solved on the coarse grid and updated approximations of the global solutions are computed. These global solutions are again used for multiscale basis function computation after possible post-processing with smoothers if needed. A convergence criterion based on the difference of consecutive approximate MsFEM solutions can be used to stop the iterations. In order to avoid using the same space of multiscale basis functions, one can use different sizes of oversampling domains in the computation of basis functions.

An algorithm with a similar concept was introduced in [93]. In [93], the authors proposed the use of a MsFVEM solution as a global solution. Numerical results show that one can achieve substantial improvement when small oversampling is employed in computing the global solutions. Moreover, one can apply this approach iteratively, by re-computing the MsFVEM solution. This iterative procedure converges in two to three iterations for heterogeneous permeability fields such as SPE 10. In general, the correction to the multiscale solution via iterations can be very useful in many practical applications. Indeed, computing a global solution each time when heterogeneities or flow fields change can be expensive. On the other hand, the iterative approaches that can compute the approximate global solution by updating a few

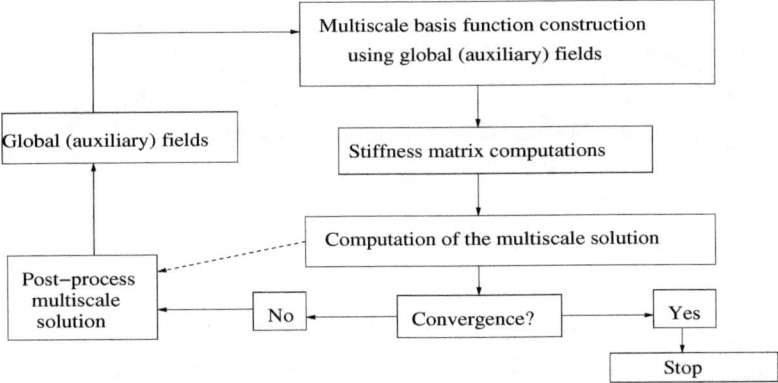

**Fig. 4.22.** An outline of iterative MsFEM.

multiscale basis functions can be very useful in fast flow simulations. These approaches share some similarities with domain decomposition methods (e.g., [137]), although there are important differences. The approach proposed in [93] iteratively computes multiscale basis functions that can be re-used for different source terms and boundary conditions, and domain decomposition methods correct the solution in the iterations. One can also keep multiscale basis functions the same during the iterations and compute the corrections to the solution in the iterations.

### 4.4.2 The use of approximate global information

Another possible approximation of global solutions can be obtained by removing the small-scale details that can be localized. This way, one will compute only important nonlocal features of the global fields by homogenizing some of the small-scale features that can be recovered in the basis function construction (see Figure 4.23 for the illustration). This can provide CPU savings in global solution computations because not all small-scale features are resolved. To demonstrate this concept, we assume that the coefficients are described by $k_{\delta>,\epsilon}(x)$, where $\delta^>$ refers to the hierarchy of scales that are larger than $\delta$, and the parameter $\epsilon$ ($\epsilon \ll \delta$) refers to the small scale that can be homogenized. Denote the partially homogenized coefficients by $k_{\delta>}^*(x)$. By homogenizing $\epsilon$ scales, one can use $k_{\delta>}^*(x)$ to compute the global fields. The use of nonuniform coarsening will allow us to discretize the equation with the coefficients $k_{\delta>}^*(x)$ on a coarser grid compared to the equation with the coefficients $k_{\delta>,\epsilon}(x)$. Indeed, many fine-scale features are due to $\epsilon$ scales. This will provide CPU savings in the computation of auxiliary global fields. Moreover, one can use smaller regions (RVE) as in Section 2.6 in the computation of $k_{\delta>}^*(x)$. In [108], we use the global solutions computed with the coefficients $k_{\delta>}^*(x)$ to construct multiscale basis functions and investigate the convergence of the method.

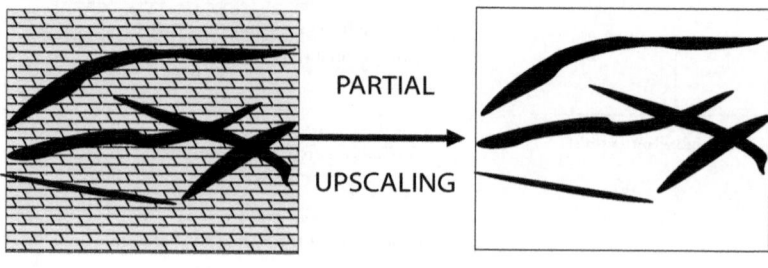

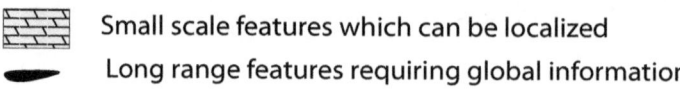

Small scale features which can be localized

Long range features requiring global information

**Fig. 4.23.** Schematic description of partial upscaling.

## 4.5 Discussions

One of the theoretical works on using limited global information in MsFEMs is by Owhadi and Zhang [218]. In this work, the authors show that the solution is smooth in a harmonic coordinate system. These results are shown under some suitable assumptions for the case $d = 2$ and more restrictive assumptions for the case $d = 3$. The use of harmonic coordinates in homogenization is not new. In [169], the author used harmonic coordinates to transform the elliptic equations with random coefficients into the elliptic equations in non-divergence forms (without lower-order terms). The homogenization of elliptic equations in non-divergence form is carried out by using spatial averaging. Harmonic coordinates in [218] consist of directional solutions of the single-phase flow equation. This suggests that one can solve the flow equation with an arbitrary right-hand side or smooth mobility $\lambda(x)$ in a harmonic coordinate system using standard finite element basis functions on a coarse grid. In the original (physical) coordinate system, this method entails solving the flow equations with multiscale basis functions that span the global solutions and, perhaps, constant or low-order polynomials. Moreover, the coarse grid in the original coordinate system is the image of the regular coarse-grid block in the harmonic coordinate system. This image is taken under the inverse of harmonic coordinate transformation. These coarse-grid blocks are usually highly distorted [218].

The multiscale methods using limited global information usually perform well (numerically) for high-contrast media. However, rigorous analysis for general high-contrast media is still an open question. Some results along this direction have been obtained recently in [137, 76, 67]. In [76], piecewise constant heterogeneities are considered. The authors show that by constructing some appropriate multiscale boundary conditions for the basis functions that take into account the local geometric property of the solution, the MsFEM converges with an optimal convergence rate independent of the aspect ratio of the heterogeneous coefficient. This is perhaps the first result in which one can obtain an optimal convergence rate independent of the high-contrast of the coefficients for a finite element method which does not require alignment of the finite element mesh with the interface boundary. In [67], the high contrast problem is formulated as an interface problem without a high contrast. Then the basis functions for the interface on the coarse grid are computed in addition to regular multiscale basis functions.

The question of whether local changes in the permeability can be treated by modifying multiscale basis functions locally is addressed in [77]. The authors consider the mobility functions $\lambda(S)$ in (4.1) which are discontinuous functions. It is shown that by changing the basis functions only near the discontinuities, one can achieve a convergent method for problems without scale separation.

The limited global information can be very useful in coarsening. In [218], the authors use the level sets of the directional solutions to generate the coarse grid. The use of level sets of the directional solutions has limitations. In previous findings, the pressure-streamline coordinates have been used in coarsening. In a recent work [93], we propose a generic algorithm, extending the main idea of [69], for performing nonuniform coarsening using a single-phase velocity field. We show that one can achieve higher accuracy with fewer degrees of freedom (compared to uniform coarsening).

As we mentioned earlier the use of limited global information in coarsening is not new. The single-phase information has been used in upscaling methods for porous media flows before. One of the main difficulties in upscaling methods when using limited global information is to recover exactly the average response of the global fine-scale information. In [140], the authors solve an optimization problem for computing the upscaled permeabilities that give nearly the same average response as the global solution. In [69], the authors propose an iterative method using global information which converges in a few iterations. The resulting upscaled coefficients give nearly the same average flow response as the global single-phase flow solution. Limited global information is also used in multiphase upscaling for upscaling of relative permeabilities.

As we mentioned above one can use multiple global solutions in computing basis functions. This is particularly useful for stochastic problems, where different realizations are used in computing basis functions, or in the situations where a priori knowledge about the change in heterogeneities or boundary conditions is known. Then, multiple global solutions can be used in constructing multiscale basis functions. In this way, one can use the same set of basis functions throughout the simulation.

We would like to note that the use of limited global information in nonlinear problems does not seem to be possible, in general. This is due to the fact that the heterogeneities depend on nonlinearities and the solution. The use of limited global information usually assumes knowledge about the spatial heterogeneities. For nonlinear problems, one does not have a priori knowledge about the heterogeneities. One approach is to identify a set of heterogeneities that will occur in nonlinear problems. For example, in nonlinear elliptic or parabolic problems, this will involve finding spatial information about $k(x, p, \nabla p)$ for all $p$. Once this has been determined, one can construct a larger set of basis functions that are capable of capturing all the long-range effects. To our best knowledge, these issues have not been addressed so far.

Finally, we mention that the multiscale methods using limited global information can be extended to other linear equations, such as wave equations [163].

# 5

# Applications of multiscale finite element methods

## 5.1 Introduction

In this chapter, we present some applications of MsFEM to fluid flows in heterogeneous porous media. We discuss multiscale methods for transport equations and their coupling to flow equations which are solved using MsFEMs. The proposed multiscale techniques for the transport equation share some similarities with nonlinear multiscale methods introduced in Chapter 3. Because of sharp interfaces, special treatment is needed near the interface. Furthermore, due to the hyperbolic nature of the transport equation, some type of limited global information is needed for constructing multiscale basis functions. These issues are discussed in Section 5.2.

In Section 5.3, we discuss the applications of MsFEMs to flows in unsaturated porous media described by Richards' equations [236]. Multiscale methods developed in Chapter 3 are applied to solve Richards' equation in heterogeneous porous formations on the coarse grid. In Section 5.4, we extend MsFEMs to solving the fluid-structure problem on the coarse grid where as a result of fluid flow in the pore region, the porous medium deforms substantially.

Applications of MsFEMs to reservoir modeling are presented using both the mixed MsFEM and MsFV in Sections 5.5 and 5.6. In these sections, more complicated porous medium equations involving compressibility, gravity, and three phases in heterogeneous reservoirs are considered. The authors address the challenging issues that arise in petroleum applications and describe the efficient use of MsFEMs in these problems.

The porous medium properties are typically described using geostatistical techniques because of uncertainties associated with prescribing permeability values to different locations. The numerical simulation of fluid flows in stochastic porous media is prohibitively expensive because the computation of each realization is CPU-demanding. In this chapter, we also consider approaches for constructing multiscale basis functions for the whole ensemble. Furthermore, the applications of MsFEMs to uncertainty quantification in inverse problems

Y. Efendiev, T.Y. Hou, *Multiscale Finite Element Methods: Theory and Applications,* 95
Surveys and Tutorials in the Applied Mathematical Sciences 4,
DOI 10.1007/978-0-387-09496-0_5, © Springer Science+Business Media LLC 2009

consisting of permeability sampling are presented. The objective here is to use MsFEMs to speedup the computations aimed at quantifying uncertainties in inverse problems.

## 5.2 Multiscale methods for transport equation

### 5.2.1 Governing equations

A prototypical example for problems studied is two-phase immiscible flow and transport in heterogeneous media. We presented the governing equations in Section 2.10 neglecting the effects of gravity, compressibility, capillary pressure and dispersion on the fine scale. We recall that the system of equations consists of the pressure equation

$$\text{div}(\lambda(S)k(x)\nabla p) = q_t, \tag{5.1}$$

where $\lambda(S)$ is the total mobility and $q_t = q_o + q_w$ is the total volumetric source term. The saturation equation has the form

$$\phi\frac{\partial S}{\partial t} + \text{div}(vf(S)) = -q_w, \tag{5.2}$$

where $f(S)$ is the fractional flow of water ($f$ is also denoted by $f_w$ often to distinguish between oil and water fractional flows), and $\phi$ is porosity. The total velocity $v$ is given by

$$v = -\lambda(S)k\nabla p. \tag{5.3}$$

In the presence of capillary effects, an additional diffusion term is present in (5.2). The above system of equations can be extended to describe the flow and transport of three-phase flow and transport (see Sections 5.5 and 5.6). In Sections 5.5 and 5.6, the applications of MsFEMs to three-phase compressible flow and transport are described.

In this section, we focus on developing multiscale methods for the transport equation described by (5.2).

### 5.2.2 Adaptive multiscale algorithm for transport equation

In this section, we present an adaptive multiscale method for solving the transport equation following [5]. The main idea of this approach is to construct multiscale basis functions similar to the construction in nonlinear MsFEMs presented in Chapter 3. Because the solution of the transport equation has sharp interfaces, a separate treatment is needed for these interfaces.

The adaptive multiscale method that we propose here consists of two parts. An adaptive criterion determines if a block is in a transient flow region. Here, by transient region, we refer to those regions with sharp saturation fronts. In these regions we use local fine-grid computations to advance the saturation solution to the next time-step. In regions with slow transients, we use a multiscale coarse-grid solver to advance the saturation solution to the next time-step. Then, instead of doing a fine-grid calculation, we map the coarse-grid solution onto a fine-grid solution using special interpolation operators.

Before we give an outline of the algorithm, we need to introduce some additional notation. First, denote the coarse grid by $\mathcal{T} = \{K_i\}$ and an underlying fine grid by $\mathcal{K} = \{\tau_i\}$. The grids used here need not coincide with the coarse and fine grids for multiscale methods used for the pressure equation and can be unstructured. In this particular application, we use a mixed MsFEM.

We introduce now the upstream fractional flow function for $\gamma_{ij} = \partial K_i \cap \partial K_j$:

$$V_{ij}(S) = f(S_i) \max\{v_{ij}, 0\} + f(S_i) \min\{v_{ij}, 0\}, \tag{5.4}$$

where $v_{ij}$ is the Darcy flux across $\gamma_{ij}$ that we get from the mixed MsFEM solution. Next, let $\bar{S}_i^n$ be the coarse-grid saturation in $K_i$ at time $t_n$, and denote by $T_{\text{tr}}^n$ the family of grid blocks that are identified to be in a transient flow region at time $t_n$. One can use various criteria based on coarse-scale saturation values or their gradients to identify transient regions. In this section, the following criteria are used to identify transient flow regions:

$$K_i \in T_{\text{tr}}^n \text{ if } \max\{|\bar{S}_i^n - \bar{S}_j^n| : |\partial K_i \cap \partial K_j| > 0\} \geq \alpha_i. \tag{5.5}$$

For each $K_i \in T_{\text{tr}}^n$, we define

$$K_i^E = K_i \cup \{\tau \in \mathcal{K} : |\partial \tau \cap \partial K_i| > 0\}.$$

Hence, $K_i^E$ consists of grid cells that are either contained in $K_i$, or that share a common interface with a cell in $K_i$. Finally, we introduce a family of operators $\{I_K : K \in \mathcal{T}\}$ that map coarse-grid saturations onto fine-grid saturation fields inside the respective blocks. The adaptive multiscale method is now outlined in Algorithm 5.2.1.

Next, we briefly describe the algorithm. In this algorithm, first, the fine-grid saturations in the transient flow regions are updated. This update involves solving the local transport equation on the fine grid in the transient region. Coarse-grid saturations in nontransient regions are updated using (5.7). The equation (5.7) is obtained by averaging the transport equation over the coarse-grid block $K$ and describes the update for the coarse-scale saturation field. Once the coarse-scale saturation field is updated, it is mapped onto the fine grid with the coarse-to-fine grid interpolation operators. This step is similar to nonlinear MsFEMs as described in Section 3.1. In particular, the basis functions are computed for different levels of average saturation within the coarse grid block and, then, interpolated. In the algorithm, implicit time integration methods are used. There are no constraints on the time-steps $\Delta t$, but they should be chosen small enough to avoid an excessive numerical diffusion.

---

**Algorithm 5.2.1** Adaptive multiscale algorithm for modeling flow in porous media

---

For each $K \in \mathcal{T}_{\mathrm{tr}}^n$, do
- For $\tau_i \subset K^E$, compute

$$S_i^{n+1/2} = S_i^n + \frac{\Delta t}{\int_{\tau_i} \phi \, dx} \left[ \int_{\tau_i} -q_w(S^{n+1/2}) dx - \sum_{j \neq i} V_{ij}^* \right], \quad (5.6)$$

where $V_{ij}^* = \begin{cases} V_{ij}(S^n) & \text{if } \gamma_{ij} \subset \partial K^E \text{ and } v_{ij} < 0. \\ V_{ij}(S^{n+1/2}) & \text{otherwise.} \end{cases}$

- Set $S^{n+1}|_K = S^{n+1/2}|_K$.

For each $K \notin \mathcal{T}_{\mathrm{tr}}^n$, do
- Set $S^{n+1}|_K = S^n|_K$.
- While $\sum_j \triangle_j t \leq \Delta t$, compute

$$\bar{S}_K^{n+1} = \bar{S}_K^n + \frac{\triangle_j t}{\int_K \phi \, dx} \left[ \int_K -q_w(S^{n+1}) \, dx - \sum_{\gamma_{ij} \subset \partial K} V_{ij}(S^{n+1}) \right], \quad (5.7)$$

and set $S^{n+1}|_K = I_K(\bar{S}_K^{n+1})$.

---

The fractional function $f$ is in general a nonlinear function of saturation. We therefore solve the fine-grid equations (5.6) using a Newton–Raphson method. Here saturation from the previous time-step is used to determine boundary conditions along the inflow boundary on $\partial K^E$. This gives rise to a mass–balance error because the inflow on grid block boundaries corresponding to the saturation from the previous time-step will not match exactly the inflow on grid block boundaries corresponding to the saturation at the current time-step. In our numerical simulations, we observed that this mass–balance error is usually very small, and generally insignificant. Note also that if $\mathcal{T}_{\mathrm{tr}} = \emptyset$, and the coarse-to-fine grid interpolation conserves mass locally, then (5.7) ensures that mass is conserved, also globally. Thus, under the assumption that the coarse-to-fine grid interpolation conserves mass locally, the latter part of the adaptive multiscale algorithm is mass conservative on both coarse and fine grids.

Next, observe that fluxes across coarse-grid interfaces in (5.7) are evaluated on fine-grid interfaces $\gamma_{ij} \subset \partial K$. Thus, rather than using a flux function that models the total flux across coarse-grid interfaces as a function of the net saturation in the upstream block, we evaluate the term $fv$ in (5.2) on the scale of the fine grid. This requires that we have fine-grid saturation values in all cells adjacent to grid block boundaries. The coarse-to-fine grid interpolation operators $\{I_K\}$ are therefore not just tools to get better resolution. In addition to improving the global accuracy of Algorithm 5.2.1 by providing a better approximation to flow across coarse-grid interfaces, they provide initial fine-grid saturation values for (5.6) in the transition when a block is identified as being part of a transient flow region. Without the interpolation, the initial saturation field for (5.6) would be constant in $K$, and the fractional flow across the coarse-grid interfaces would have to be based on the net grid block saturations only, as pseudo-functions generally do [171].

We remark that the proposed adaptive multiscale method has some similarities to the multiscale framework developed for nonlinear equations in which multiscale basis functions are constructed by mapping the coarse dimensional space defined over the entire region. Furthermore, this map is used in the global coarse-grid formulation of the fine-scale problem to compute the coarse-scale solution. In our multiscale approach, the basis functions are constructed as a function of average saturation in each coarse block, and then used in the global formulation of the problem. In both approaches, the main task is to determine an accurate and efficient multiscale map that improves the global coarse-grid formulation of the problem.

### 5.2.3 The coarse-to-fine grid interpolation operator

In the following we attempt to construct operators that map each coarse-grid saturation field onto a fine-scale saturation profile that is close to the corresponding profile that one would get by solving the saturation equation on the global fine grid. The basic idea is to approximate the fine-scale saturation in $K_i$ as a linear combination of two basis functions $\Phi_i^k$ and $\Phi_i^{k+1}$ with $\int_{K_i} \Phi_i^k \phi\, dx \leq \bar{S}_i^n \int_{K_i} \phi\, dx < \int_{K_i} \Phi_i^{k+1} \phi\, dx$:

$$I_{K_i}(\bar{S}_i^n) = \eta\Phi_i^k + (1-\eta)\Phi_i^{k+1}. \tag{5.8}$$

Here $\eta \in [0,1]$ is chosen such that the interpolation preserves mass, that is such that

$$\int_{K_i} I_{K_i}(\bar{S}_i^n)\phi\, dx = \bar{S}_i^n \int_{K_i} \phi\, dx. \tag{5.9}$$

This condition states that the fluid contained in $K_i$ is distributed inside $K_i$ in such a way that the total fluid volume in $K_i$ is conserved. The basis functions $\Phi_i^k = s_i(x, \tau_k)$ represent snapshots of the solution of the following equation:

$$\phi\frac{\partial s_i}{\partial t} + \mathrm{div}(f(s_i)v) = -q_w \quad \text{in } K_i. \tag{5.10}$$

For the local problem (5.10) to be well defined, we need to specify initial conditions and boundary conditions, and provide a possibly time-varying velocity field in $K_i$. Unfortunately, we do not know a priori what the velocity will be during the simulation, nor what boundary conditions to impose. Assumptions must therefore be made as to how the velocity and saturation approximately evolve. We describe below an approach that is local in terms of boundary and initial conditions, however, one can naturally incorporate global information into this approach. The proposed approach assumes that global boundary conditions for the pressure equation (5.1) are not changed, and that the source terms are fixed. We assume also that an upstream method is used to solve the local equations (5.10). Thus, we need only specify boundary conditions on the inflow boundaries $\Gamma_K^{in} = \{\gamma_{jl} \subset \partial K : \tau_l \subset K,\ v_{jl} < 0\}$.

For fixed flow conditions, the fine-scale velocity features will generally not change significantly during a flow simulation. This is discussed in [5]. One option is therefore to solve the pressure equation (5.1) at the initial time with the mixed MsFEM, use $v = v(x, t_0)|_K$ in (5.10), and the same initial data as for the global problem (5.2). A local way of generating saturation basis functions based on this approach requires that sensible boundary conditions for (5.10) can be imposed for each block independently. In our numerical simulations, we impose $s_i = 1$ on the inflow boundary $\Gamma_T^{in}$, although other boundary conditions can be imposed (see discussions in Section 5.2.7).

An approach that is often used in practice for upscaling the saturation equation entails the use of so-called pseudo-relative permeabilities $(k_{rj}^*)_i = (k_{rj}^*)_{K_i}$ in place of the fine-scale $k_{rj}$. Because the fine-scale $k_{rj}$ are typically functions only of saturation $S$, pseudo-relative permeabilities, or pseudo-functions for brevity, are commonly assumed to depend only on the coarse-grid saturation $\overline{S}$, though the curves can vary between coarse grid blocks. The proposed technique shares some similarities with pseudo-function approaches although there are some important differences. The proposed approach allows recovering fine-scale features of the saturation field and can be used for accurate upscaling. The relation between proposed methods and pseudo-function approaches is discussed in [5].

### 5.2.4 Numerical results

We now use the proposed methodology to model incompressible and immiscible two-phase flow on test cases with permeability and porosity from SPE 10 [78]. This model was discussed before and consists of a Tarbert formation on top of a fluvial upper Ness formation. Although both formations are very heterogeneous, the upper Ness formation gives rise to more complex flow. We employ here mostly data modeling parts of the fluvial upper Ness formation. Because fluvial formations are particularly hard to upscale, the upper Ness formation should serve as an appropriate model for testing and validation of the proposed multiscale method. The upper Ness model is Cartesian and consists of $60 \times 220 \times 50 = 6.6 \cdot 10^5$ grid cells.

We assume that the reservoir is initially fully oil-saturated, and inject water at a constant rate in grid cells penetrated by a vertical well at the center of the domain. We then produce at the producers which are vertical wells located at each of the four corners. The water and oil mobilities are defined by

$$\lambda_w(S) = \frac{S^2}{\mu_w} \text{ and } \lambda_o(S) = \frac{(1-S)^2}{\mu_o}, \tag{5.11}$$

where the water and oil viscosities are assumed to be equal: $\mu_w = \mu_o = 0.003$ cp.

To measure the overall accuracy of a saturation solution we compute the error in the fine- and coarse-grid saturation profiles relative to a reference solution,

$$e(S, S_{\text{ref}}, t) = \frac{\|\phi S_{\text{ref}}(\cdot, t) - \phi S(\cdot, t)\|_{L^2}}{\|\phi S_{\text{ref}}(\cdot, t) - \phi S_{\text{ref}}(\cdot, 0)\|_{L^2}}.$$

Here time is measured in dimensionless time PVI, that is time measures the fraction of the total accessible pore volume in $\Omega$ that has been injected into $\Omega$.

For all test cases, we use Cartesian coarse grids, and assume that the fine-grid cells coincide with grid cells in the original Cartesian grid. The reference solution $S_{\text{ref}}$ is computed using an implicit upstream method on the fine grid, and a corresponding coarse-grid solution is computed using the same method on a coarse grid. Moreover, note that although we use a fixed set of basis functions for the mixed MsFEM, we solve the pressure equation repeatedly to account for mobility variations. Thus, the velocity fields in the simulations will differ from the velocity field used to generate the saturation basis functions. However, to assess the accuracy of solutions obtained using the adaptive multiscale algorithm (AMsA), we compute, at each pressure time-step, the velocity field corresponding to the reference solution for saturation, and use this velocity field in AMsA, and to compute the coarse-grid solution. This allows us to monitor the error that stems from AMsA only.

### 5.2.5 Results for a two-dimensional test case

We consider first a test case representing the bottom layer of the SPE model. The coarse grid is defined so that each grid block contains $10 \times 10$ grid cells. The saturation plots in Figure 5.1 show that the solutions obtained using AMsA with $\alpha = 0$, $\alpha = 0.1$, and $\alpha = 0.2$ (the same threshold is used in all grid blocks, see (5.5)) are very similar to the reference solution. We recall that $\alpha = 0$ corresponds to the case when the saturation update is performed in all coarse blocks and $\alpha = 1$ corresponds to the case when no saturation update is performed. The solution obtained using $\alpha = 1$ looks quite different compared to the cases with other values of $\alpha$. The sharp edges that we see in this plot are due to the fact that the boundary conditions used to generate the saturation basis functions overestimate the inflow. We therefore get too much saturation along the inflow part (with respect to the initial velocity field) of each grid block boundary. This indicates that without the adaptive component, AMsA is not able to provide plausible fine-grid saturation profiles. To achieve this, one has to build more information about the global flow problem into the saturation basis functions by specifying appropriate coarse grid blocks using global information or appropriate dynamic boundary conditions for (5.10).

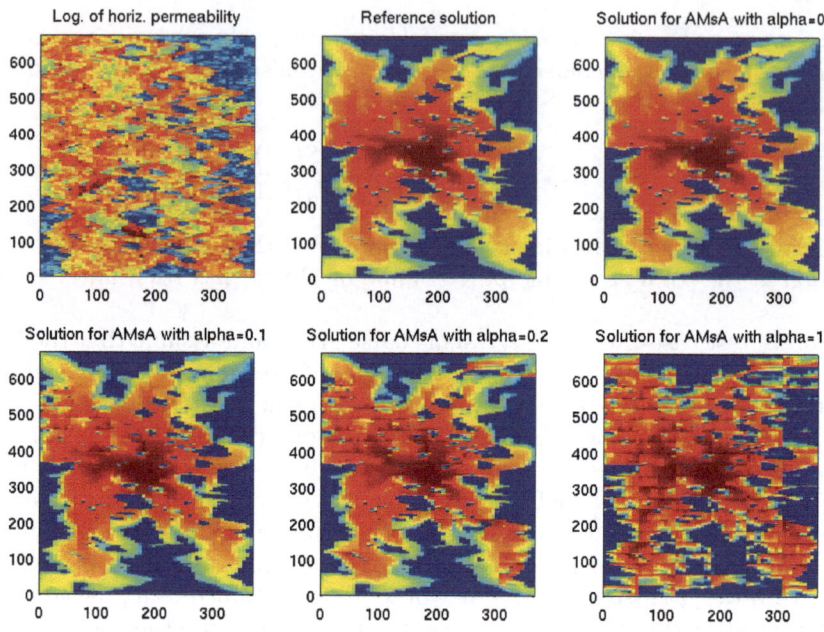

**Fig. 5.1.** Saturation profiles at $\sim 0.7$ PVI for simulations on the bottom layer.

Figure 5.2 shows that the accuracy of AMsA decays with increasing $\alpha$. However, for all $\alpha$, AMsA gives a significantly more accurate solution on the coarse grid than the standard upstream method on the coarse grid gives.

### Computational efficiency

Except for $\alpha = 1$, for which local problems are not solved during the course of a flow simulation, the computational cost of AMsA is dominated by the cost of solving the local equations (5.6). In particular, for small $\alpha$ the computational cost $C(\alpha)$ of solving (5.2) using AMsA scales roughly as

$$C(\alpha) \sim F_u(\alpha) N_t C(0),$$

where $N_t$ is the total number of time-steps and $F_u(\alpha)$ is the average fraction of blocks that belong to a transient flow region. Note that $C(0)$ is the cost when the transient region is the entire domain.

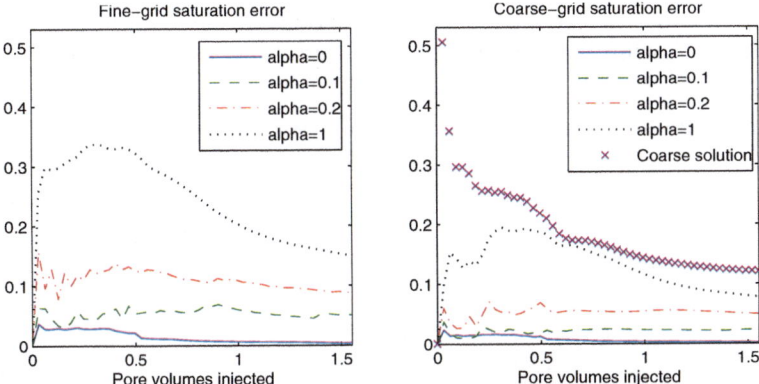

**Fig. 5.2.** Saturation errors for saturation solutions obtained from simulations on the bottom layer of the upper Ness formation. The fine-grid curves measure the error with $e(S, S_{\mathrm{ref}}, t)$ on the fine grid relative to the reference solution, and the coarse-grid curves measure the error on a coarse grid with $e(\bar{S}, \bar{S}_{\mathrm{ref}}, t)$ relative to the projection of the reference solution onto the coarse grid.

Clearly, $F_u$ is a decreasing function of $\alpha$. Hence, there is a trade-off between high accuracy and low computational cost. Note also that, in addition to $\alpha$, $F_u$ depends implicitly on various factors (e.g., the coarse grid, the criteria used to identify transient flow regions, the fluid parameters, the heterogeneous structures, etc.). In particular, AMsA is in general more efficient (and accurate) for spatially correlated variogram-based permeability models than for models with fluvial heterogeneity, as is illustrated in Figure 5.3. Whereas, on average, 73% and 55% of the blocks in the upper Ness model are identified as belonging to transient flow regions for $\alpha = 0.1$ and $\alpha = 0.2$ respectively, the corresponding numbers for the Tarbert model are 46 and 27. The potential efficiency of AMsA is therefore highly dependent on the type of model to which it is applied. Relative to AMsA with $\alpha = 0$, we may expect good accuracy on both coarse and fine grids, with a speed-up factor about two for models with fluvial heterogeneity, and a speed-up factor three or four for models with smoother heterogeneity. The speed-up strongly depends on the adaptivity criteria which can be adjusted for a particular problem. In our simulations, the criteria based on gradients of the coarse-scale saturation are used. We have observed an increase in speed-up when the criteria based on saturation values are used. Without the adaptive component, the computational complexity of AMsA is comparable to the complexity of coarse-grid simulations using pseudo-functions. As we mentioned earlier, the accuracy of AMsA can be improved by choosing adaptive coarse gridding. This procedure will also enhance the efficiency of AMsA, because it localizes sharp fronts. Finally, we note that the purpose of the interpolator is not primarily to get the fine-scale details correct, but rather to introduce a flexible mechanism that allows us to capture the subgrid transport effects on a coarse scale.

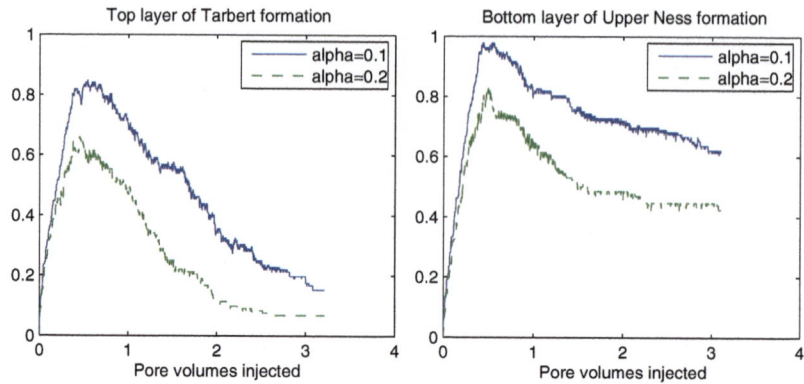

**Fig. 5.3.** Fraction of blocks that are identified to belong to transient flow regions during the course of two-phase flow simulations on the top layer of the Tarbert formation (left) and the bottom layer of the upper Ness formation (right).

### 5.2.6 Three-dimensional test cases

In this section we want to examine the accuracy of AMsA when applied to two-phase flow simulations on three-dimensional models from the upper Ness formation. Here we consider only AMsA using $\alpha = 0$, $\alpha = 0.1$, and $\alpha = 1$ in all blocks. The case $\alpha = 0$ is referred to as the domain decomposition (DD) algorithm, the case $\alpha = 0.1$ is referred to as the adaptive algorithm, and the case $\alpha = 1$ is called the multiscale algorithm.

In order for AMsA to provide a valuable tool in reservoir simulation, it should, in addition to being significantly more accurate than the coarse-grid solution, capture fine-scale characteristics of the reference solution at well locations. This is demonstrated by comparing water-cut curves (fraction of water in the produced fluid) for AMsA with water-cut curves for the reference solution. To get accurate production characteristics, it is essential that high-flow channels are resolved adequately because high-flow channels often carry the majority of the flow that reaches the producers. Thus, if AMsA can be used to model these regions properly, then they should provide a more robust alternative to reservoir simulation on upscaled models.

Consider first the ten bottom layers of the upper Ness formation, and define the coarse grid so that each grid block in the coarse grid consists of $10 \times 10 \times 5$ grid cells. Figures 5.4 and 5.5 demonstrate that all AMsAs give significantly more accurate results than the solution obtained by solving the saturation equation on the coarse grid with the implicit upstream method. We notice, in particular, that the water-cut curves for the multiscale algorithm are much more accurate than the corresponding water-cut curves for the coarse-grid solution. This indicates that AMsA is more capable of resolving high-flow regions adequately, also without the local fine-grid computations.

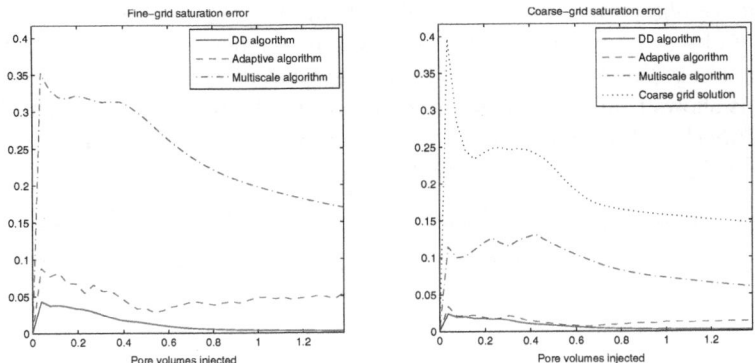

**Fig. 5.4.** Saturation errors for simulations on the bottom ten layers of the upper Ness formation.

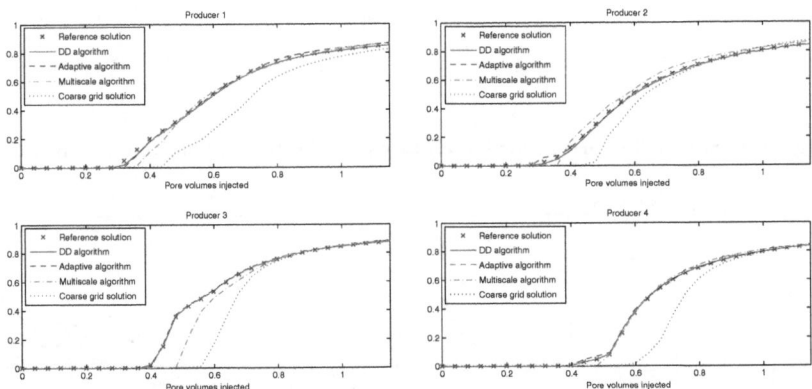

**Fig. 5.5.** Water-cut curves for simulations on the bottom ten layers of the upper Ness formation.

We turn now to the full three-dimensional model of the upper Ness formation. The previous examples showed that the DD algorithm seems to produce solutions that very closely match the reference solution, and it is computationally very expensive to compute a solution on the full upper Ness model using the implicit upstream method on the fine grid, therefore we use here the solution obtained using the DD algorithm as the reference solution. Again we let the coarse grid be defined so that each grid block in the grid consists of $10 \times 10 \times 5$ grid cells.

Figure 5.6 demonstrates that the errors are approximately the same as in the previous example. We observe also that the saturation error on the coarse grid for the multiscale algorithm is less than half of the corresponding error for the coarse-grid solution. Furthermore, the water-cut curves for the multiscale algorithm depicted in Figure 5.7 closely match the water-cut curves for the adaptive algorithm and the DD algorithm, except possibly for producer 4 where we observe a mismatch. In contrast, the coarse-grid solution continues to overestimate the breakthrough times, and thus overpredicts the oil production. This shows that the multiscale method may be used as an alternative to pseudo-functions for enhancing the accuracy of coarse-grid simulations.

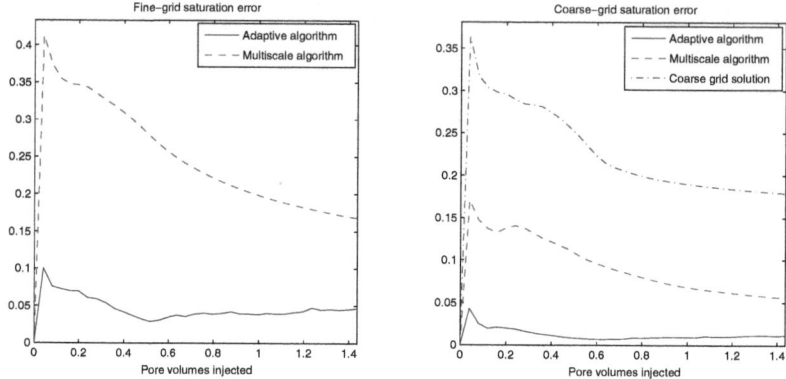

**Fig. 5.6.** Saturation errors for simulations on the full upper Ness formation.

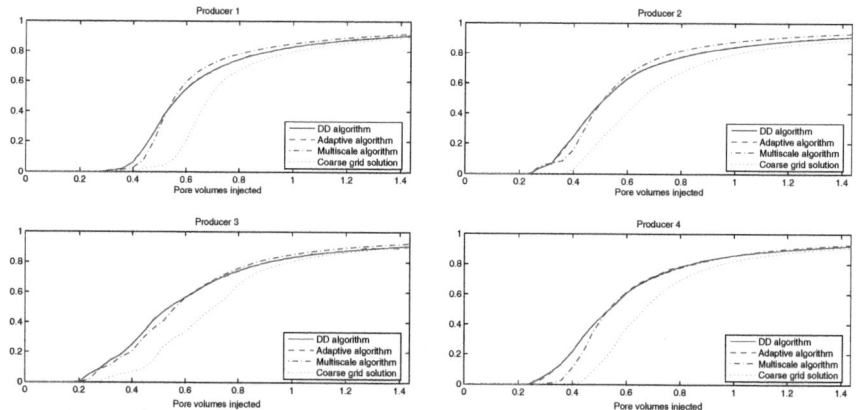

**Fig. 5.7.** Water-cut curves for simulations on the full model of the upper Ness formation.

### 5.2.7 Discussion on local boundary conditions

In our numerical simulations, we imposed $s_i = 1$ on the inflow boundary $\Gamma_T^{in}$ (see (5.10)). For general coarse grids, these boundary conditions may seem a bit crude. Indeed, these boundary conditions are exact only if there is a sharp front in the global solution that, for each block, hits the whole inflow boundary at approximately the same instant. It should be emphasized that the purpose of the interpolator is not primarily to get the fine-scale details correct, but rather to introduce a flexible mechanism that allows us to model the flow on a coarse scale more correctly.

To get accurate solutions, also on fine grids, one must either use an adaptive component to improve the solution in transient flow regions, or build more information into the interpolator. For instance, note that the inherent flexibility with respect to coarse grids allows us to reduce the error associated with this type of boundary condition by using flow-based, non-Cartesian grids. In particular, by using coarse blocks with boundaries aligned with level sets of time-of-flight function, one can achieve higher accuracy compared to the approaches where Cartesian coarse blocks (or coarse blocks selected independent of global flow features) are used. This option is discussed in [5]. We note that our numerical results show that the multiscale approach using Cartesian coarse blocks still provides a good overall accuracy. One can also use limited global information, such as the time-of-flight function, in constructing coarse blocks.

### 5.2.8 Other approaches for coarsening the transport equation

There are a number of other techniques for coarsening the saturation equation that can be coupled to the pressure equation. Next, we describe a few of these approaches very briefly without detailed numerical studies which can be found in the literature.

### A macrodispersion model for transport equation

The approach entails using a macrodispersion formulation for the coarse-scale saturation equation. We consider the upscaling of the saturation equation using perturbation techniques following, for example, [102, 101]. We omit the details of the derivation of the upscaled model. We first consider the case $\lambda(S) = 1$ and $f(S) = S$ in (5.1) and (5.2) (with $q_w = 0$ in (2.41)). The upscaled model was derived using perturbation arguments for (5.2), in which the saturation $S$ and the velocity $v$ on the fine scale are assumed to be the sum of their volume-averaged and fluctuating components,

$$v = \overline{v} + v', \quad S = \overline{S} + S'. \tag{5.12}$$

Here the overbar quantities designate the volume average of fine-scale quantities over coarse blocks. For simplicity, one can assume that the coarse blocks are rectangular which allows stating $\overline{\nabla F} = \nabla \overline{F}$, if averages are taken over dual volume. In general, one can also perform the perturbation technique directly on the target coarse block as done in [102]. In this case, the averages of divergences can be written over the boundaries of the coarse blocks. Substituting (5.12) into the saturation equation for single-phase and averaging over coarse blocks we obtain

$$\frac{\partial \overline{S}}{\partial t} + \overline{v} \cdot \nabla \overline{S} + \overline{v' \cdot \nabla S'} = 0. \tag{5.13}$$

The term $\overline{v' \cdot \nabla S'}$ represents subgrid effects due to the heterogeneities of convection. This term can be modeled using the equation for $S'$ that is derived by subtracting (5.13) from the fine-scale equation (5.2),

$$\frac{\partial S'}{\partial t} + \overline{v} \cdot \nabla S' + v' \cdot \nabla \overline{S} + v' \cdot \nabla S' = \overline{v' \cdot \nabla S'}.$$

This equation can be solved along the characteristics $dx/dt = \overline{v}$ by neglecting higher-order terms. Carrying out the calculations in an analogous manner to the ones performed in [102] we can easily obtain the following coarse-scale saturation equation

$$\frac{\partial \overline{S}}{\partial t} + \overline{v} \cdot \nabla \overline{S} = \mathrm{div}(D(x,t)\nabla \overline{S}(x,t)), \tag{5.14}$$

where $D(x,t)$ is the dispersive matrix coefficient, whose entries are written as

$$D_{ij}(x,t) = \left[\int_0^t \overline{v_i'(x)v_j'(x(\tau))}d\tau\right]. \tag{5.15}$$

Next it can be easily shown that the diffusion coefficient can be approximated up to the first order by

$$D_{ij}(x,t) = \overline{v_i'(x)L_j^D}$$

where $L_j^D$ is the displacement of the particle in the $j$ direction that starts at the point $x$ and travels with velocity $-v$. The diffusion term in the coarse model for the saturation field (5.14) represents the effects of the small scales on the large ones. Note that the diffusion coefficient is a correlation between the velocity perturbation and the displacement. This is different from [102] where the diffusion is taken to be proportional to the length of the coarse-scale trajectory. Using MsFEMs for the pressure equation we can recover the small-scale features of the velocity field that allow us to compute the fine-scale displacement.

For the nonlinear flux $f(S)$, we can use a similar argument by expanding $f(S) = f(\overline{S}) + f_S(\overline{S})S' + \cdots$. In this expansion we take into account only linear terms and assume that the flux is nearly linear. This case is similar to the linear case and the analysis can be carried out in an analogous manner. The resulting coarse-scale equation has the form

$$\frac{\partial \overline{S}}{\partial t} + \overline{v} \cdot \nabla \overline{S} = \mathrm{div}(f_S(\overline{S})^2 D(x,t) \nabla \overline{S}(x,t)), \tag{5.16}$$

where $D(x,t)$ is the macrodiffusion corresponding to the linear flow. This formulation has been derived within the stochastic framework in [173]. We note that the higher-order terms in the expansion of $f(S)$ may result in other effects that have not been studied extensively to our best knowledge. In [101] the authors use a similar formulation although their implementation is different from ours. Numerical results can be found in [102, 101].

### Coarsening in a flow-based coordinate system

In [106, 247], a flow-based coordinate system is used to coarsen the saturation equation. A flow-based coordinate system consists of single-phase pressure and the corresponding streamfunction fields. The use of global information can improve the multiscale finite element method. In particular, the solution of the pressure equation at the initial time is used to construct the boundary conditions for the basis functions. It is interesting to note that the multiscale finite element methods that employ limited global information reduce to the standard multiscale finite element method in a flow-based coordinate system. This can be verified directly and the reason behind it is that we have already employed limited global information in a flow-based coordinate system.

To achieve a high degree of speedup in two-phase flow computations, we consider the upscaling of the transport equation in a flow-based coordinate system. Flow-based coordinate systems simplify the scale interaction and allow us to perform upscaling of the transport equation. In particular, in a flow-based coordinate system, the saturation equation becomes one-dimensional with a varying velocity field along the streamlines. This allows us to use the perturbation approach and perform upscaling using macrodispersion models.

Extensive numerical studies are presented in [247, 106]. These numerical tests use the MsFVEM for two-phase flow. Note that global information is already incorporated into the multiscale basis functions and the standard MsFVEM is equivalent to the MsFVEM using limited global information introduced earlier. In our simulations, a moving mesh is used to concentrate the points of computation near the sharp front. Because the saturation equation is one-dimensional in the pressure–streamline coordinates, the implementation of the moving mesh is straightforward and efficient. We have presented the numerical results for different types of heterogeneities. All numerical results show that one can achieve accurate results with low computational cost.

### Multiscale analysis for convection dominated equations

In this section, we consider a systematic upscaling framework for the transport equation based on multiscale homogenization. In [144], Hou, Westhead, and Yang introduced a novel multiscale analysis for the two-phase immiscible flows in heterogeneous porous media. In particular they derived the homogenized equations by projecting the fluctuation of saturation onto a suitable subspace. Furthermore, they demonstrated by extensive numerical experiments that the upscaling method can accurately capture the multiscale solution of the two-phase flow. Very recently, Hou and Liang [142] further improved the multiscale analysis of Hou et al. [144] and developed a systematic multiscale analysis to upscale convection-dominated transport equations.

To demonstrate the main idea, we consider the following transport equation which contains a strong convection term and a weak diffusion term

$$\frac{\partial S_\epsilon}{\partial t} + v(x, \frac{x}{\epsilon}, t) \cdot \nabla S_\epsilon = \epsilon^m \text{div}(D(x, \frac{x}{\epsilon}, t) \nabla S_\epsilon),$$

where $S_\epsilon|_{t=0} = S_I(x)$, $m \in [2, \infty]$ is an integer, $v(x, y, t)$ and $D(x, y, t)$ are assumed to be periodic in $y = x/\epsilon$, and $\epsilon$ characterizes the small scale in the media, Moreover, we assume that $v$ is oscillatory divergence-free with respect to the fast variable $y$; that is $div_y(v) = 0$. The local Peclet number is of order $O(\epsilon^{-m+1})$.

Next, we define a null space $\mathcal{N}$, $\mathcal{N} = \{f \in H_Y^1, v \cdot \nabla_y f = 0, \forall y \in Y\} \subset L_Y^2$, where $L_Y^2$ is the $L^2$ space of periodic functions. This functional space plays an important role in our multiscale analysis. We also introduce a range space $\mathcal{W}$, $\mathcal{W} = \{v \cdot \nabla_y \theta : \theta \in H_Y^1\}$. In [144], the authors have shown that $\mathcal{N}$ and $\mathcal{W}$ form an orthogonal decomposition of $L_Y^2$; that is

$$L_Y^2 = \mathcal{N} \oplus \overline{\mathcal{W}}.$$

Let $\mathcal{P}$ be the projection $H_Y^1 \to \mathcal{N}$. Define the projection $\mathcal{Q}$: $L_Y^2 \to \mathcal{W}$ as

$$\|g - \mathcal{Q}(g)\| = \min_{\theta \in H_Y^1} \|g - v \cdot \nabla_y \theta\|.$$

As pointed out in [144], $\mathcal{P}$ is related to $\mathcal{Q}$ via $\mathcal{P}(g) = g - \mathcal{Q}(g)$, and $\mathcal{Q}$ can be computed by $\mathcal{Q}(g) = v \cdot \nabla_y \theta$, where $\theta$ is the solution of

$$\text{div}_y(E \nabla_y \theta) = v \cdot \nabla_y g, \quad y \in Y, \tag{5.17}$$

with periodic boundary condition and the matrix is defined by $E = v^T v$ whose $(i, j)$ entry is given by $v_i v_j$, where $v = (v_1, v_2, v_3)$. Moreover, the projection operator $\mathcal{P}$ is equivalent to the streamline averaging projection operator [144, 142].

Guided by our multiscale analysis, we look for a multiscale expansion of the concentration in the form

$$S_\epsilon(x, t) = S_0(x, x/\epsilon, t) + \epsilon S_1(x, x/\epsilon, t) + O(\epsilon^2),$$

where $S_j$ $(j = 0, 1)$ are periodic functions of $y$.

In [142], we showed that the leading-order approximation $S_0$ satisfies the following homogenized equations

$$v \cdot \nabla_y S_0 = 0, \tag{5.18}$$

$$\frac{\partial S_0}{\partial t} + v \cdot \nabla_x S_0 + v \cdot \nabla_y w = 0, \tag{5.19}$$

where $w \in \mathcal{W}$, and the initial condition is given by $S_0|_{t=0} = S_I(x)$.

Note that there are two equations for $S_0$ given by (5.18) and (5.19), but there is no evolution equation for $w$. The equation for $w$ can be derived by imposing the algebraic constraint (5.18) for $S_0$. The role of $w$ is to enforce $v \cdot \nabla_y S_0 = 0$, which is similar to the role that the pressure plays in the incompressible Navier–Stokes equations. The solution $w$ can be obtained by solving (5.17). In [144], an effective iterative method was introduced to solve the degenerate elliptic equation (5.17).

One of the main contributions of [142] is to show that the homogenized equations (5.18) and (5.19) are well-posed and obtain an optimal error estimate

$$\|S_\epsilon(x,t) - S_0(x, \frac{x}{\epsilon}, t)\|_{L^2} \leq C\epsilon.$$

We now decompose $c_0$ and $v$ into the sum of their average and fluctuation, $S_0(x,y,t) = \overline{S_0}(x,t) + S_0'(x,y,t)$, $v(x,y,t) = \overline{v}(x,t) + v'(x,y,t)$, where $\overline{f}(x,t) = \int_Y f(x,y,t)dy$. It is easy to show that $\overline{S_0}$ and $S_0'$ satisfy the following equations

$$\frac{\partial \overline{S_0}}{\partial t} + \overline{v} \cdot \nabla_x \overline{S_0} + \overline{v' \cdot \nabla_x S_0'} = 0, \tag{5.20}$$

$$\frac{\partial S_0'}{\partial t} + \overline{v} \cdot \nabla_x S_0' + v' \cdot \nabla_x \overline{S_0} + v' \cdot \nabla_x S_0' - \overline{v' \cdot \nabla_x S_0'} + v \cdot \nabla_y w = 0.$$

We remark that the term $\overline{v' \cdot \nabla_x S_0'}$ in (5.20) plays a role similar to the Reynolds stress term in turbulence modeling. This is the term that introduces the nonlocal memory effect into the average equation.

The above multiscale analysis has been applied to upscale the saturation in the two-phase flow in [144]. To solve the coupled elliptic equation for pressure and the transport equation for saturation, we can use the IMPES method, where the pressure equation is solved using MsFVEM and then the velocity approximation is used for upscaling of the transport equation. In [144], the authors presented many numerical experiments for the immiscible flows in porous media based on a multiscale analysis similar to the one described here. They showed that their upscaling method captures both the average and the small-scale fluctuation very well for permeability fields described using two-point correlation functions. By using a new reparameterization technique introduced in [151], we have applied this upscaling method to simulate more realistic heterogeneous porous media without scale separation or periodic structure in [144].

### 5.2.9 Summary

In summary, the main purpose of this section has been to introduce a new (adaptive) multiscale method for solving the transport equation that arises in immiscible two-phase flow in porous media. The basic idea is to compute the global flow on a coarse grid, and map the averaged grid block saturations onto plausible saturation profiles on a finer subgrid. To enhance the accuracy of the coarse-grid saturation profile, while at the same time avoiding an upscaling phase involving, for example, the construction of pseudo-relative permeability functions, we introduce a numerical scheme for solving the transport equation on a coarse grid that honors fine-scale structures in the velocity field in a mathematically consistent manner. Moreover, to capture rapid transitions in saturation values near propagating saturation fronts accurately, we propose to include an adaptive component in the algorithm. In the adaptive algorithm, we solve the saturation locally on a fine grid in transient flow regions. The proposed (adaptive) multiscale method has been analyzed and tested on models with complex heterogeneous structures. We have also extended and implemented multiscale methods for transport equations on unstructured corner-point grids (see [6]). In this section, we also discussed a few other approaches for coarsening transport equations.

## 5.3 Applications to Richards' equation

### 5.3.1 Problem statement

In this section we consider the applications of MsFEMs to Richards' equation ([236]), which describes the infiltration of water flow into porous media whose pore space is filled with air and some water. The equation describing Richards' equation under some assumptions is given by

$$\frac{\partial}{\partial t}\theta(p) - \mathrm{div}(k(x,p)\nabla(p + x_3)) = 0 \ \ \mathrm{in} \ \Omega, \qquad (5.21)$$

where $\theta(p)$ is the volumetric water content and $p$ is the pressure. The following are assumed ([236]) for (5.21): (1) the porous media and water are incompressible; (2) the temporal variation of the water saturation is significantly larger than the temporal variation of the water pressure; (3) the air phase is infinitely mobile so that the air pressure remains constant (in this case it is atmospheric pressure which equals zero); and (4) neglect the source/sink terms.

Constitutive relations between $\theta$ and $p$ and between $k(x,p)$ and $p$ are developed appropriately, which consequently gives nonlinearity behavior in (5.21). The relation between the water content and pressure is referred to as the moisture retention function. The equation written in (5.21) is called the coupled-form of Richards' equation. This equation is also called the mixed form of Richards' equation, due to the fact that there are two variables involved in it, namely, the water content $\theta$ and the pressure head $p$. Taking advantage of the differentiability of the soil retention function, one may rewrite (5.21) as follows

$$C(p)\frac{\partial}{\partial t}p - \text{div}(k(x,p)\nabla(p+x_3)) = 0 \text{ in } \Omega, \tag{5.22}$$

where $C(p) = d\theta/dp$ is the specific moisture capacity. This version is referred to as the head-form ($h$-form) of Richards' equation. Another formulation of the Richards' equation is based on the water content $\theta$,

$$\frac{\partial}{\partial t}\theta - \text{div}(D(x,\theta)\nabla\theta) - \frac{\partial k}{\partial x_3} = 0 \text{ in } \Omega, \tag{5.23}$$

where $D(\theta) = k(\theta)/(d\theta/dp)$ defines the diffusivity. This form is called the $\theta$-form of Richards' equation.

The sources of nonlinearity of Richards' equation come from the moisture retention and relative hydraulic conductivity functions, $\theta(p)$ and $k(x,p)$, respectively. Reliable approximations of these relations are in general tedious to develop and thus also challenging. Field measurements or laboratory experiments to gather the parameters are relatively expensive, and furthermore, even if one can come up with such relations from these works, they will be somehow limited to the particular cases under consideration.

Perhaps the most widely used empirical constitutive relations for the moisture content and hydraulic conductivity is due to the work of van Genuchten [131]. He proposed a method of determining the functional relation of relative hydraulic conductivity to the pressure head by using the field observation knowledge of the moisture retention. In turn, the procedure would require curve-fitting to the proposed moisture retention function with the experimental/observational data to establish certain parameters inherent to the resulting hydraulic conductivity model. There are several widely known formulations of the constitutive relations: the Haverkamp model

$$\theta(p) = \frac{\alpha(\theta_s - \theta_r)}{\alpha + |p|^\beta} + \theta_r, \quad k(x,p) = k_s(x)\frac{A}{A + |p|^\gamma};$$

van Genuchten model [131]

$$\theta(p) = \frac{\alpha(\theta_s - \theta_r)}{[1 + (\alpha|p|)^n]^m} + \theta_r, \quad k(x,p) = k_s(x)\frac{\left\{1 - (\alpha|p|)^{n-1}[1 + (\alpha|p|)^n]^{-m}\right\}^2}{[1 + (\alpha|p|)^n]^{m/2}};$$

exponential model [268]

$$\theta(p) = \theta_s\, e^{\beta p}, \quad k(x,p) = k_s(x)\, e^{\alpha p}.$$

## 5.3.2 MsFVEM for Richards' equations

The spatial field $k_s(x)$ in the above models is also known as the saturated hydraulic conductivity. It has been observed that the hydraulic conductivity has a broad range of values, which together with the functional forms presented above confirm the nonlinear behavior of the process. Furthermore, the water content and hydraulic conductivity approach zero as the pressure head goes to very large negative values. In other words, the Richards' equation has a tendency to degenerate in a very dry condition, that is conditions with a large negative pressure. Because we are interested in mass conservative schemes, finite volume formulation (3.13) of the global problem instead of finite element formulation is used. For (5.21), it is to find $p_h \in W_h$ such that

$$\int_{V_z} (\theta(\eta^{p_h}) - \theta^{n-1}) \, dx - \Delta t \int_{\partial V_z} k(x, \eta^{p_h}) \nabla p_{r,h} \cdot n \, ds = 0, \quad \forall z \in Z_h^0, \quad (5.24)$$

where $\theta^{n-1}$ is the value of $\theta(\eta^{p_h})$ evaluated at time-step $n - 1$, and $p_{r,h} \in \mathcal{P}_h$ is a function that satisfies the boundary value problem:

$$-\mathrm{div}(k(x, \eta^{p_h}) \nabla p_{r,h}) = 0 \quad \text{in } K,$$
$$p_{r,h} = p_h \quad \text{on } \partial K.$$

Here $V_z$ is the control volume surrounding the vertex $z \in Z_h^0$ and $Z_h^0$ is the collection of all vertices that do not belong to the Dirichlet boundary (see Section 3.2).

MsFEM (or MsFVEM) offers a great advantage when the nonlinearity and heterogeneity of $k(x, p)$ are separable; that is

$$k(x, p) = k_s(x) \, k_r(p). \quad (5.25)$$

In this case, as we discussed earlier, the local problems become linear and the corresponding $\mathcal{P}_h$ is a linear space; that is we may construct a set of basis functions $\{\phi_z\}_{z \in Z_h^0}$ (as before) such that they satisfy

$$-\mathrm{div}(k_s(x) \nabla \phi_z) = 0 \quad \text{in } K,$$
$$\phi_z = \phi_z^0 \quad \text{on } \partial K,$$

where $\phi_z^0$ is a piecewise linear function. We note that if $p_h$ has a discontinuity or a sharp front region, then the multiscale basis functions need to be updated only in that region. The latter is similar to the use of MsFEM in two-phase flow applications. In this case the basis functions are only updated along the front. Now, we may formulate the finite-dimensional problem. We want to seek $p_{r,h} \in \mathcal{P}_h$ with $p_{r,h} = \sum_{z \in Z_h^0} p_z \phi_z$ such that

$$\int_{V_z} (\theta(\eta^{p_h}) - \theta^{n-1}) \, dx - \Delta t \int_{\partial V_z} k_s(x) \, k_r(\eta^{p_h}) \, \nabla p_{r,h} \cdot n \, ds = 0,$$

for every control volume $V_z \subset \Omega$. To this equation we can directly apply the linearization procedure described in [133]. Let us denote

$$r^m = p_{r,h}^m - p_{r,h}^{m-1}, \quad m = 1, 2, 3, ..., \quad (5.26)$$

where $p_{r,h}^m$ is the iterate of $p_{r,h}$ at the iteration level $m$. Thus, we would like to find $r^m = \sum_{z \in Z_h^0} r_z^m \phi_z$ such that for $m = 1, 2, 3, \ldots$ $\|r^m\| \leq \delta$ with $\delta$ being some pre-determined error tolerance

$$\int_{V_z} C(\eta^{p_h^{m-1}}) r^m \, dx - \Delta t \int_{\partial V_z} k_s(x) \, k_r(\eta^{p_h^{m-1}}) \, \nabla r^m \cdot n \, ds = R^{h,m-1},$$

with

$$R^{h,m-1} = -\int_{V_z} (\theta(\eta^{p_h^{m-1}}) - \theta^{n-1}) \, dx + \Delta t \int_{\partial V_z} k_s(x) \, k_r(\eta^{p_h^{m-1}}) \, \nabla p_{r,h}^{m-1} \cdot n \, ds.$$

$$(5.27)$$

The superscript $m$ at each of the functions means that the corresponding functions are evaluated at an iteration level $m$.

### 5.3.3 Numerical results

We present several numerical experiments that demonstrate the ability of the coarse models presented in the previous subsections. The coarse models are compared with the fine model solved on a fine mesh. We have employed a finite volume difference to solve the fine-scale equations. This solution serves as a reference for the proposed coarse models. The problems that we consider are typical water infiltration into an initially dry soil. The porous media that we consider is a rectangle of size $L_1 \times L_2$ (see Figure 5.8). The fine model uses $256 \times 256$ rectangular elements, and the coarse model uses $32 \times 32$ rectangular elements.

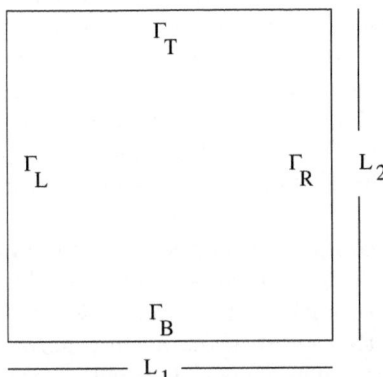

**Fig. 5.8.** Rectangular layout of porous media.

A realization of the hydraulic conductivity field $k_s(x)$ is generated using geostatistical package GSLIB ([85]). We have used a spherical variogram with prescribed correlation lengths $(l_1, l_2)$ and the variance $(\sigma)$ for this purpose. All examples use $\sigma = 1.5$.

The first problem is a soil infiltration, which was first analyzed by Haverkamp (cf. [64]). The porous media dimension is $L_1 = 40$ and $L_2 = 40$. The boundary conditions are as follows. $\Gamma_L$ and $\Gamma_R$ are impermeable, and Dirichlet conditions are imposed on $\Gamma_B$ and $\Gamma_T$, namely $p_T = -21.7$ in $\Gamma_T$, and $p_B = -61.5$ in $\Gamma_B$. The initial pressure is $p_0 = -61.5$. We use the Haverkamp model to construct the constitutive relations. The related parameters are $\alpha = 1.611 \times 10^6$, $\theta_s = 0.287$, $\theta_r = 0.075$, $\beta = 3.96$, $A = 1.175 \times 10^6$, and $\gamma = 4.74$. For this problem we assume that the nonlinearity and heterogeneity are separable, where the latter comes from $k_s(x)$ with $\overline{k_s} = 0.00944$. We assume that appropriate units for these parameters hold. There are two cases that we consider for this problem, namely the isotropic heterogeneity with $l_1 = l_2 = 0.1$, and the anisotropic heterogeneity with $l_1 = 0.01$ and $l_2 = 0.20$. For the backward Euler scheme, we use $\Delta t = 10$. Note that the large value of $\Delta t$ is due to the smallness of $\overline{k_s}$ (average magnitude of the diffusion). The comparison is shown in Figures 5.9 and 5.10, where the solutions are plotted at $t = 360$.

The second problem is a soil infiltration through porous media whose dimension is $L_1 = 1$ and $L_2 = 1$. The boundary conditions are as follows. $\Gamma_L$ and $\Gamma_R$ are impermeable. Dirichlet conditions are imposed on $\Gamma_B$ with $p_B = -10$. The boundary $\Gamma_T$ is divided into three parts. On the middle part, a zero Dirichlet condition is imposed, and the rest are impermeable. We use the exponential model to construct the constitutive relations. with the following related parameters: $\beta = 0.01$, $\theta_s = 1$, $\overline{k_s} = 1$, and $\overline{\alpha} = 0.01$. The heterogeneity comes from $k_s(x)$ and $\alpha(x)$. Clearly, for this problem the nonlinearity and heterogeneity are not separable. Again, isotropic and anisotropic heterogeneities are considered with $l_1 = l_2 = 0.1$ and $l_1 = 0.20$, $l_2 = 0.01$, respectively. For the backward Euler scheme, we use $\Delta t = 2$. The comparison is shown in Figures 5.11 and 5.12, where the solutions are plotted at $t = 10$.

We note that the problems that we have considered are vertical infiltration on the porous media. Hence, it is also useful to compare the cross-sectional vertical velocity that will be plotted against the depth $z$. Here, the cross-sectional vertical velocity is obtained by taking an average over the horizontal direction ($x$-axis).

Figure 5.13 shows comparison of the cross-sectional vertical velocity for the Haverkamp model. The average is taken over the entire horizontal span because the boundary condition on $\Gamma_T$ (and also on $\Gamma_B$) is all Dirichlet condition. Both plots in this figure show a close agreement between the fine and coarse models. For the exponential model, as we have described above, there are three different segments for the boundary condition on $\Gamma_T$; that is a Neumann condition on the first and third part, and a Dirichlet condition on the second/middle part of $\Gamma_T$. Thus, we compare the cross-sectional vertical velocity in each of these segments separately. Figure 5.14 shows the comparison for one of these segments. The agreement between the coarse-grid and fine-grid calculations is excellent.

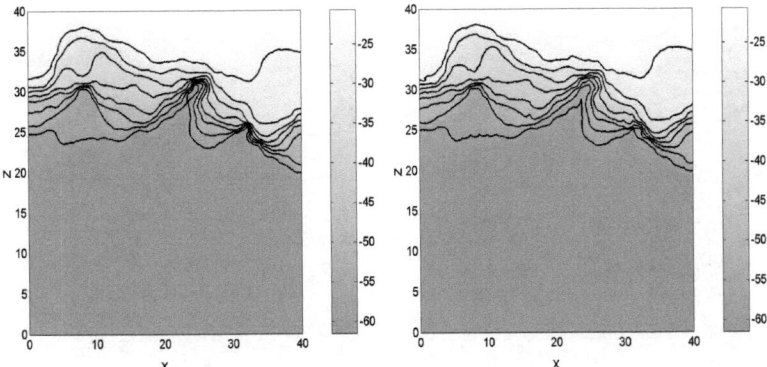

**Fig. 5.9.** Haverkamp model with isotropic heterogeneity. Comparison of water pressure between the fine model (left) and the coarse model (right).

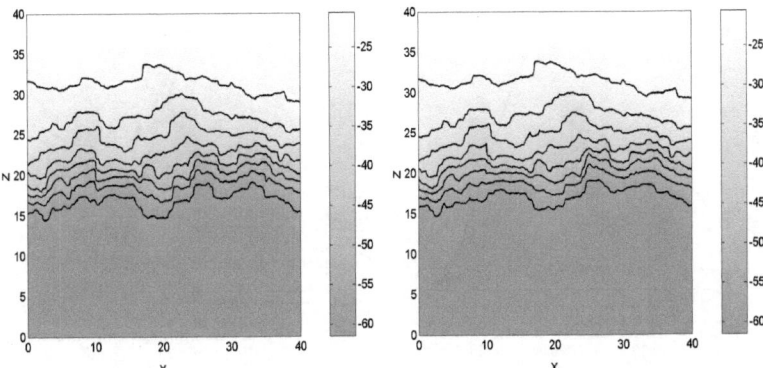

**Fig. 5.10.** Haverkamp model with anisotropic heterogeneity. Comparison of water pressure between the fine model (left) and the coarse model (right).

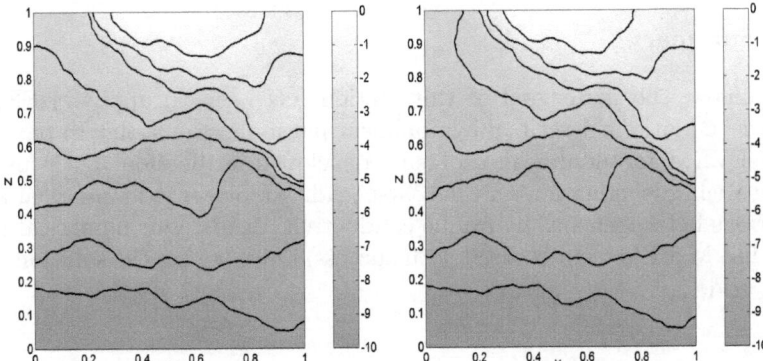

**Fig. 5.11.** Exponential model with isotropic heterogeneity. Comparison of water pressure between the fine model (left) and the coarse model (right).

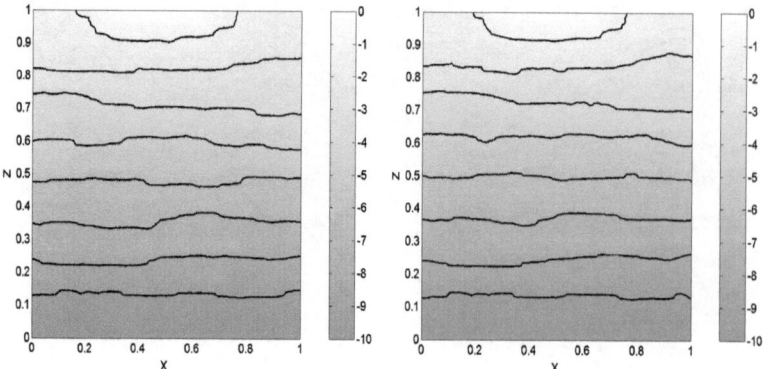

**Fig. 5.12.** Exponential model with anisotropic heterogeneity. Comparison of water pressure between the fine model (left) and the coarse model (right).

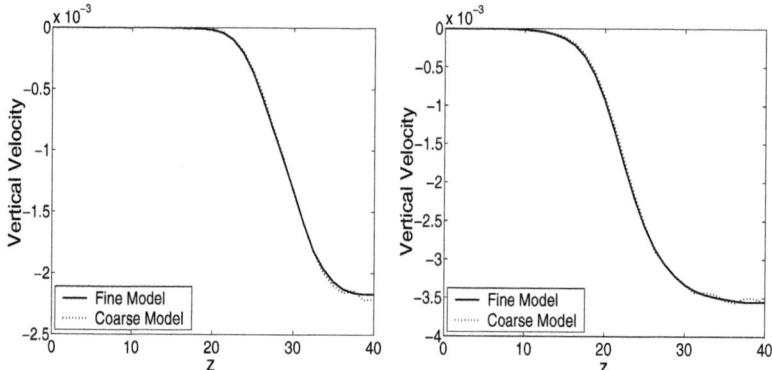

**Fig. 5.13.** Comparison of vertical velocity on the coarse grid for Haverkamp model: isotropic heterogeneity (left) and anisotropic heterogeneity (right).

## 5.3.4 Summary

In summary, the main goal of this section has been to apply MsFEMs to Richards' equations described by nonlinear parabolic equations. In particular, the MsFVEM for nonlinear problems developed in Section 3.2 is used for solving Richards' equation on the coarse grid. We presented numerical results for various heterogeneous hydraulic conductivity fields. Our numerical results show that MsFEMs can be used with success in predicting the solution on the coarse grid.

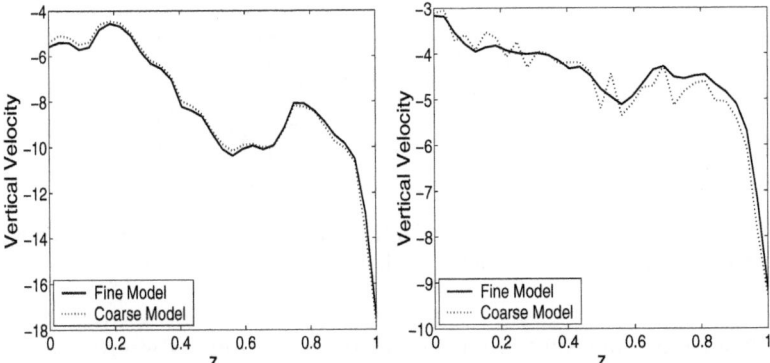

**Fig. 5.14.** Comparison of vertical velocity on the coarse grid for exponential model: isotropic heterogeneity (left) and anisotropic heterogeneity (right). The average is taken over the second third of the domain.

## 5.4 Applications to fluid–structure interaction

### 5.4.1 Problem statement

MsFEMs can also be used to solve complex multiphysics problems. In this section, we extend the MsFEM to solving a fluid–structure interaction (FSI) problem on the coarse grid. At the fine scale, we consider Stokes flow past an elastic skeleton. Thus, our domain $\Omega$ has two parts: a fluid domain $\Omega_0^f$ and a solid domain $\Omega_0^s$. The subscript 0 indicates that these are the domains of the two constituents (solid and fluid) at rest. As macroscopic boundary conditions are applied, the fluid starts to flow, thus exerting forces on the solid, causing them to deform. As a steady state is achieved the fluid flows in a domain $\Omega^f = \Omega \setminus \Omega^s$, and the forces that the fluid exerts on the solid at their interface $\Gamma = \left(\partial\Omega^f \cap \partial\Omega^s\right) \setminus \partial\Omega$ are balanced by the elastic stresses inside the solid. The precise formulation of the FSI problem is:

$$\Gamma = \{X + u(X)|\forall X \in \Gamma_0\}, \tag{5.28}$$

$$-\mu\Delta v + \nabla p = b \text{ in } \Omega^f, \quad \operatorname{div}(v) = 0 \text{ in } \Omega^f, \quad v = 0 \text{ on } \Gamma, \tag{5.29}$$

$$-\operatorname{div}(S(E)) = b_0 \text{ in } \Omega_0^s, \tag{5.30}$$

$$\det(\nabla u + I)(-pI + 2\mu D(x(X)))(\nabla u + I)^{-T} n_0 = S(E)n_0 \text{ on } \Gamma_0. \tag{5.31}$$

Note that $\Gamma$ is the set of points $X + u(X)$. The above equation utilizes the standard notation from continuum mechanics. The deformation gradient $F(X) = \nabla x(X)$, the displacements in the solid $u(X) = x(X) - X$, the infinitesimal strain $E(X) = \frac{1}{2}\left(\nabla u(X) + \nabla u(X)^T\right)$, the fluid velocity $v(x)$ and, finally, the symmetric part of its gradient $D(x) = \frac{1}{2}\left(\nabla v(x) + \nabla v(x)^T\right)$. Furthermore, the usual Cauchy stress tensor is denoted by $T(x)$, which is the

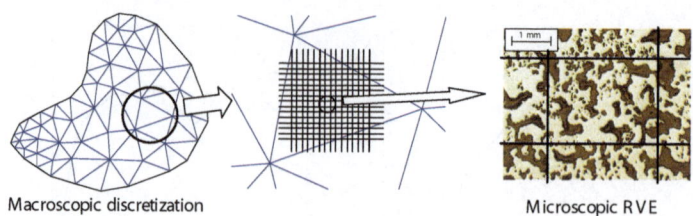

Macroscopic discretization                                Microscopic RVE

**Fig. 5.15.** Schematic of nonlinear MsFEM for FSI. The microstructure shown is an actual sample of porous shape memory alloy [172].

convenient stress measure when describing the fluid. Observe that Cauchy stress tensor is a spatial field, defined on the deformed configuration of the body. For the solid part, the first Piola-Kirchhoff stress tensor $S(X)$ is more appropriate as it gives a description of the stresses in Lagrangian coordinates. The two are related by the identity (see, e.g., [157]):

$$S(X) = \det(F(X))T(x(X))F^{-T}(X). \tag{5.32}$$

In the above formulation of the FSI system the constitutive equation for the Piola-Kirchoff stress $S$ is left unspecified. It has to be specified taking into account the particular solid at hand (see [157] for details). In our numerical examples, the linear elasticity model will be used; that is $S(E) = \mathcal{C} : E$. We refer, for example, to [158, 232] for full details on deriving the FSI problem.

Observe that the position of the interface is a part of the boundary value problem, and the solid–fluid coupling term (5.31) is nonlinear in $u$. Therefore, the FSI problem is nonlinear, even when the constitutive equation for the solid is a linear one.

### 5.4.2 Multiscale numerical formulation

The mapping $E^{MsFEM}$, which couples the coarse scale pressure $p_0$ and displacements $u_0$ to the fine-scale fluid velocity $v$, pressure $p$ and displacements $u$ is defined through the fine-scale FSI problem (5.28)-(5.31). In our problem, we use RVE for local computations (see Figure 5.15 for the illustration[1]). Note that $E^{MsFEM}$ defines a map from a coarse-scale solution $\{p_h, u_h\}$ with given $\Gamma_0$ to a fine-scale approximation $\{p_{r,h}, v_{r,h}, u_{r,h}\}$ via the local solution of (5.28)-(5.31). Various boundary conditions can be chosen for local problems. In our simulations, we use periodic boundary conditions such that the spatial averages of $p_{r,h}$ and $u_{r,h}$ are the same as those for $p_h$ and $u_h$. In general, one can also take $p_h$ and $u_h$ as boundary conditions. In the computation of the local FSI solution with given $p_h$, $u_h$ and the reference interface $\Gamma_0$, one solves an

---

[1] The right figure is the courtesy of the Shape Memory Alloy Research Team (SMART) at Texas A & M University

iterative problem. We assume that this iterative problem converges and provides a unique fine-scale solution $\{p_{r,h}, v_{r,h}, u_{r,h}\}$. This condition guarantees that $E^{MsFEM}$ is a single-valued map.

Next, we discuss the coarse-scale formulation of the problem. In coarse-scale simulations, our goal is to find approximations of $p_0$ and $u_0$, denoted by $p_h$ and $u_h$. When substituting $(p_{r,h}, v_{r,h}, u_{r,h})$ (given $p_h$ and $u_h$) into the fine-scale equations, one needs to solve the resulting system on the coarse-dimensional space. There are various approaches as discussed earlier (see Section 2.4). In particular, one can multiply the fine-scale residual by coarse-scale test functions, or minimize the residual at some coarse points, or use coarse-scale equations when available. These procedures result in a nonlinear equation for finding $(p_h, u_h)$

$$G(p_h, u_h) = 0, \tag{5.33}$$

where $G$ is the reduced variational formulation. This equation is solved via a fixed-point iteration. Here, we consider a simple, physically intuitive, iterative method. In particular, we assume that the coarse-scale equation for the pressure is given by the Darcy equation (see (1.1)) and the coarse-scale equation for the elasticity has the same form as the underlying fine-scale equations, but with upscaled elastic properties that are computed based on local RVE computations. We carry out numerical simulations iteratively. Given $p_h^n$ and $u_h^n$ at the $n$th iteration, $p_{r,h}^n$, $v_{r,h}^n$, and $u_{r,h}^n$ are computed. This is done by using a local problem in RVE as described above. This step involves the solution of the elasticity problem and yields new pore geometry based on the deformations. Furthermore, taking into account local geometry of the pore space, the permeabilities $k^n$ and upscaled elastic properties are computed via standard cell problems (e.g., [240, 42]). Once the permeabilities are computed, the global problem

$$\text{div}(k^n(x)\nabla p^{n+1}) = f$$

is solved and $p_h^{n+1}$ (finite element projection of $p^{n+1}$) is calculated. Similarly, $u_h^{n+1}$ is computed by solving elasticity equation with upscaled elastic properties (e.g., $\mathcal{C}^*$ for linear solids). This iterative procedure can be summarized in Algorithm 5.4.1. Modifications of this algorithm are presented in [232].

---

**Algorithm 5.4.1** Iterative homogenization of strongly coupled FSI problem

---

- Initialize all micro- and macro-fields to zero.
- Project $p_h^n$ and $u_h^n$ using $E^{MsFEM}$.
- Evaluate the permeability and elastic properties in a coarse-grid block that involves the computation of the deformed pore geometry.
- Compute macroscopic quantities $p_h^{n+1}$ and $u_h^{n+1}$.
- Check for convergence and, if necessary, return to Step 2.

---

In the numerical examples, we consider flow past elastic obstacles. The fluid surrounds the obstacles and the obstacles are supported rigidly in their center. Note that the rigid support is necessary, otherwise the flow will move them. Observe also that in 2D either the fluid or the solid domains can be connected, but not both. Therefore, to study upscaling of deformable porous media the solid domain has to be disconnected, so that the fluid can flow throughout the domain and interact with the solid. This simplification allows us to formulate a coarse-scale equation for the macroscopic pressure $p_0$ only, and thus, $E^{MsFEM}$ is defined for a given $p_h$.

### 5.4.3 Numerical examples

In the numerical examples, we consider flow past a 2D periodic arrangement of elastic obstacles (Figure 5.16). The obstacles are, in the reference configuration, circular and centered in the middle of a square unit cell (Figure 5.16(a)). The macroscopic domain is assumed periodic (with the period size $\epsilon$) in the reference configuration and we consider a series of macroscopic domains with $\epsilon^{-1} = 4, 8, 16, ....$ The case $\epsilon = 1/16$ is shown in Figure 5.16(b).

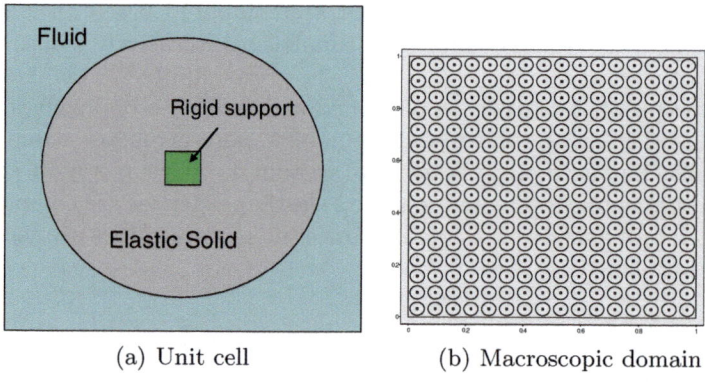

(a) Unit cell                    (b) Macroscopic domain

**Fig. 5.16.** The unit cell (a) consists of circular linear elastic material, surrounded by the fluid. The elastic media is supported rigidly in the center. The unit cell is arranged periodically to form the macroscopic domain. (b) Macroscopic domain with a $16 \times 16$ periodic arrangement of the unit cell (a).

The elastic material under consideration is linear and isotropic with Young's modulus $E = 1.44$ and Poisson's ratio $\nu = 0.1$. The fluid has viscosity 0.1. These non-dimensional properties are selected such that a pressure in the range $0.1 - 0.5$ will produce a sizeable deformation in the solid and lead to strongly coupled FSI problems.

The main objectives of this example are to demonstrate the behavior of the iterative Algorithm 5.4.1. There are two main questions that need to be illuminated: first, whether the nonlinear iteration converges, and second, the approximation that Algorithm 5.4.1 provides to the fine-scale solution of the FSI problem needs to be investigated with respect to the scale parameter $\epsilon$.

The boundary value problem is thus designed to meet both of these goals. The macroscopic domain is the unit square (Figure 5.16(b)) and a uniform pressure $P_l$ is applied at the left side of the domain. The pressure at the right side is 0 and no-flow boundary conditions are considered at the top and bottom sides of the domain. These boundary conditions imply the fine-scale solution is periodic in the $x_2$-direction with the period being one horizontal strip of $(1/\epsilon)$ unit cells. Also, the averaged macroscopic quantities are essentially one-dimensional. This very simple boundary value problem is selected to allow direct numerical simulations (DNS) of the fine-scale solution to the FSI problem. A DNS is computationally very intensive both in memory consumption and CPU time. However, with the selected boundary conditions, a DNS can be performed on a single strip of unit cells and then periodically repeated in the $x_2$-direction. This leads to a factor of $1/\epsilon$ reduction in computational effort and allows us to compute the DNS solution on a series of domains with $\epsilon^{-1} = 4, 8, 16, 32, 64$.

We perform a series of computations with $P_l = 0.1$ and $P_l = 0.2$. The first observation is that Algorithm 5.4.1 in fact behaves as a contraction operator and converges. The approximate upscaled pressure is plotted in Figure 5.17. Because, as already discussed, the upscaled pressure does not vary in the $x_2$ due the boundary conditions, the plot is a cross-section of the upscaled pressure at a fixed location $x_2 = \text{const}$. The number of iterations it took for Algorithm 5.4.1 to reach a relative accuracy of $\times 10^{-6}$ is reported in Table 5.1.

Based on the results it is seen that the algorithm behaves as a contraction

**Table 5.1.** Performance of Algorithm 5.4.1. Listed are the iteration number it took Algorithm 5.4.1 to converge as well as the error between the "exact" DNS and the MsFEM (fine-scale) solution for fine-scale displacements.

| $\epsilon$ | $P_l = 0.1$ | | | | |
|---|---|---|---|---|---|
| | Iterations | $L^\infty$ Error | $L^\infty$ Rel. Error | $L^2$ Error | $L^2$ Rel. Error |
| 1/4 | 6 | $1.23 \times 10^{-3}$ | 0.18 | $2.48 \times 10^{-4}$ | 0.23 |
| 1/8 | 6 | $3.18 \times 10^{-4}$ | 0.10 | $4.39 \times 10^{-5}$ | 0.13 |
| 1/16 | 6 | $8.07 \times 10^{-5}$ | 0.053 | $7.75 \times 10^{-6}$ | 0.069 |
| 1/32 | 6 | $2.03 \times 10^{-5}$ | 0.027 | $1.37 \times 10^{-6}$ | 0.0351 |
| | $P_l = 0.2$ | | | | |
| | Iterations | $L^\infty$ Error | $L^\infty$ Rel. Error | $L^2$ Error | $L^2$ Rel. Error |
| 1/4 | 8 | $2.96 \times 10^{-3}$ | 0.22 | $4.93 \times 10^{-4}$ | 0.22 |
| 1/8 | 8 | $7.94 \times 10^{-4}$ | 0.126 | $8.78 \times 10^{-5}$ | 0.127 |
| 1/16 | 8 | $2.06 \times 10^{-4}$ | 0.068 | $1.56 \times 10^{-5}$ | 0.067 |
| 1/32 | 8 | $5.25 \times 10^{-5}$ | 0.035 | $2.75 \times 10^{-6}$ | 0.034 |

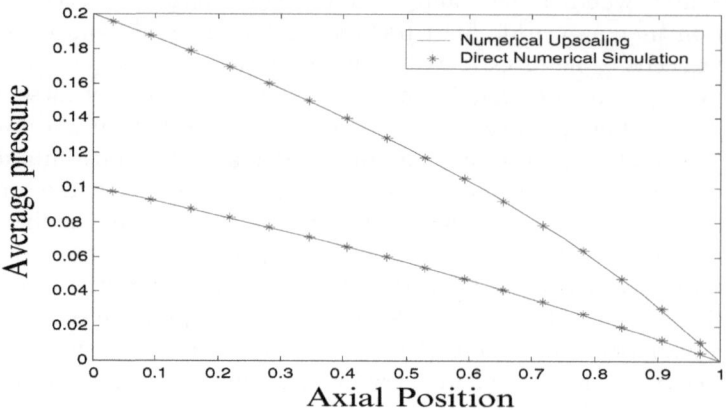

**Fig. 5.17.** Comparison of coarse-scale pressure profiles at $x_2 = 0.5$.

operator and the number of iterations is independent of the length scale $\epsilon$. The same table also lists the comparisons of the DNS solution with the projected fine-scale displacements (via the mapping $E^{MsFEM}$). Based on the error between the DNS displacements and the fine-scale displacements obtained via the MsFEM, it is seen that the method is convergent in terms of $\epsilon$. The actual convergence rate requires a detailed theoretical analysis which is reported in [232]. [232].

### 5.4.4 Discussions

In this section, the application of nonlinear MsFEM to complex multiphysics problems was studied. Here, our goal was simply to discuss an application of the MsFEM to FSI problems and we did not discuss many other existing methods (e.g., [141, 123, 123, 174]). Note that the governing equations do not have elliptic or parabolic forms such as those discussed in Chapter 3, but the general concept of MsFEMs (see Section 2.4) can be applied for solving such systems.

## 5.5 Applications of mixed MsFEMs to reservoir modeling and simulation (by J. E. Aarnes)

Reservoir simulation — the modeling of flow and transport of hydrocarbons in oil and gas reservoirs — is perhaps the most widely considered application in the literature on numerical models for porous media flow. In fact, numerical reservoir simulation has a history that goes back to the early days of the computer. Due to constraints on computational capability, reservoir simulation has been performed on very coarse models with limited spatial

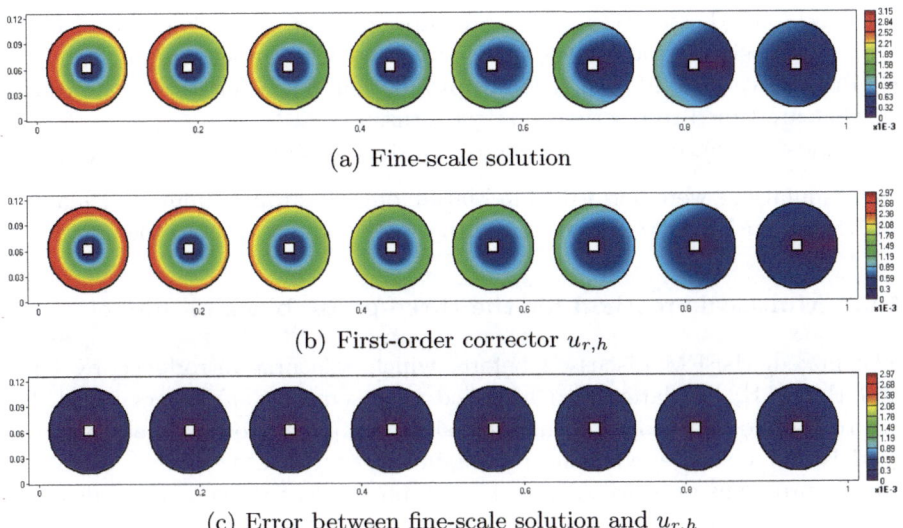

(a) Fine-scale solution

(b) First-order corrector $u_{r,h}$

(c) Error between fine-scale solution and $u_{r,h}$

**Fig. 5.18.** Displacements in a typical solution to the model problem. The macroscopic domain has $8 \times 8$ unit cells, and due to periodicity in the $x_2$-direction, only one horizontal row of unit cells is shown. The exact fine-scale displacements (a) can be compared with the first-order corrector (b). The difference between the two is shown in (c).

resolution. However, the current trend in geomodeling – the process of developing a conceptual geological description of the reservoir – is to build detailed high-resolution models that match as closely as possible the geologists' perception of the reservoir. As a result, there is a steadily increasing gap between the size of the geological model built by geologists, and the model used for reservoir simulation. The reservoir simulation model is normally obtained by coarsening or upscaling the geological model.

As an alternative to upscaling it has been suggested that multiscale methods can be used to run simulations directly on geological models. To this end it is generally assumed that the fine and coarse grids overlap such that each block in the coarse grid simply consists of a number of cells from the underlying fine grid. This means that one can perform the coarsening of the grid in index space rather than in physical space, and thereby significantly simplify the process of generating the coarse grid. In particular, one avoids the practical problems of resampling nonoverlapping cells/blocks in the fine/coarse grid that are traditionally associated with upscaling.

In this section we make an effort to demonstrate some applications where multiscale flow solvers used in combination with various methods for fast computation of fluid transport may spur new ways of using flow information as part of reservoir planning and management. In particular, we demonstrate how the multiscale methods can be used to

- Accelerate the solution of the pressure equation in three-phase black oil reservoir simulation models (and retain the solution accuracy).
- Provide very rapid estimation of production characteristics on flow-grids that are tuned to reservoir flow patterns.
- Almost instantly estimate injector–producer pairs and swept volumes.

The simulations were performed in Matlab on a desktop computer with a dual AMD Athlon X2 4400+ processor with 1 MB cache and 2 GB memory.

### 5.5.1 Multiscale method for the three-phase black oil model

The mixed MsFEM discussed before, which was first introduced by Chen and Hou [71], has later been modified in a sequence of papers [1, 11, 13, 12] to handle the geometric and physical complexity of real-field reservoir models. For instance, whereas the original method was developed for solving elliptic problems on Cartesian grids, the most recent version [12] is designed for solving the parabolic pressure equation of three-phase black oil models on real-field corner-point grids with faults. The three-phase black oil model describes the flow of an aqueous phase $(a)$, usually water, a liquid phase $(l)$ containing oil and liquefied gas, and a vapor phase $(v)$ containing gas and vaporized oil. The pressure equation for the three-phase black oil model may be expressed on the following form:

$$\left( \frac{\partial \phi_{\text{por}}}{dp_l} + \phi_{\text{por}} \sum_j c_j S_j \right) \frac{\partial p_l}{dt} + \nabla \cdot \left( \sum_j v_j \right) + \sum_j c_j v_j \cdot \nabla p_l = q, \quad (5.34)$$

where $p_l$ is liquid pressure, $\phi_{\text{por}}$ is porosity, $v_j$, $c_j$ and $S_j$ are phase velocities, compressibilities, and saturations, respectively, and $q$ is a volumetric source term. The phase velocities are related to the phase pressures $p_j$ through Darcy's law:

$$v_j = -\frac{k k_{rj}}{\mu_j} \left( \nabla p_j + \rho_j g e_3 \right), \quad j = a, l, v. \quad (5.35)$$

Here $\rho_j$ is the density of phase $j$, $g$ is the magnitude of acceleration of gravity, $e_3$ is the unit normal pointing vertically upwards, $k$ is the absolute permeability, and $k_{rj}$ and $\mu_j$ are the relative permeability and viscosity of phase $j$, respectively. See [12] for the definition of the phase compressibilities.

When applied to the three-phase black oil model the mixed MsFEM approximates the liquid pressure $p_l$ and the total velocity $v = \sum_j v_j$ in finite-dimensional subspaces defined over the coarse grid. Recall that the pressure is approximated in a regular mixed finite-element space consisting of functions that are constant on each coarse block, and the velocity is approximated in a special multiscale space spanned by special multiscale basis functions that correspond to localized solutions of the pressure equation with a prescribed direction of flow; see [1, 13]. Given these basis functions, the mixed MsFEM

finds the best linear superposition (in a certain sense) under the constraint
that the velocity field is mass conservative on the coarse scale. Moreover, if
the local flow solutions are mass conservative on the fine grid, then so will the
global mixed MsFEM solution be.

To perform a reservoir simulation using the mixed MsFEM one proceeds
as follows.

1. Introduce a coarse grid, for instance, by partitioning in index space as
   seen in Figure 2.10.
2. Detect all pairs of adjacent blocks.
3. For each pair, compute a velocity basis function.
4. Start simulation, for each time-step, do
   a) Assemble and solve the coarse-grid system.
   b) Recover fine-grid velocities/fluxes.
   c) Solve the fluid-transport equations.

For increased stability, one may iterate on solving the pressure and transport
equations before advancing to the next time-step and thereby obtain a fully
implicit method [183]. Similarly, for cases with strong displacement fronts, one
may also update a few basis functions throughout the simulation to account
more accurately for a strong saturation dependence; see [1, 167].

To illustrate the accuracy of the multiscale solutions, we consider a two-
dimensional test-case modeling layer 68 from model 2 of the SPE comparative
solution project [78], henceforth called the SPE 10 model. This particular
layer is known to be a very difficult model, (see, e.g., [167]). The simulations
start with 0.4 PVI of gas injection followed by 0.6 PVI of water injection.
The reservoir is initially filled with 5% gas and 95% oil, four injection wells
constrained to inject at 300 bar are located at each corner, and one rate-
constrained production well is located in the middle.

We consider both the mixed MsFEM in [13] and the corresponding method
using limited global information to define the multiscale basis functions. The
fine grid is a 60-by-220 Cartesian grid and the coarse grid for the mixed
MsFEM is defined to be a 5-by-11 Cartesian grid. Accuracy of the mixed
MsFEM solutions is assessed by comparing the water-cut (fraction of water in
produced fluid) and gas-cut (fraction of gas in produced fluid) curves obtained
using a mixed MsFEM with the corresponding curves obtained by solving the
pressure equation on the fine grid using a mimetic finite difference method
(FDM) [177, 41]. The latter solution is referred to as the reference solution.
In all simulations the saturation equations are solved on the fine grid.

Figure 5.19 shows the logarithm of permeability, magnitude of velocity at
initial time, and water-cut and gas-cut as functions of PVI (pore volume in-
jected). Although there are certain differences between $\log |v|$ computed using
a mixed MsFEM (without limited global information) and the fine-grid solution depicted in Figure 5.19(b), we see that the mismatch has limited influence
on the production curves. Indeed, even with a coarse grid with only 55 blocks

the mixed MsFEM produces water-cut and gas-cut curves that match the reference solution closely. This demonstrates that the mixed MsFEM captures the main flow characteristics.

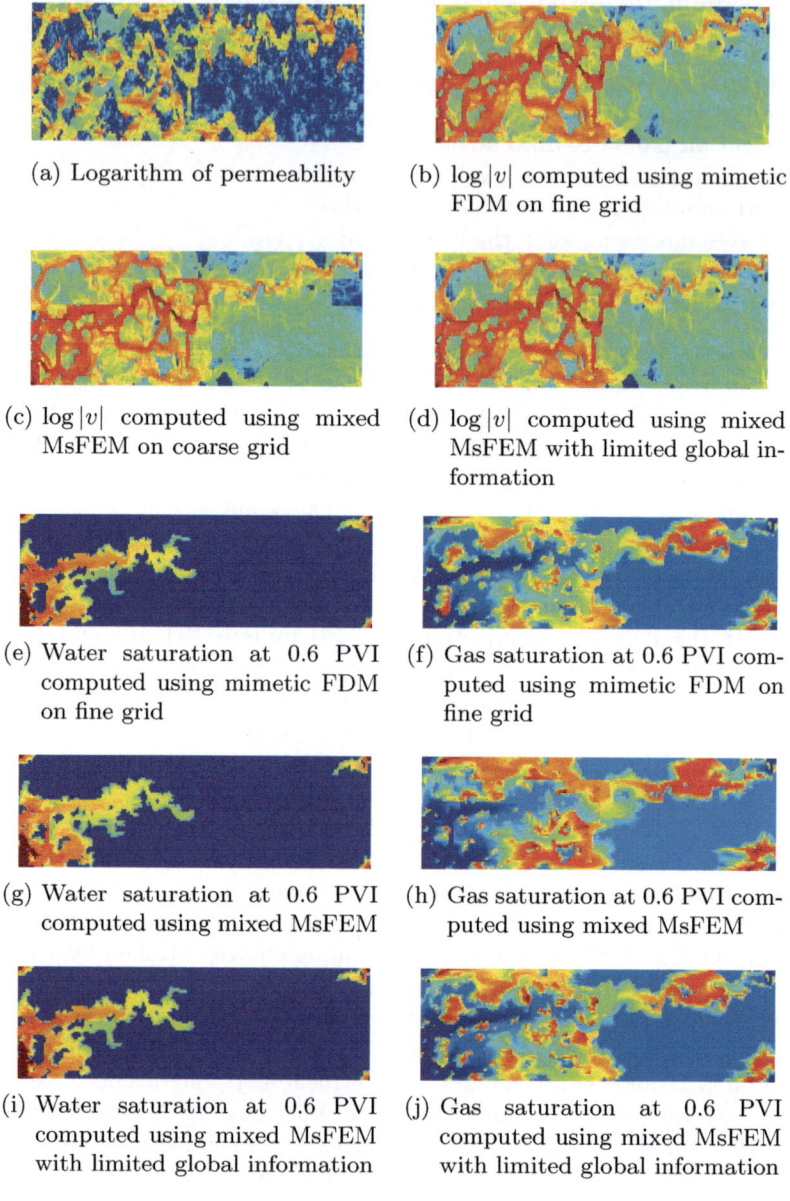

(a) Logarithm of permeability

(b) $\log|v|$ computed using mimetic FDM on fine grid

(c) $\log|v|$ computed using mixed MsFEM on coarse grid

(d) $\log|v|$ computed using mixed MsFEM with limited global information

(e) Water saturation at 0.6 PVI computed using mimetic FDM on fine grid

(f) Gas saturation at 0.6 PVI computed using mimetic FDM on fine grid

(g) Water saturation at 0.6 PVI computed using mixed MsFEM

(h) Gas saturation at 0.6 PVI computed using mixed MsFEM

(i) Water saturation at 0.6 PVI computed using mixed MsFEM with limited global information

(j) Gas saturation at 0.6 PVI computed using mixed MsFEM with limited global information

**Fig. 5.19.** Velocity solutions at initial time and saturation profiles at 0.6 PVI.

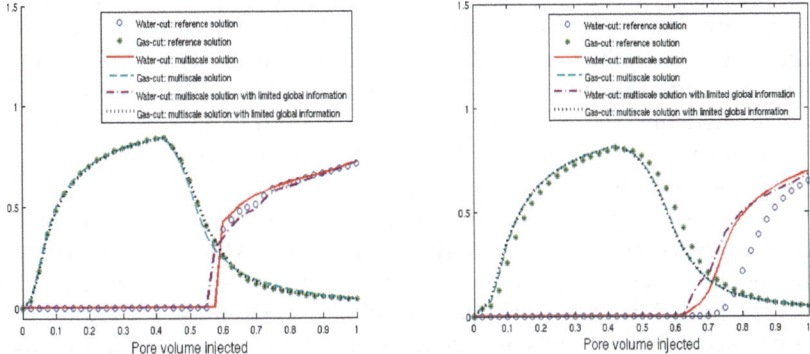

(a) Simulations on layer 68 from the SPE10 model

(b) Simulations on layers 81–85 from the SPE10 model

**Fig. 5.20.** Water-cut and gas-cut for simulations on a two-dimensional model (layer 68 from SPE 10) and a three-dimensional model (layers 81–85 from SPE10).

When solving the pressure equation on the fine grid using the mimetic FDM the time spent on solving the pressure equation stands for 86% of the computation time. In the multiscale simulations, on the other hand, the time spent on solving the saturation equations dominates the computation time (54%). With our current Matlab implementation the multiscale simulations run seven times faster than the fine-grid simulation. On larger models the difference will generally be more substantial because the computational complexity of the mixed MsFEM scales linearly with the model size. Moreover, the mixed MsFEM is very easily parallelized: the assembly of the coarse-grid system, which accounts for nearly 100% of the computation time, is called embarrassingly parallel, and perfectly suited for the multicore computers and distributed memory computing platforms. A further reduction in the computation time spent on solving the pressure equation, or alternatively an increase in model size, can therefore easily be achieved with parallel computing resources. However, to reduce the total computation time further one should also consider alternative strategies for solving the saturation equations.

Our numerical results show that mixed MsFEMs using limited global information give two-fold improvement in water and gas saturation errors when single-phase flow information is used in the construction of multiscale basis functions as discussed in Section 4.2. These results will be reported elsewhere.

### 5.5.2 Adaptive coarsening of the saturation equations

For large problems solving the saturation equations on the fine grid with a finite difference method may not be feasible, or may become a bottleneck. An alternative is to employ streamline methods [83] that advect the fluid phases along one-dimensional trajectories tangential to the velocity. These methods are generally very fast provided large time-steps between each pressure step can be taken. But it is also natural to ask if it is possible to exploit fine-grid velocity resolution in a multiscale type approach for the saturation equation. However, modeling the flow and transport accurately on coarse grids is difficult due to the dynamic nature of coarse-grid relative permeability functions [37] and the need to capture sharp propagating fronts. Fortunately, recent work [9] shows that one can model the main flow characteristics on relatively coarse grids without using pseudo-functions provided the coarse grid adapts to the local heterogeneity and resolves the dominant features in the velocity field (e.g., high-flow channels). In the following, we present an approach for generating such coarse grids and demonstrate how these grids can be used to get accurate production data.

Assume that the velocity is modeled on a high-resolution model (e.g., using the subresolution in mixed MsFEMs), and that it is prohibitively computationally expensive to solve the saturation equations on the same grid. Thus, we propose creating an upscaled model only for the saturation equation. This is done by generating a coarse grid that resolves underlying flow patterns more accurately than traditional coarse grids used in reservoir simulation. These grids allow us to capture more accurately flow quantities of interest, such as production characteristics, without resorting to multiphase upscaling.

As for a mixed MsFEM, we use the term *block* to denote a cell in the coarse grid to distinguish it from a cell in the fine grid. The coarsening strategy presented in [9], henceforth called the nonuniform coarsening algorithm, is essentially based on grouping cells according to flow magnitude. The algorithm involves two parameters that determine the degree of coarsening: a lower bound $V_{\min}$ on the volume of each block and an upper bound $G_{\max}$ on total amount of flow through each block. These parameters are selected to give the desired resolution of the saturation. A general rule for how to select the parameters is given in [9].

The steps in the nonuniform coarsening algorithm are as follows:

1. Use the logarithm of the velocity magnitude in each cell to segment the cells in the fine grid into ten different bins; that is, each cell $c$ is assigned a number $n(c) = 1, \ldots, 10$ by upper-integer interpolation in the range of $g(c) = \frac{10(\log |v(c)| - \min \log |v|)}{\max \log |v| - \min \log |v|}$.
2. Create an initial coarse grid with one block assigned to each connected collection of cells with the same value of $n(c)$.
3. Merge each block with less volume than $V_{\min}$ with a neighboring block.
4. Refine each block that has more flow than $G_{\max}$.
5. Repeat Step 3 and terminate.

Note that only the saturation equations are discretized on this grid. To this end, we employ a backward Euler method where the spatial discretization is a finite volume method that is upstream weighted at the fine-grid level; see [9].

This implies that we utilize the fine-grid resolution in the velocity field when solving for saturation on the coarse grid.

Figures 5.21(b)–5.21(d) show the logarithm of a velocity field as a piecewise constant function on the fine grid, on a coarse Cartesian grid with 240 blocks, and on a nonuniformly coarsened grid with 236 blocks. If we denote the reservoir by $\Omega$, and $N$ is the number of cells in the fine grid, then the nonuniform coarse grid is generated under the constraint that each block $B$ satisfies

$$\int_B dx \geq \frac{15}{N} \int_\Omega dx \quad \text{and} \quad \int_B \log|v| dx \leq \frac{75}{N} \int_\Omega \log|v|\, dx.$$

We clearly see that the nonuniformly coarsened grid adapts to underlying flow patterns. In contrast, the channels with high velocity are almost impossible to detect in Figure 5.21(c). The fact that the coarse grid is capable of resolving the main flow trends leads to improved accuracy in modeled production characteristics. This is illustrated in Figure 5.21(e) which shows water-cut curves obtained on the nonuniform coarse grid are closer to the water-cut curve obtained on the fine grid than the water-cut curve obtained on the Cartesian coarse grid.

The robustness and accuracy of the nonuniform coarsening approach relative to modeling saturation on uniformly coarsened grids is demonstrated in a series of numerical examples in [9]. Instead of including further numerical results here, we only state the main conclusions from [9]:

- Nonuniformly coarsened grids give significantly more accurate water-cut curves than one obtains using uniformly coarsened grids with a similar number of blocks.
- It is very easy to select parameters $V_{min}$ and $G_{max}$ to give a desired level of upscaling. Moreover, the accuracy of water-cut curves obtained on nonuniform coarse grids is nearly insensitive to the degree of upscaling.
- Although the nonuniform coarsening algorithm employs an initial velocity field, the coarse grid does not have to be regenerated if the flow field changes significantly, for example, if new wells are opened, or choke settings are altered. This is due to the fact that the nonuniform coarse grid essentially adapts to high permeable regions with good connectivity.
- The grid needs to be regenerated if the geology is altered significantly. However, the time spent on generating the coarse grid is usually small relative to the simulation time. The nonuniform coarsening algorithm therefore allows grids to be generated at run-time.

Hence, in combination with mixed MsFEM the nonuniform coarsening approach provides a foundation for a simulation technology that is capable of selecting grids at run-time and performing simulations in a matter of minutes, rather than hours or days. This type of simulation time may open up for using reservoir simulation for operational decision support. In the next section we discuss how the mixed MsFEM alone may be used to provide flow-based information that can be used in operational reservoir management workflows.

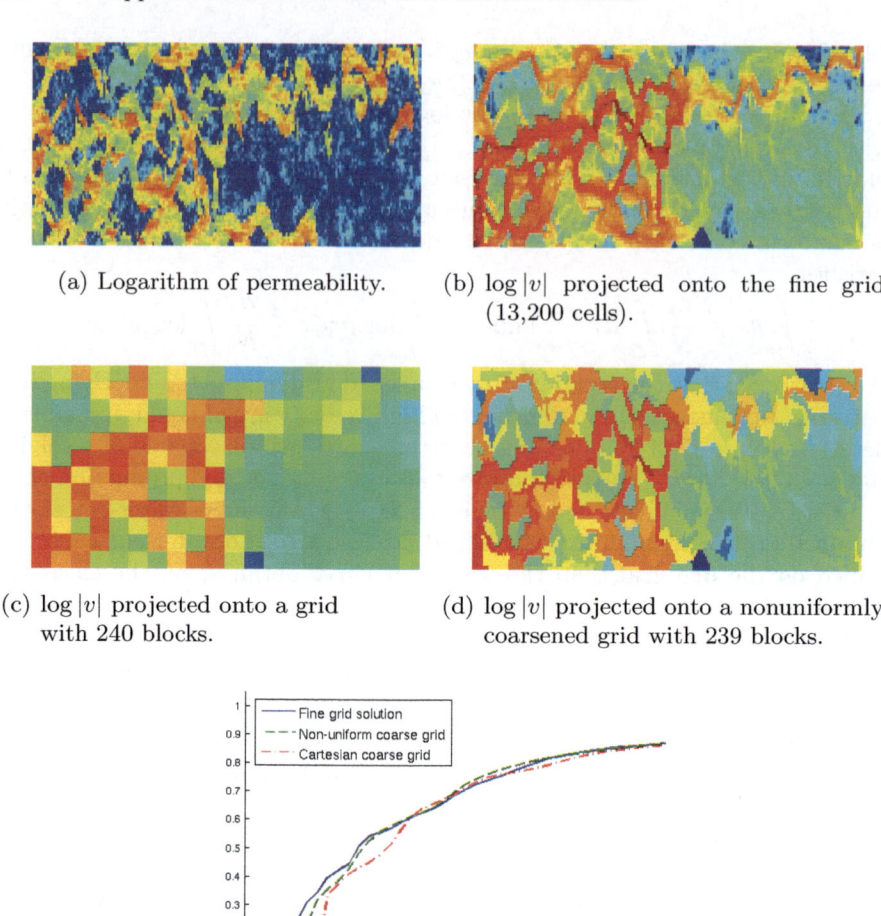

(a) Logarithm of permeability.

(b) $\log|v|$ projected onto the fine grid (13,200 cells).

(c) $\log|v|$ projected onto a grid with 240 blocks.

(d) $\log|v|$ projected onto a nonuniformly coarsened grid with 239 blocks.

(e) Water-cut curves

**Fig. 5.21.** Illustration of the nonuniform coarsening algorithm ability to generate grids that resolve flow patterns and produce accurate production estimates.

### 5.5.3 Utilization of multiscale methods for operational decision support

Reservoir simulation is used today as part of long term strategic planning (e.g., to predict production, quantify uncertainty, and evaluate the objective function in history matching). But to our knowledge, it is not common to utilize reservoir simulations or static flow-based information (for instance, a snapshot of the reservoir pressure and velocity fields) for operational decision support. A reason may be that traditional reservoir simulators are built as black-box tools targeting applications where only the phase saturations and production data are needed or used. Another reason may be that reservoir simulation is generally time-consuming and limited to low-resolution models, or localized high-resolution models. Hence, using reservoir simulation for decision support is not regarded as an option when decisions need to be made on a daily, hour-by-hour, or minute-by-minute basis.

With state-of-the-art multiscale techniques it is possible to evaluate flow responses of suggested well locations almost instantly. These techniques can of course also be used to run very fast reservoir simulations, but to release the full potential of multiscale methods one should not see them only as tools to accelerate simulations. Indeed, because multiscale methods can provide (accurate) information about flow patterns almost instantly, also on large-scale high-resolution models, they may have a huge potential for improving current decision support tools and work processes where flow information is not used, for example, due to too long response-time of conventional simulators. Using a multiscale solver will be particularly efficient if the flow field needs to be updated due to small or localized changes in the reservoir parameters, well configuration, and so on. Then, all that is needed is to update a few local basis functions to reflect changes in reservoir properties and so on, before the global flow can be solved very efficiently on a relatively coarse grid in (less than) a few seconds.

In the following we discuss various ways of using a snapshot of the reservoir velocity field to extract information that we believe can be valuable in operational reservoir management. Examples of information that can be extracted from a snapshot of the velocity field include:

- Injector–producer dependence
- Estimated well-sweep, that is, regions flooded by each injector
- How the flow changes by altering choke settings or inserting new wells
- How the flow is affected by perturbing the geology

One option for utilizing a velocity solution for these applications is to map streamlines (lines tangential to the velocity). This option is available today with commercial streamline simulators (e.g., FrontSim and 3DSL) or certain geomodeling tools (e.g., IRAP). Although streamline tracing scales very well with model sizes, current methods for solving the flow field do not. Utilizing a multiscale pressure solver will improve the scaling dramatically and open up significantly faster response times or larger model sizes.

For complex reservoirs with strong heterogeneity, many wells, and/or a large number of faults, streamlines typically form complex intertwined bundles and it can be difficult for the naked eye to distinguish the different well-sweep regions. In addition, accurate tracing of streamlines on models with complex grid geometry is nontrivial. For such cases, it may be more advantageous to provide information about reservoir partitioning and communication patterns in terms of volumetric objects that are bounded by surface patches or consist of a collection of grid cells. In the following, we present means to provide much of the same information one can extract from a streamline map directly on the physical grid in a way that is easy to compute and visualize.

Consider two equations of the same form: the time-of-flight equation

$$v \cdot \nabla \tau = \phi_{\text{por}}, \quad \tau(\cup w_i^{\text{in}}) = 0, \tag{5.36}$$

where $w_i^{\text{in}}$ denotes injection well $i$, and the stationary tracer equation

$$v \cdot \nabla c_i = 0, \quad c_i(w_i^{\text{in}}) = 1. \tag{5.37}$$

Here $c_i$ models the eventual concentration of a tracer if released continually from injection well $i$, that is, if the injected substance is a unique tracer.

Presuming now that the velocity $v$ is known, the time-of-flight equation (5.36) and the tracer equations (5.37) can be solved efficiently using an upstream-weighted discontinuous Galerkin (dG) method [206]. For instance, to compute the time-of-flight $\tau$ using a first-order upstream weighted dG method, we solve the following system of equations,

$$\int_{\partial T_i} \tau^+ v \cdot n \, ds = \int_{T_i} \phi_{\text{por}} \, dx, \tag{5.38}$$

for all cells $T_i$. Here $\tau$ is a cellwise constant function, $n$ is the unit normal on $\partial T_i$ pointing outward, and $\tau^+$ is $\tau$ evaluated on the upstream side of each interface; that is

$$\int_{\partial T_i} \tau^+ v \cdot n \, ds = \sum_j \left( \tau(T_i) \max\{v_{ij}, 0\} + \tau(T_j) \min\{v_{ij}, 0\} \right),$$

where $v_{ij}$ is the flux from $T_i$ to $T_j$. Using an optimal reordering of the cells, the discretized system can be cast as a block-triangular system that can be solved hyperfast [206].

Figure 5.22 shows time-of-flight and stationary tracer distribution for a case with ten pressure-constrained injection wells - eight along the perimeter and two in the middle — and six rate-constrained production wells. In the tracer profile plots the color of a cell corresponds to the tracer with the highest concentration. We see here that the reservoir is neatly divided into separate regions. Combining the time-of-flight information with the tracer data (i.e.,

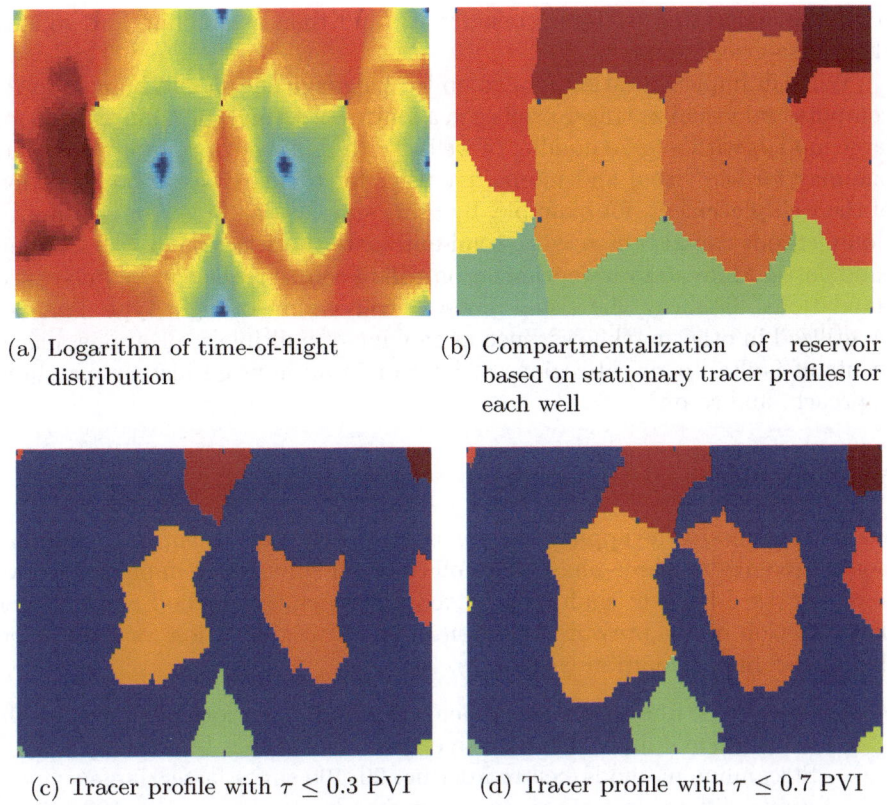

(a) Logarithm of time-of-flight distribution

(b) Compartmentalization of reservoir based on stationary tracer profiles for each well

(c) Tracer profile with $\tau \leq 0.3$ PVI

(d) Tracer profile with $\tau \leq 0.7$ PVI

**Fig. 5.22.** Example of how time-of-flight and stationary tracer profiles can be used to give a visual picture of flooded regions.

by only coloring cells with $\tau \leq T$), we can easily estimate and visualize the regions that are expected to be flooded at time $t = T$ by a unit displacement front arising from each injector at time $t = 0$.

One area where plots of synthetic tracer profiles can be valuable is planning of new wells, where tracer/time-of-flight data may be used to visually inspect how adding a new or moving an existing well affects the injector-producer coupling, breakthrough times, and flooded/drained regions. It is also easy to add a mathematical measure indicating the quality of a well location. This may be useful for reservoirs with many wells where it can be difficult to assess the quality of a potential well location visually. When a new well is added or a well is moved, the flow field needs to be updated before the tracer distribution can be computed. Using mixed MsFEM with precomputed basis functions, this can be performed very efficiently by only updating the basis functions affected by the change in the given well (one basis function for each block the associated well-bore penetrates) and then assembling and solving the coarse-grid system.

With an implementation in a compiled language one should be able to re-compute and visualize tracer profiles in a matter of seconds or minutes, also for large models with a large number of wells. With this type of computation time one may foresee visual and interactive ways for the early-stage optimization of the well placement, for example, by using simple trial and error. Optimization methods can also be used to semi-automate the selection of well location candidates. More accurate optimization will of course require more fine-tuned simulations. Similarly, the visual power of tracer/time-of-flight type data can be utilized in other workflows, such as ranking of multiple realizations, placements of faults, to reveal regions of interest in an assisted history-matching approach, and so on.

### 5.5.4 Summary

We have discussed the application of a multiscale finite element-based simulation technology for three-phase black oil reservoir simulation. In particular, we have discussed how the multiscale mixed finite element method allows faster discretization of the pressure equation in reservoir simulation, or simulation directly on high resolution geomodels.

Computation time/model size when saturation equations are solved with
- Conventional finite difference method: 10–30 times faster/larger
- Streamline methods/cell reordering: 20–100 times faster/larger
- Finite difference method on nonuniformly coarsened grid: 100– times faster/larger

We have also discussed how using MsFEMs may open up for using flow-based information for operational decision support, for instance by allowing almost instant computation and visualization of well-sweep and injector-producer pairs. Providing tools for rapid computation of this type of flow-based information can be instrumental in increasing the interactivity and reduce the turnaround time for various reservoir management workflows. In particular, to bridge the gap between the geomodel and the simulation model, it may be necessary that simulation grids and suitable simulation technology can be selected in a semi-automated manner at run-time to fit response-time requirements and available resources.

## 5.6 Multiscale finite volume method for black oil systems (by S. H. Lee, C. Wolfsteiner and H. A. Tchelepi)

Most practical reservoir simulation studies are performed using the so-called black oil model, in which the phase behavior is represented using solubilities and formation volume factors. We extend the multiscale finite volume (MsFV) method to deal with nonlinear immiscible three-phase compressible flow in the presence of gravity and capillary forces (i.e., black oil model). Consistent with the MsFV framework, flow and transport are treated separately and differently using a sequential implicit algorithm. A multiscale operator splitting strategy is used to solve the overall mass balance (i.e., the pressure equation). The black oil pressure equation, which is nonlinear and parabolic, is decomposed into three parts. The first is a homogeneous elliptic equation, for which the original MsFV method is used to compute the dual basis functions and the coarse-scale transmissibilities. The second equation accounts for gravity and capillary effects; the third equation accounts for mass accumulation and sources/sinks (wells). With the basis functions of the elliptic part, the coarse-scale operator can be assembled. The gravity/capillary pressure part is made up of an elliptic part and a correction term, which is computed using solutions of gravity-driven local problems. A particular solution represents accumulation and wells. The reconstructed fine-scale pressure is used to compute the fine-scale phase fluxes, which are then used to solve the nonlinear saturation equations. For this purpose, a Schwarz iterative scheme is used on the primal coarse grid. The framework is demonstrated using challenging black oil examples of nonlinear compressible multiphase flow in strongly heterogeneous formations.

### 5.6.1 Governing equations and discretized formulation

The standard black oil model has two hydrocarbon phases (i.e., oil and gas) and one aqueous phase (water) with rock and fluid compressibility, gravity effects, and capillary pressure. The thermodynamic equilibrium between the hydrocarbon phases is modeled via the solubility of the gas pseudo-component in the oil phase. The conservation equations are nonlinear due to the strong nonlinear character of the relative permeability and capillary pressure relations, the large gas compressibility, phase appearance and disappearance effects, and large density and viscosity differences.

The governing equations for the black oil formulation [29] are:

$$\frac{\partial}{\partial t}\left(\phi_{\mathrm{por}} b_o S_o\right) = \operatorname{div}\left(b_o \lambda_o \left(\nabla p_o - g\rho_o e_3\right)\right) - q_o, \tag{5.39}$$

$$\frac{\partial}{\partial t}\left(\phi_{\mathrm{por}} b_w S_w\right) = \operatorname{div}\left(b_w \lambda_w \left(\nabla p_w - g\rho_w e_3\right)\right) - q_w, \tag{5.40}$$

$$\frac{\partial}{\partial t}\left(\phi_{\mathrm{por}}\left(b_g S_g + R_s b_o S_o\right)\right) = \operatorname{div}\left(b_g \lambda_g \left(\nabla p_g - g\rho_g e_3\right)\right) - q_g \tag{5.41}$$
$$+ \quad \operatorname{div}\left(R_s b_o \lambda_o \left(\nabla p_o - g\rho_o e_3\right)\right) - R_s q_o,$$

on the domain $\Omega$, with boundary conditions on $\partial\Omega$. Here, $\lambda_l = k(x) k_{r_l}/\mu_l$ is the mobility of phase $l$, where $l = o, w, g$ (i.e., oil, water, and gas); $b_l = 1/B_l$

where $B_l$ is the formation volume factor (i.e., ratio of volume at reservoir conditions to volume at standard conditions). $S_l$, $k_{r_l}$, $\mu_l$, $\rho_l$ denote, respectively, the saturation, relative permeability, viscosity, and density of phase $l$. The well volumetric flow rate is $q_l$. The tensor $k$ describes the permeability field, which is usually represented as a complex multiscale function of space. Porosity is denoted by $\phi_{\text{por}}$, $p_l$ is the phase pressure, $g$ is gravitational acceleration, $e_3$ denotes the unit vector along the reservoir depth, and $R_s$ is the solubility of gas in oil. In general, $\mu_l$, $\rho_l$, $B_l$, $R_s$, and $\phi_{\text{por}}$ are functions of pressure. The relative permeabilities, $k_{r_l}$, are functions of saturation.

Saturations are constrained by $1 = S_o + S_w + S_g$, and the three phase pressures $p_w$, $p_o$, and $p_g$ are related by two independent capillary pressure functions:

$$p_w - p_o = p_{cwo}(S_o, S_g, S_w), \quad p_g - p_o = p_{cgo}(S_o, S_g, S_w).$$

We choose the oil phase pressure as the primary variable, $p = p_o$. Multiplication of the semi-discretized equations of (5.39) to (5.41) with

$$\alpha_o = \frac{1}{b_o^{n+1}} - \frac{R_s^{n+1}}{b_g^{n+1}}, \quad \alpha_w = \frac{1}{b_w^{n+1}}, \text{ and } \alpha_g = \frac{1}{b_g^{n+1}},$$

respectively, and summation of the resulting equations gives the pressure equation:

$$L_{BO}\, p^{\nu+1} = -\frac{C_w}{\Delta t}(p^{\nu+1} - p^\nu) + RHS1 + RHS2,$$

where the operator for black oil is defined by

$$L_{BO} \equiv -\sum_\ell \alpha_\ell \operatorname{div}\left(\lambda'^{\nu}_\ell \nabla\right)$$

and the right-hand sides are given by

$$RHS1 = -\sum_\ell \alpha_\ell \operatorname{div}\left(g\rho_\ell \lambda'^{\nu}_\ell \nabla z\right) + \sum_{\ell=w,g} \alpha_\ell \operatorname{div}\left(\lambda'_\ell \cdot \nabla p_{c\ell o}\right)^\nu$$

$$RHS2 = \frac{\phi^n_{\text{por}}}{\Delta t}\left(\sum_\ell \alpha_\ell b^n_\ell S^n_\ell + \alpha_g R^n_s b^n_o S^n_o\right) - \frac{\phi^\nu_{\text{por}}}{\Delta t} - \sum_\ell \alpha_\ell q^\nu_\ell - \alpha_g (R_s q_o)^\nu.$$

The $C_w$ is a weak function of pressure defined in Lee et al. [175].

### 5.6.2 Multiscale finite volume formulation

In the multiscale finite volume (MsFV) algorithm introduced in [159, 160, 161], the global (fine-scale) problem is partitioned into primal and dual coarse volumes as illustrated in Section 2.5.1. A set of basis functions is computed for each dual volume, and the coarse-scale problem is assembled. Using the coarse-scale system, the coarse-scale pressure is computed. The same basis functions allow for local reconstruction of the fine-scale pressure from the coarse solution.

The original MsFV algorithm [159] was designed to solve the (elliptic) pressure equation of incompressible flow in highly heterogeneous formations. The black oil model, which accounts for compressibility and capillarity, yields a nonlinear parabolic pressure equation. However, these effects are, in general, local in nature, and the pressure equation usually exhibits near-elliptic behavior. We construct a multiscale algorithm that takes advantage of this characteristic.

A multiscale, operator splitting approach is used to solve the nonlinear parabolic overall mass balance equation for the pressure field. Specifically, the black oil pressure equation is decomposed into three equations, one homogeneous and two inhomogeneous. The homogeneous (elliptic) equation is used to compute the dual basis functions and the coarse-scale transmissibilities. The first inhomogeneous part, $p_g$, accounts for gravity and capillarity. The second inhomogeneous equation is solved for the particular solution $p_p$, which accounts for accumulation (i.e., rock and fluid compressibility) and sink/source terms. Specifically, the black oil pressure equation is decomposed as follows.

$$L_{BO}\, p_h^{\nu+1} = 0, \tag{5.42}$$

$$L_{BO}\, p_g^{\nu+1} = RHS1, \tag{5.43}$$

$$L_{BO}\, p_p^{\nu+1} = -\frac{C_w}{\Delta t}[(p_h + p_g + p_p)^{\nu+1} - (p_h + p_g + p_p)^{\nu}] + RHS2. \tag{5.44}$$

**Homogeneous pressure solution**

The original MsFV method [159] employs locally computed basis functions (on the fine scale) and a pressure operator on a coarse grid. The fine-scale pressure field can then be obtained via a reconstruction step. Recently, Lunati and Jenny [186] presented a MsFV method for compressible multiphase flow. Their third proposed algorithm is somewhat similar to the scheme presented in this section; however, we do not use explicitly computed coarse-scale formation volume factors.

A conforming coarse grid with $N$ nodes and $M$ cells is constructed on the original fine grid. Each coarse cell $K_c^i$ with $i \in \{1, ..., M\}$ is composed of multiple fine cells. A dual coarse grid is constructed such that each dual coarse cell $K_d^j$, $j \in \{1, ..., N\}$ contains exactly one coarse node. The coarse dual grid has $M$ nodes, $x_i$ ($i \in \{1, ..., M\}$), each in the interior of a coarse cell $K_d^i$. Each dual grid has $N_c$ corners (for a Cartesian grid, four in two dimensions and eight in three dimensions). A set of dual basis functions, $\phi_j^i$, is constructed, one for each corner $i$ of each dual coarse cell $K_d^j$.

The dual basis functions are used to assemble the coarse-scale transmissibility field for computation of the coarse-scale pressure $p_i^c$. The dual basis function $\phi_j^i$, for example, is the local solution of (5.42):

$$\alpha_o \operatorname{div}\left(\lambda''_o\nabla\phi_j^i\right) + \alpha_w \operatorname{div}\left(\lambda''_w\nabla\phi_j^i\right) + \alpha_g \operatorname{div}\left(\lambda''_g\nabla\phi_j^i\right) = 0 \text{ on } K_d^j,$$

where properties from the underlying fine grid (e.g., total mobility) are used. The boundary conditions are obtained by solving the reduced problems [175], although one can easily use different boundary conditions as discussed earlier. Finally, given a coarse-scale solution $p_i^c$, the phase transmissibilities of the coarse grid can be readily computed from the fluxes across the coarse grid interface [175].

### Inhomogeneous solution: Gravity and capillary pressure

As shown in (5.43), the inhomogeneous solution $p_g$ accounts for gravity and capillary forces. Due to the complexity of the fractional flow function in the presence of gravity, the potential field cannot be represented by a simple superposition of the basis functions. Lunati and Jenny [187] proposed a method where $p_g$ is split into two parts. The first part is represented by the original dual basis functions; the second part is a locally computed correction term that accounts for buoyancy effects. Following their approach, $p_g$ can be written as

$$p_g = p_g^a + p_g^b = \sum_i \phi_j^i p_g^{c,i} + p_g^b \text{ in } K_d^j. \tag{5.45}$$

Note that within a dual coarse grid, $p_g^a$ is represented by a linear combination of basis functions, weighted by the coarse-scale pressures.

The additional correction term $p_g^b$ is obtained using (5.43),

$$-\alpha_o \operatorname{div}\left(\lambda''_o\nabla p_g^b\right) - \alpha_w \operatorname{div}\left(\lambda''_w\nabla p_g^b\right) - \alpha_g \operatorname{div}\left(\lambda''_g\nabla p_g^b\right)$$
$$= RHS1 \text{ in } K_d, \tag{5.46}$$

where solutions of reduced problems consistent with (5.46) serve as boundary conditions. Note that the correction term $p_g^b$ is computed with the simple boundary conditions for the reduced system that is independent of the global pressure distribution. This particular localization assumption to compute $p_g^b$ is analogous to the one used to construct the dual basis function in the absence of gravity effects. Lunati and Jenny [187] showed its effectiveness in resolving the fine-scale structures of complex heterogeneous problems, when buoyancy plays an important role.

Substitution of (5.45) in (5.43) and applying Green's theorem to the coarse operator [187], one can readily show that $p_g^b$ acts as an additional source/sink term in the coarse-scale pressure system.

### Particular solution: Mass accumulation and wells

The particular solution $p_p$, governed by (5.44), is used to model sources and sinks and the effects of compressibility (i.e., fluid accumulation). Accurate modeling of wells is crucial for any practical reservoir simulation problem. A treatment of wells specifically designed for the MsFV method has recently been proposed by Wolfsteiner, Lee and Tchelepi [259]. The framework allows for modeling wells that penetrate one or multiple fine cells, and accommodates fixed-rate or fixed-pressure operating conditions. In their approach, the near-singular pressure distribution around the well is removed by a change of variables. The well effects are approximated using special basis functions that are then added to a smoothly varying background solution computed using the standard MsFV method. Here, we employ a very simple model, where wells are represented only on the coarse grid. The corresponding fine cells receive source terms of equal strength [159].

Once the coarse-scale pressure is computed, the fine-grid pressure in the dual grid can be obtained using the basis functions.:

$$p_p(x) + p_g(x) = \sum_{i=1}^{N_c} \phi_j^i(x)(p_{p,i}^c + p_{g,i}^c) + p_g^b(x), \quad \text{for } x \in K_d^j. \qquad (5.47)$$

The pressure from the particular solution and the linear gravity part are interpolated using the dual basis functions, and then the gravity correction term, $p_g^b$, is added.

Jenny, Lee and Tchelepi [159, 160] found that the fine-scale velocity field computed directly from the reconstructed pressure (i.e., using the coarse-grid solution and the dual basis functions) suffered from local mass balance errors along the dual coarse cell boundaries. As a remedy, they proposed a second set of (primal) basis functions that guarantee a conservative fine-scale velocity field. In doing so, it is critical to honor the fine-scale fluxes from the overlapping dual basis functions as boundary conditions. That approach can be expensive, however. This is because the number of primal basis functions is large, and even if they need to be recomputed occasionally, the cost can be significant.

Here, we do not use this second set of bases. Instead we solve local problems on the primal coarse grid as follows. The reconstructed fine-scale pressure at the boundaries of a primal coarse cell is used to compute the fine-scale fluxes, which then serve as boundary conditions for local problems on the primal coarse grid. These local problems solve the nonlinear black oil equations, which may include compressibility, capillarity, and solubility effects. In essence the multiscale pressure approximation is used to prescribe flux boundary conditions for the full black oil equation set on the local primal coarse-cell level. Our experience is that the fine-scale pressure solution obtained from these local Neumann problems is quite accurate (i.e., locally consistent with the velocity field) when compared to the pressure that is reconstructed using the dual basis.

### 5.6.3 Sequential fully implicit coupling and adaptive computation

In the algorithm presented in the previous sections, flow and transport are solved sequentially. First, a fine-scale pressure field together with a compatible (and conservative) fine-scale velocity field is computed using the black oil MsFV method. Then, the transport problem is solved on local fine-grid domains with an implicit upwind scheme. A Schwarz overlap method is used with saturation at the boundaries from the previous iteration, which has been found to be very efficient for the saturation equations. The updated saturation distribution defines a new total mobility field for the subsequent elliptic problem (i.e., the next Newton iteration). Note that, in general, some of the basis functions have to be recomputed. These steps can be iterated until convergence of all variables at the current time level.

The MsFV approach can be easily adapted to a sequential fully implicit treatment [161]. The MsFV implementation allows for performing an IMPES, traditional sequential [29], or a fully implicit scheme. Here, the full nonlinear transmissibility terms at the new time-step level are retained so that stability is guaranteed [160]. The converged solution using this sequential approach should be identical to the solution obtained using the simultaneous solution strategy, which is usually used to deal with coupled fully implicit systems.

The MsFV approach is well suited for adaptive computation, which can lead to significant efficiency gains. The most expensive part of the algorithm is computation of the dual basis functions. In general, this is performed every iteration due to changes in the saturation (mobility) field. As discussed in Jenny et al. [160], an adaptive scheme can be used to update the dual basis functions. Because the basis functions are constructed with local support, the change of the total mobility is used to decide when and where to update the basis functions in the domain. For compressible fluid, we employ an effective total mobility change criterion for adaptable computation of the pressure field [175].

### 5.6.4 Numerical examples

**Waterflood in linear geometries**

This test case is a two-dimensional problem with $220 \times 60$ fine cells. A uniform coarse grid of $22 \times 6$ is used for the multiscale run. The permeability description is taken from the first layer of the Tenth SPE Comparative Solution Project [78]. As shown in Figure 5.23(a), a highly correlated area of low permeability is found on the left-hand-side of the model, and a high-permeability area is present on the right end of the model.

The black oil model includes three compressible fluid phases (i.e., oil, water, and gas). The pressure dependence of the densities is described using formation volume factors, and the phase equilibrium between the oil and gas phases is described using the solution gas–oil ratio [79]. Typical black oil properties are listed in Lee, Wolfsteiner, and Tchelepi [175]. For this example, we did not consider gravity effects, and we used a high oil compressibility, namely, $4.8 \cdot 10^{-4}$ for $p < p_b$ ($p_b$ is the bubble point pressure) for a stringent test of compressibility. As the pressure decreases, some solution gas is liberated from the oil phase and forms a free immiscible gas phase. Moreover, the oil-phase volume decreases as the pressure decreases below the bubble point, $p < p_b$. The solution-gas is constant above the bubble point pressure, and the oil-phase volume decreases as the pressure increases (i.e., $p > p_b$).

The model is initialized with oil ($S_o = 1$) and constant pressure (4000 psia). At $t = 0$, water is injected at a constant pressure of 5000 psia from the left side; the right boundary is maintained at 2000 psia. This numerical example is a challenging test due to the large pressure drop across the model and the large variations in permeability. A constant time-step size of 1 day is used. In Figure 5.23, the results from the black oil MsFV simulator and fine-scale reference simulations are shown at 50 days. For example, Figure 5.23(c)-(d) indicate that the water and gas distributions obtained from the MsFV approach are in excellent agreement with the reference fine-scale solutions.

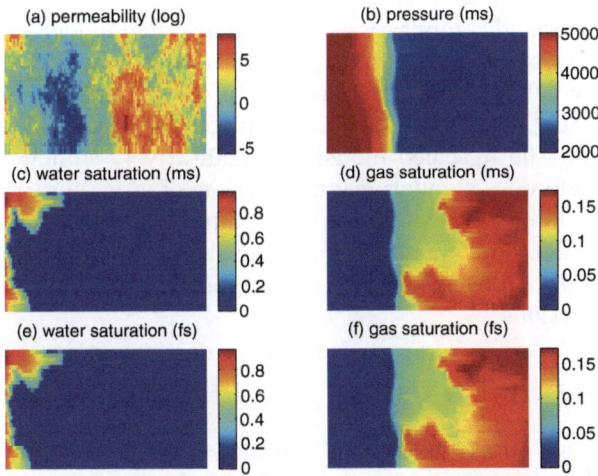

**Fig. 5.23.** Depletion with constant pressure boundary conditions: at time = 50days: (a) logarithm of permeability; (b) pressure from MsFV; (c) $S_w$ from MsFV,(d) $S_g$ from MsFV; (e) $S_w$ from fine-scale simulation; (f) $S_g$ from fine-scale simulation.

## A three-dimensional heterogeneous model with two wells

This example employs a three-dimensional model with two wells at two opposite corners and a heterogeneous permeability field. The permeability distribution is generated by the sequential Gaussian simulation method [85]. The logarithm of permeability has a Gaussian histogram with mean and standard deviation of 50 md and 1.5, respectively. The variogram is spherical with ranges of 30 m and 15 m in directions that are at 45 and 135 degrees with respect to the horizontal, and 7.5 m in the vertical direction. The permeability is shown in Figure 5.24(a). The model is $150 \times 150 \times 48$ m in size and is uniformly discretized using $45 \times 45 \times 30$ fine cells. The uniform $9 \times 9 \times 6$ coarse grid is used in the MsFV computations.

The permeability distribution is shown in Figure 5.24(a). The fluid properties for the first example are also employed. The reservoir is initially at gravitational equilibrium with 4000 psia at the bottom of the model. Water is injected at a constant rate from the bottom left corner (i.e., coarse cell 1,1,1) displacing the oil toward the producer located at the top right corner (cell 9,9,6). Figure 5.24(b) shows the oil saturation distribution at water breakthrough. The pressure around the production well is below the bubble point, and a free gas phase is present. In Figure 5.25, the production rates from MsFV are compared with those from the fine-scale reference simulation. The comparison shows that the black oil MsFV approach is able to model difficult multiphase flow problems in heterogeneous media when strong gravity and compressibility effects are present.

We also performed computations using an upscaled model for this problem. We used the basis functions to compute an upscaled (effective coarse-scale) transmissibility field, and we computed the pressure and saturation using the coarse-scale model. The results are also depicted in Figure 5.25. Even though the results from the upscaled model are qualitatively similar to those from the fine-scale reference simulation, the presence of large numerical dispersion in the upscaled model gives less accurate production rates compared with the multiscale method. This numerical example shows that reconstruction of the fine-scale information by the MsFV is an important step in obtaining accurate transport predictions.

### 5.6.5 Remarks

We developed a multiscale finite volume (MsFV) method for the black oil formulation of multiphase flow and transport in heterogeneous porous media. The black oil formulation, which involves immiscible three-phase flow with compressibility, gravity, capillary, and mass transfer, in the form of gas solubility, is widely used in practical field-scale simulations.

Our approach extends the sequential implicit MsFV method [161, 256] to the nonlinear black oil model. An operator-splitting multiscale algorithm is devised to compute the fine-scale pressure field, which is used to compute the fine-scale velocity field. The nonlinear saturation equations of the black oil model are solved on the local primal coarse grid using the fine-scale velocity field. The black oil MsFV method extends our ability to deal with large-scale problems of practical interest. The treatment ensures that the nonlinearity due to rock and fluid compressibility, gravity, and capillarity can be resolved

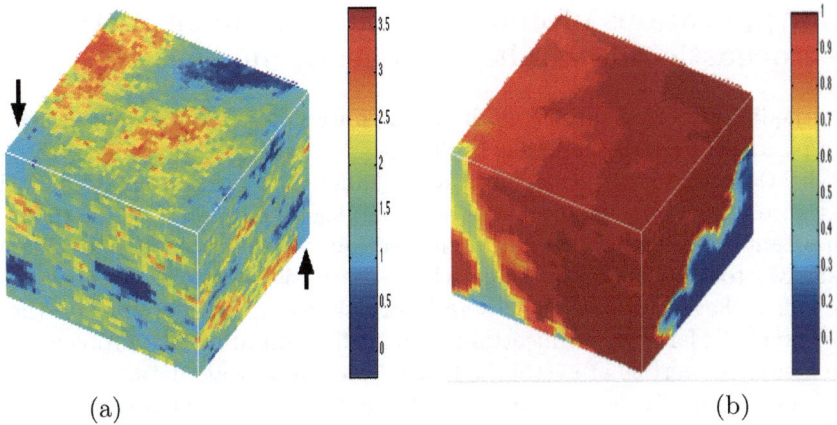

(a)                                                              (b)

**Fig. 5.24.** A heterogeneous model with two wells (Example 3): (a) log-permeability distribution, (b) oil saturation just after breakthrough.

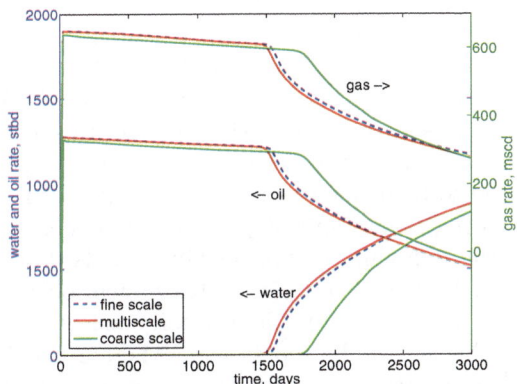

**Fig. 5.25.** Comparison of production rates for Example 3: multiscale (solid lines) fine-scale reference (dashed).

by solving specially constructed local boundary value problems. The methodology is demonstrated using several numerical examples. These examples show clearly that the MsFV scheme yields results that are in excellent agreement with reference fine-grid solutions.

Although the numerical efficiency of this new black oil MsFV simulator has not been fully examined, the numerical efficiency gains shown in references [160, 161, 256] are expected to hold (e.g., 10 ∼ 20 times faster than the conventional finite difference method). This is because all the nonlinearities due to the presence of compressibility, gravity, and capillary pressure are resolved locally.

## 5.7 Applications of multiscale finite element methods to stochastic flows in heterogeneous media

The media properties often contain uncertainties. These uncertainties are usually parameterized and one has to deal with a large set of permeability fields (realizations). This brings an additional challenge to the fine-scale simulations and necessitates the use of coarse-scale models. The multiscale methods are important for such problems. In this section, we describe the extensions of MsFEMs to stochastic equations where the basis functions are constructed such that they span both spaces and uncertainties. We also consider the applications of MsFEMs to uncertainty quantification in inverse problems when the media properties are estimated based on coarse-scale data.

First, we briefly discuss stochastic flow equations from an application point of view. Assume that the media properties are random and denoted by $k(x, \omega)$, where $\omega$ refers to a realization. Then, the solution of the flow equation is given by $p(x, \omega)$ for each realization $\omega$.

One of the commonly used stochastic descriptions of spatial fields is based on a two-point correlation function of log-permeability. To describe it, we denote by $Y(x, \omega) = \log[k(x, \omega)]$. For permeability fields described with the two-point correlation function, it is assumed that $R(x, y) = E[Y(x, \omega)Y(y, \omega)]$ is known, where $E[\cdot]$ refers to the expectation (i.e., average over all realizations) and $x, y$ are points in the spatial domain. In applications, the permeability fields are considered to be defined on a discrete grid. In this case, $R(x, y)$ is a square matrix with $N_{dof}$ rows and $N_{dof}$ columns, where $N_{dof}$ is the number of grid blocks in the domain. For permeability fields described by the two-point correlation function, one can use the Karhunen–Loève expansion (KLE) [182, 271] to obtain a permeability field description with possibly fewer degrees of freedom. This is done by representing the permeability field in terms of an optimal $L^2$ basis. By truncating the expansion, we can represent the permeability matrix by a small number of random parameters.

We briefly recall some properties of the KLE. For simplicity, we assume that $E[Y(x, \omega)] = 0$. Suppose $Y(x, \omega)$ is a second-order stochastic process with $E \int_{\Omega} Y^2(x, \omega)dx < \infty$. Given an orthonormal basis $\{\Phi_i\}$ in $L^2(\Omega)$, we can expand $Y(x, \omega)$ as a general Fourier series

$$Y(x, \omega) = \sum_{i=1}^{\infty} Y_i(\omega)\Phi_i(x), \qquad Y_i(\omega) = \int_{\Omega} Y(x, \omega)\Phi_i(x)dx.$$

We are interested in the special $L^2$ basis $\{\Phi_i\}$ that makes the random variables $Y_i$ uncorrelated. That is, $E(Y_iY_j) = 0$ for all $i \neq j$. The basis functions $\{\Phi_i\}$ satisfy

$$E[Y_iY_j] = \int_{\Omega} \Phi_i(x)dx \int_{\Omega} R(x, y)\Phi_j(y)dy = 0, \quad i \neq j.$$

Because $\{\Phi_i\}$ is a complete basis in $L^2(\Omega)$, it follows that $\Phi_i(x)$ are eigenfunctions of $R(x, y)$:

$$\int_{\Omega} R(x,y)\Phi_i(y)dy = \lambda_i\Phi_i(x), \quad i = 1, 2, \ldots, \tag{5.48}$$

where $\lambda_i = E[Y_i^2] > 0$. Furthermore, we have

$$R(x,y) = \sum_{i=1}^{\infty} \lambda_i\Phi_i(x)\Phi_i(y). \tag{5.49}$$

Denote $\theta_i = Y_i/\sqrt{\lambda_i}$; then $\theta_i$ satisfy $E(\theta_i) = 0$ and $E(\theta_i\theta_j) = \delta_{ij}$. It follows that

$$Y(x,\omega) = \sum_{i=1}^{\infty} \sqrt{\lambda_i}\theta_i(\omega)\Phi_i(x), \tag{5.50}$$

where $\Phi_i$ and $\lambda_i$ satisfy (5.48). We assume that the eigenvalues $\lambda_i$ are ordered as $\lambda_1 \geq \lambda_2 \geq \cdots$. The expansion (5.50) is called the Karhunen–Loève expansion. In the KLE (5.50), the $L^2$ basis functions $\Phi_i(x)$ are deterministic and resolve the spatial dependence of the permeability field. The randomness is represented by the scalar random variables $\theta_i$. After we discretize the domain $\Omega$ by a rectangular mesh, the continuous KLE (5.50) is reduced to finite terms and $\Phi_i(x)$ are discrete fields. Generally, we only need to keep the leading order terms (quantified by the magnitude of $\lambda_i$) and still capture most of the energy of the stochastic process $Y(x,\omega)$. For an $N$-term KLE approximation $Y_N = \sum_{i=1}^{N} \sqrt{\lambda_i}\theta_i\Phi_i$, define the energy ratio of the approximation as

$$e(N) := \frac{E\|Y_N\|^2}{E\|Y\|^2} = \frac{\sum_{i=1}^{N} \lambda_i}{\sum_{i=1}^{\infty} \lambda_i}.$$

If $\lambda_i, i = 1, 2, \ldots$, decay very fast, then the truncated KLE would be a good approximation of the stochastic process in the $L^2$ sense.

Next, we discuss some example cases. Suppose the permeability field $k(x,\omega)$ is a log-normal homogeneous stochastic process; then $Y(x,\omega) = \log(k(x,\omega))$ is a Gaussian process, and $\theta_i$ are independent standard Gaussian random variables. In this case, the covariance function of $Y(x,\omega)$ has the form

$$R(x,y) = \sigma^2 \exp\left(-\frac{|x_1 - y_1|^2}{2l_1^2} - \frac{|x_2 - y_2|^2}{2l_2^2}\right). \tag{5.51}$$

In the above formula, $l_1$ and $l_2$ are the correlation lengths in each dimension, and $\sigma^2 = E(Y^2)$ is the variance. We first solve the eigenvalue problem (5.48) numerically on the rectangular mesh and obtain the eigenpairs $\{\lambda_i, \Phi_i\}$. We put 8 points per correlation length in our numerical simulations. Because the eigenvalues decay fast, the truncated KLE approximates the stochastic process $Y(x,\omega)$ fairly well in the $L^2$ sense. Therefore, we can sample $Y(x,\omega)$ from the truncated KLE (5.50) by generating Gaussian random variables $\theta_i$. In Figure 5.26, we plot eigenvalues and three eigenvectors corresponding to eigenvalues (in decreasing order) 1, 6, and 15. In particular, we plot eigenvalues for the log-normal permeability field described by (5.51) as well as by

$$R(x, y) = \sigma^2 \exp\left(-\frac{|x_1 - y_1|}{l_1} - \frac{|x_2 - y_2|}{l_2}\right). \tag{5.52}$$

As we see from these figures the eigenvalues decay quickly for log-normal permeability fields compared to log-permeability fields described by (5.52). Moreover, the eigenvectors corresponding to smaller (in value) eigenvectors contain finer-scale features of the media.

For some simplified cases, one can derive formulas for eigenvalues and eigenvectors (e.g., [277]). In the one-dimensional case, $R(x, y) = \sigma^2 \exp(-|x_1 - y_1|/l_1)$, the eigenvalues have the form

$$\lambda_n = \frac{2l_1\sigma^2}{l_1^2\zeta_n^2 + 1}$$

and

$$\Phi_n(x) = \frac{1}{\sqrt{(l_1^2\zeta_n^2 + 1)L/2 + l_1}}(l_1\zeta_n\cos(\zeta_n x) + \sin(\zeta_n x)),$$

where $L$ is the length of the domain and $\zeta_n$ are positive roots of the characteristic equation

$$(\zeta^2 l_1^2 - 1)\sin(\zeta L) = 2\zeta l_1 \cos(\zeta L). \tag{5.53}$$

For problems in a multidimension, if the covariance function is in the form $R(x, y) = \sigma^2 \exp(-|x_1 - y_1|/l_1 - |x_2 - y_2|/l_2)$, the eigenvalues have the form

$$\lambda_{ij} = \frac{4l_1 l_2 \sigma^2}{(l_1^2(\zeta_i^1)^2 + 1)(l_2^2(\zeta_j^2)^2 + 1)}$$

and

$$\Phi_{ij}(x) = \Phi_i(x_1)\Phi_j(x_2),$$

where $\zeta_i^1$ and $\zeta_j^2$ are positive roots of (5.53) using parameters $(L_1, l_1)$ and $(L_2, l_2)$, respectively, with $L_1$ and $L_2$ being the lengths of the whole domain in the $x_1-$ and $x_2-$directions.

## 5.7.1 Multiscale methods for stochastic equations

In this section, we present a multiscale approach for solving stochastic flow equations. The main idea of the proposed approaches is to construct multiscale basis functions that capture the small-scale information across the realizations of stochastic equations. Once the basis functions are constructed, the solution is projected into the finite-dimensional space spanned by the multiscale basis functions. The pre-computed basis functions are constructed based on selected realizations of the stochastic permeability field and the method can be regarded as an extension of MsFEMs to stochastic porous media equations. The proposed approaches, although they do not require any interpolation in stochastic space, can be combined with interpolation-based approaches to

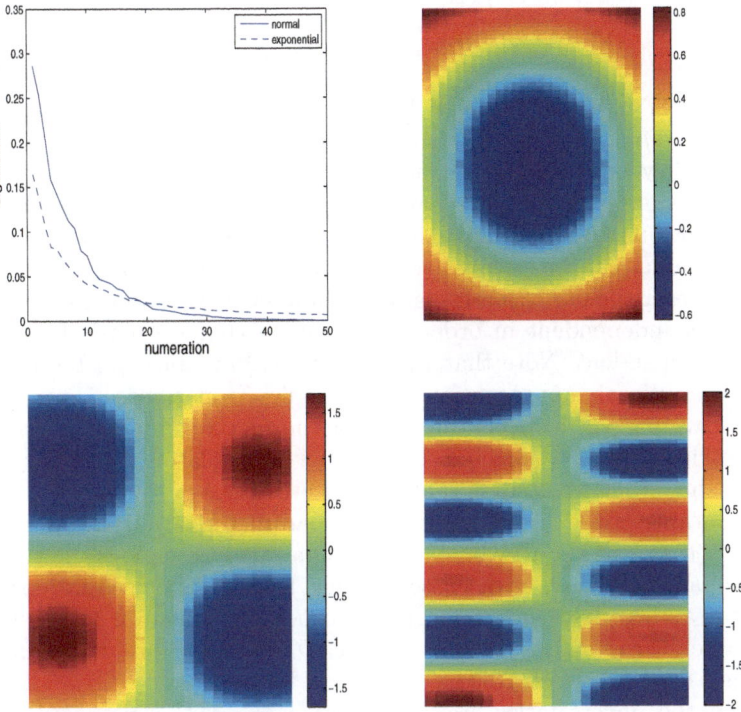

**Fig. 5.26.** Top left: Eigenvalue distribution. Top right: 1st eigenvector. Bottom left: 6th eigenvector. Bottom right: 15th eigenvector.

predict the solution on the coarse grid. The permeability fields under consideration do not have scale separation. For this reason, we employ multiscale methods using limited global information in our simulations. The main idea of these approaches is to use some global fields that contain nonlocal information as discussed in Chapter 4. We use the mixed MsFEM framework here, although other global couplings can also be used.

To present the approach, we consider realizations of permeability fields $k_i$ sampled from a stochastic distribution. For each $k_i$, let $p_i$ and $v_i$ be a solution obtained by solving the flow equation on a fine grid using a suitable mass conservative numerical method. Then, we define $\mathcal{V}_h(k_i)$ the space spanned by mixed multiscale basis functions, $\psi_{il}^K$, defined via (4.6) that have the following boundary conditions

$$\psi_{il}^K \cdot n = \frac{v_i \cdot n}{\int_{e_l} v_i \cdot n\, ds},$$

on $e_l$ for each $K$. We employ $N$ realizations of the permeability field and define a finite-dimensional space that consists of a direct sum of mixed multiscale finite element basis functions corresponding to all realizations:

$$\mathcal{V}_h = \oplus_{i=1}^{N} \mathcal{V}_h(k_i).$$

Hence, in this case we obtain $N$ basis functions for each edge (face) in the coarse grid. Once the basis functions are constructed, the solution of the stochastic flow equation for an arbitrary realization is projected onto this finite-dimensional space. Note that this approach does not require any interpolation formula in uncertainty space, although interpolation techniques, if easily available, can be used to reduce the size of $\mathcal{V}_h$ locally in uncertainty space (see later the use of interpolation techniques). We assume that $v_1, ..., v_N$ are linearly independent in order to guarantee that the basis functions are linearly independent. Note that the local basis functions can be used in the proposed multiscale approach for stochastic flow equations.

Next, we present a formal analysis of the method under the assumption that the chosen realizations can be used to interpolate an arbitrary realization. To show this, we assume that the uncertainties of the permeability field can be parameterized. As a result of this parameterization, the permeability is expressed as $k = k(x, \theta)$ where $\theta \in \mathbb{R}^L$. One such example is the Karhunen–Loève expansion (KLE) as discussed earlier. KLE can be used in representing the permeability fields given via the two-point correlation function, where $k(x, \theta) = \exp(Y(x, \theta))$, $Y(x, \theta) = \sum_{i=1}^{L} \Theta_i \Phi_i(x)$, $\Phi_i(x)$ pre-determined functions, and $\theta = (\Theta_1, ..., \Theta_L)$.

When the uncertainties are parameterized and $L$ is not large, one can employ sparse interpolation techniques in $\mathbb{R}^L$ (e.g., [272]), where the solution is computed for some values of $\theta = (\Theta_1, ..., \Theta_L)$, denoted by $\theta_k$, and then interpolated for an arbitrary $\theta \in \mathbb{R}^L$. Assuming that $k(x, \theta)$ smoothly depends on $\theta$, we can approximate the solution for an arbitrary $\theta$ as

$$p(x, \theta) \approx \sum_i p(x, \theta_i) \beta_i(\theta), \qquad (5.54)$$

where $\beta_i(\theta)$ are the corresponding weights which are in general difficult to obtain. We note that the interpolation error depends on the choice of interpolation points and the smoothness of $p(x, \theta)$ with respect to $\theta$. Denoting the velocity field for two-phase flow by $v$, we have

$$v(x, \theta) \approx \sum_i v(x, \theta_i) \beta_i(\theta). \qquad (5.55)$$

Equation (5.54) shows that the solution of the stochastic flow equation can be approximated if we provide approximations of $p(x, \theta_i)$ for each $\theta_i$. Because the solution for each selected realization can be approximated using corresponding global fields, we have

$$v(x, \theta_i) \approx \sum_j c_{ij}^*(x) v_j(x, \theta_i).$$

In our numerical simulations, we use single-phase velocity fields following previous discussions in Section 4.2 (see also [1, 3]). One can, in general, use directional flows as proposed in a more general setting in [218]. We note that in our multiscale simulations, the basis functions are constructed using $v_j(x, \theta_i)$. One can show the convergence of the proposed approach following, for example, [8].

We note that the proposed method can be applied in a local region of the uncertainty space by selecting realizations that correspond to this region. The latter is useful when one would like to perform uncertainty quantification in a subregion of the uncertainty space. One can use the localization in the uncertainty space for more accurate probabilistic estimations by partitioning the uncertainty space. To our best knowledge, the idea of local partitioning of uncertainty space in the context of stochastic PDEs was first investigated in [267] where the authors introduced a multi-element generalized polynomial chaos approach. In our approaches, we can borrow this idea and combine it with MsFEMs. To describe the procedure, we denote by $U$ the uncertainty space and assume that $U$ is partitioned into $U_i$. In each region $U_i$, we choose selected realizations $\theta_j^i$ representing these local regions. Then, the basis functions are defined as before for these selected realizations in each $U_i$. This approach is an implementation of the earlier proposed technique simply in local regions of uncertainty space. In particular, the multiscale basis functions are constructed as before although with local support both in spatial and uncertainty spaces. When performing simulations for a particular (arbitrary) realization, the multiscale basis functions from the local uncertainty region that contains this particular realization will be used. This will provide high accuracy and reduce the computational cost.

We note that the proposed method can be applied in local regions of uncertainty space and, consequently, the support of basis functions can be localized in uncertainty space. To describe the procedure, we denote by $U$ the uncertainty space and assume that $U$ is partitioned into $U_i$. Here $U_i$ can be regions larger than the characteristic length scale in uncertainty space. Furthermore, in each region $U_i$, we choose realizations $\theta_j^i$ representing these local regions. Then, the basis functions are defined as before for these realizations in each $U_i$. This approach is an implementation of the earlier proposed technique simply in local regions of uncertainty space. In particular, the multiscale basis functions are constructed as before although with local support. One can draw a parallel between this approach and a general multiscale approach where the coefficients strongly vary with respect to spatial variables and uncertainties. In particular, we would like to construct multiscale basis functions for permeability fields $k(x, \theta)$ over a coarse region that is larger than spatial and uncertainty heterogeneities. In this case, one needs to construct the local spatial basis functions for each $\theta_j^i$ in $U_i$. Because the dependence on $\theta$ is parametric, one needs to capture the spatial heterogeneities for all values of $\theta$ in $U_i$. In this case, the basis functions are derived from the solution of

$$\text{div}(k(x, \theta)\nabla w_i^K(x, \theta)) = 0.$$

These basis functions, which are smooth with respect to $\theta$, can be approximated by choosing appropriate realizations. Thus, the proposed approach can be regarded as an extension of the mixed MsFEM to problems with uncertainties. When using local patches in uncertainty space, one needs to determine a partition to which a particular realization belongs. We note that pre-computed multiscale basis functions can be repeatedly used for different boundary conditions/source terms and for dynamic two-phase flow and transport simulations.

The main practical advantage of the proposed mixed MsFEM is that one does not need interpolation formulas. Indeed, when an approximation space consists of a union of subspaces generated using the solutions corresponding to different permeability realizations, one is actually projecting the true solution onto this enriched approximation space. Thus, the velocity solution will be a superposition of basis functions corresponding to each of the sample fields, but the interpolation weights are determined automatically from the projection property of the mixed MsFEM. In particular, the interpolation weights will vary throughout the uncertainty domain. This approach is interpolation-free, easy to use, and provides a computationally cost-efficient methodology for performing multiple simulations, for instance, to quantify uncertainty. We also note that when an interpolation formula is easily available, one can interpolate the set of pre-computed multiscale basis functions to calculate the basis functions for a particular realization. However, the nature of this interpolation (pointwise or $L^2$ or so on) will be pre-determined. Our proposed approach chooses the best interpolation both in spatial and stochastic space. Finally, we would like to note that in upscaling approaches, to our best knowledge, one cannot avoid interpolation techniques.

### Numerical results

*Experimental setup.* In our simulations below, we take $k_{rw}(S) = S^2$, $\mu_w = 0.1$, $k_{ro}(S) = (1 - S)^2$, and $\mu_o = 1$ in two-phase flow and transport simulations (see (5.1), (5.2)). The log-permeability field $Y(x)$ is given on a $100 \times 100$ fine Cartesian grid. This grid is then coarsened to form a uniform $5 \times 5$ Cartesian grid so that each block in the coarse grid contains a $20 \times 20$ cell partition from the fine grid. We solve the pressure equation on the coarse grid using the mixed MsFEM and then reconstruct the fine-scale velocity field as a superposition of the multiscale basis functions. The reconstructed field is used to solve the saturation equation on the fine grid. The saturation equation is solved using an implicit upstream finite volume (discontinuous Galerkin) method. We would like to emphasize that the multiscale basis functions are constructed at time zero, that is, they are not recomputed during the simulations.

In the numerical examples that are reported below we consider a traditional quarter-of-a-five-spot problem. That is, $\Omega$ is taken to be a square domain, we inject water at the upper left corner, and produce the fluid that

reaches the producer at the lower right corner. To assess the quality of the
respective saturation solutions obtained using the mixed MsFEM, we compute for each realization a reference solution $S_{\mathrm{ref}}$ obtained by solving the
time-dependent pressure equation on the fine grid with the given permeability field (using the lowest-order Raviart–Thomas mixed finite element method
for Cartesian grids). Then, in addition to measuring the relative saturation
error in the $L^1$-norm:

$$\|S - S_{\mathrm{ref}}\|_{L^1} / \|S_{\mathrm{ref}}\|_{L^1},$$

we compare various production characteristics. We use the water-cut curve
defining the fraction of water in the produced fluid as a function of time
measured in pore volumes injected (PVI) (see (2.44)). We recall that

$$w(t) = \frac{q_w(t)}{q_w(t) + q_o(t)},$$

where $q_o$ and $q_w$ are flow rates of oil and water at the producer at time $t$.
    We monitor the following quantities

- The relative water-cut error in the $L^2$-norm:

$$\|w - w_{\mathrm{ref}}\|_{L^2} / \|w_{\mathrm{ref}}\|_{L^2}.$$

- The breakthrough time (defined as $w^{-1}(0.05)$) at the producer.
- The cumulative oil production at 0.6 PVI:

$$Q_o = -\frac{1}{\int_\Omega \phi\, dx} \int_0^{0.6PVI} \left( \int_\Omega \min(q_o(x, \tau), 0)\, dx \right) d\tau.$$

    Before we embark on the numerical experiments, we note that the Raviart–
Thomas mixed finite element discretization of the pressure equation results in
a linear system with $N_{\mathrm{fine}}^2 + 2 \times (N_{\mathrm{fine}} - 1) \times N_{\mathrm{fine}} = 29800$ unknowns, where
$N_{\mathrm{fine}} = 100$. In comparison, when using a sample of $N$ permeability fields to
generate the mixed MsFEM basis functions, the stochastic multiscale method
gives rise to a linear system with $N_{\mathrm{coarse}}^2 + 2 \times (N_{\mathrm{coarse}} - 1) \times N_{\mathrm{coarse}} \times N =
25 + 40N$ unknowns, where $N_{\mathrm{coarse}} = 5$. Hence, when using a sample size
of 25, for instance, the number of the unknowns in the fine-grid system is
roughly 30 times larger than the number of unknowns in the mixed MsFEM
system. In this section, we present our results which employ samples of 10–50
permeability fields. In other words, we compute 10–50 velocity basis functions
for each interface in the coarse grid. We note that in order for the proposed
methods to be computationally efficient one needs to use fewer basis functions in each coarse-grid block to represent the heterogeneities across space
and uncertainties. In particular, the number of basis functions needs to be
less than the number of fine-grid blocks within the target coarse-grid block.
Otherwise, one can simply use fine-scale basis functions which are the same
for an arbitrary realization. In the case of the latter, the stochasticity does
not affect the choice of the finite element function space.

*Gaussian fields.* For Gaussian fields, one can reduce the dimension of the uncertainty space dramatically due to the fast decay of eigenvalues. To sample the realizations that are used to generate the multiscale basis functions, we use the first order Smolyak collocation points $\theta_i$ in $[-3, 3]^L$ (see, e.g., [272]). That is, $\theta_0 = 0$, $\theta_{2i-1} = 3\delta_{ij}$, and $\theta_{2i} = -3\delta_{ij}$, $i = 1, ..., L$. We note that the choice of interpolation points does not affect the implementation of our approach.

Our first results are for the isotropic case with $l_1 = l_2 = 0.2$ and $\sigma^2 = 2$. In this case, we can reduce the dimension of the stochastic permeability to 10. From this stochastic model for the permeability we draw randomly 100 realizations and perform simulations on the corresponding permeability fields.

In Figure 5.27 we compare breakthrough times and cumulative oil production at 0.6 PVI. We see that there is nearly a perfect match between the results obtained with the mixed MsFEM and the corresponding results derived from the reference solutions. Next, in Figure 5.28, we plot $L^2$ errors in the saturation field for these realizations as well as the water-cut errors. It can be observed from this figure that the saturation errors are mostly below 3%. Finally, we plot in Figure 5.29 a histogram of the breakthrough times and cumulative oil production values depicted in Figure 5.27 to demonstrate that the mixed MsFEM essentially provides the same statistics as one obtains from the set of reference solutions. These results suggest that with a few precomputed basis functions in each coarse grid block we can solve two-phase flow equations on the coarse grid for an arbitrary realization and obtain nearly the same results as one obtains by doing fine-grid simulations for each realization.

We have also considered numerical results for an anisotropic Gaussian field with $l_1 = 0.5$, $l_2 = 0.1$, and $\sigma^2 = 2$ in [7]. Due to anisotropy, KLE requires 12 terms. We sample the realizations that are used to generate the multiscale basis functions using the first order Smolyak collocation points as in the isotropic case. The numerical results obtained for the anisotropic Gaussian fields are qualitatively the same as the results shown in Figure 5.27 – Figure 5.29. We include only the anisotropic equivalent of Figure 5.29. Histograms of breakthrough time and cumulative oil production at 0.6 PVI for 100 randomly chosen realizations are depicted in Figure 5.30. The histograms confirm that the multiscale method essentially provides the same breakthrough time and cumulative oil production statistics as one obtains from the set of reference solutions.

*Exponential variogram fields.* For our second set of results, we consider permeability fields with exponential covariance matrix

$$R(x, y) = \sigma^2 \exp\left(-\frac{|x_1 - y_1|}{l_1} - \frac{|x_2 - y_2|}{l_2}\right). \tag{5.56}$$

Because of the slow decay of eigenvalues, one usually needs to keep many terms in KLE and deal with a large uncertainty space. To approximate the permeability fields, KLE requires 300 to 400 eigenvectors depending on correlation lengths and variance. This is a large-dimensional problem for performing

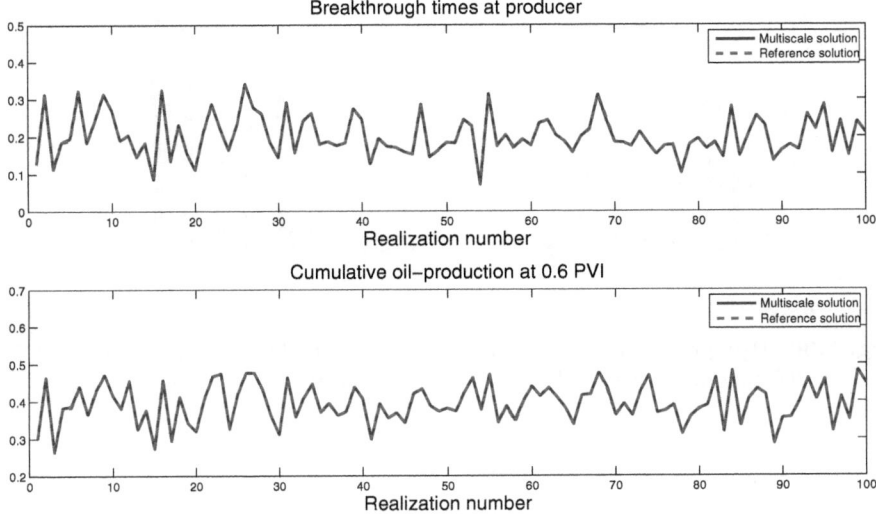

**Fig. 5.27.** Breakthrough time and cumulative oil production at 0.6 PVI for 100 random realizations from a Gaussian field with $l_1 = l_2 = 0.2$ and $\sigma^2 = 2$.

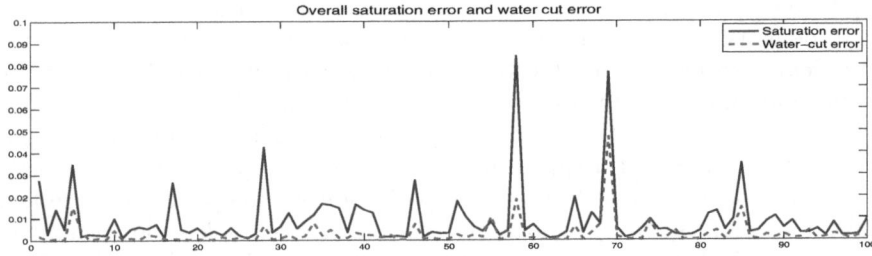

**Fig. 5.28.** $L^2$ errors of the saturation field and water-cut errors for 100 randomly chosen realizations (the number of a realization is indicated along the horizontal axis). Gaussian field with $l_1 = l_2 = 0.2$, $\sigma^2 = 2$.

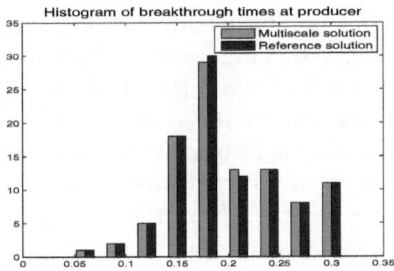 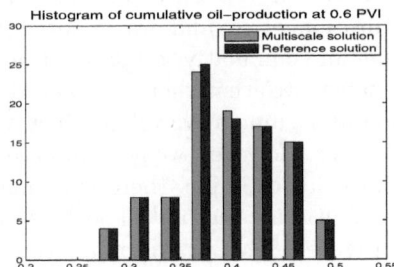

**Fig. 5.29.** Histograms of the breakthrough times and cumulative oil production values shown in Figure 5.27.

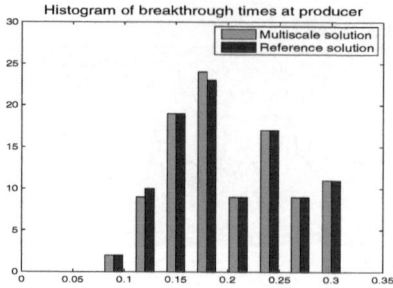 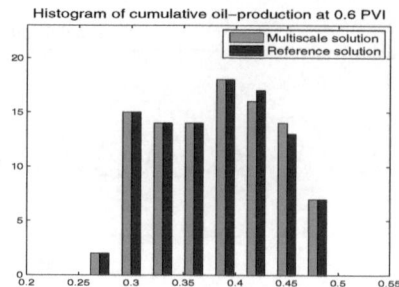

**Fig. 5.30.** Histograms of breakthrough time and cumulative oil production at 0.6 PVI for 100 random Gaussian fields with $l_1 = 0.5$ and $l_2 = 0.1$ and $\sigma^2 = 2$.

direct interpolation using multiscale basis functions. Instead we suggest using few independent realizations in constructing basis functions, and then performing statistical studies on a much larger set of realizations. We note that for independent realizations, we do not have an easily available interpolation formula. Moreover, the use of independent realizations is quite easy and one can use this technique for more general permeability fields in as much as it only requires independent samples of the permeability field.

To demonstrate the performance of the stochastic multiscale method for these fields, we present results for a case where the permeability fields are drawn from an anisotropic exponential variogram distribution with $l_1 = 0.5$, $l_2 = 0.1$, and $\sigma^2 = 2$ (the results for the isotropic case are similar, and not reported here). The KLE requires 350 eigenvectors to represent this stochastic permeability distribution. From this distribution we sample 20 independent realizations and use these realizations to generate the multiscale basis functions. Figure 5.31 displays one randomly chosen realization and corresponding saturation profiles at 0.6 PVI obtained by solving the pressure equation on the fine grid, and on the $5 \times 5$ coarse grid with the mixed MsFEM, respectively.

Figures 5.32, 5.33, and 5.34, show: breakthrough time at producer and cumulative oil production at 0.6 PVI for 100 randomly chosen realizations for both the reference solution and the multiscale solution; relative overall saturation error and water-cut error; and histograms of the breakthrough times and cumulative oil production values depicted in Figure 5.32. Figure 5.32 demonstrates that there is generally a good match between the breakthrough time and cumulative oil production curves for the reference and multiscale solutions. However, we now observe that there is a slight bias in the multiscale results, for example, there is a small time-lag in the breakthrough times for the multiscale method. The bias can also be observed from the histograms in Figure 5.34, but the magnitude of the bias is small, and the multiscale

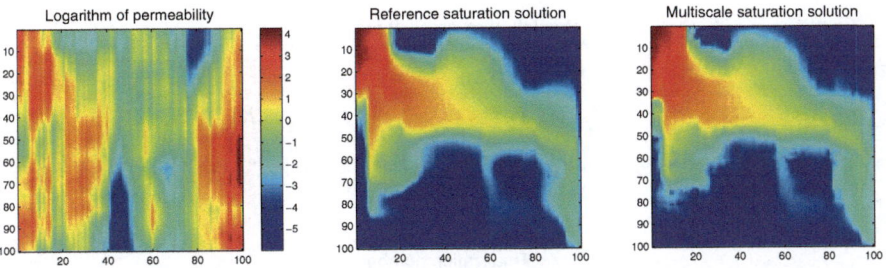

**Fig. 5.31.** An exponential variogram field with $l_1 = 0.5$, $l_2 = 0.1$, and $\sigma^2 = 2$, and a comparison of the reference saturation field at 0.6 PVI and the corresponding saturation field obtained using the stochastic multiscale method.

solutions are generally quite close to the reference solution, as is illustrated in Figures 5.31 and 5.33.

We now demonstrate that the bias in breakthrough time and cumulative oil production persists, but is efficiently reduced by increasing the number of realizations used to generate the multiscale basis functions. Figures 5.35, 5.36, and 5.37 show, respectively, the saturation and water-cut error for each of the 100 randomly selected realizations for the stochastic multiscale method with different sample sizes, the cumulative probability distribution of breakthrough times and cumulative oil production, and the corresponding histograms of the breakthrough times and the cumulative oil production values. Here, the sample size refers to the number of realizations selected in constructing multiscale basis functions. The plots show the following: the saturation and water-cut errors decay with increasing sample size; the time lag in the breakthrough times (also observed in the cumulative oil production) decays rapidly with increasing sample size, and that using 50 basis functions for each coarse-grid interface generates statistics that are nearly unbiased, and generally match the statistics derived from the set of reference solutions very well. Observe that a sample size of 50 gives rise to a linear system with 2025 unknowns, roughly 1/15 as many as in the fine-grid system.

## Summary

In conclusion, we have developed and studied the stochastic mixed multiscale finite element method. This method solves stochastic porous media flow equation on the coarse grid using a set of pre-computed basis functions. The pre-computed basis functions are constructed based on selected realizations of the stochastic permeability field, and thus span both spatial scales and uncertainties. The proposed method can be regarded as an extension of mixed MsFEM to stochastic porous media flow equations. The proposed approach does not require any interpolation in stochastic space and is capable of predicting the

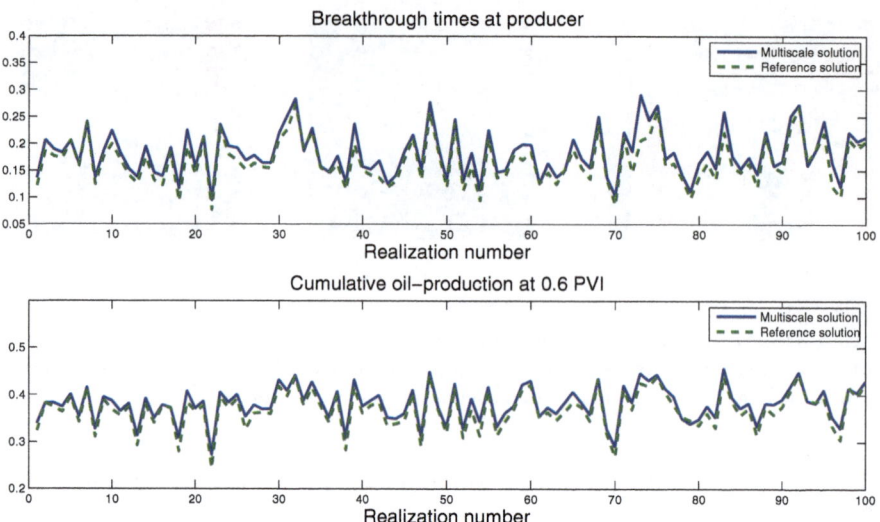

**Fig. 5.32.** Breakthrough time and cumulative oil production at 0.6 PVI for 100 random realizations from an exponential variogram field with $l_1 = 0.5$, $l_2 = 0.1$, and $\sigma^2 = 2$.

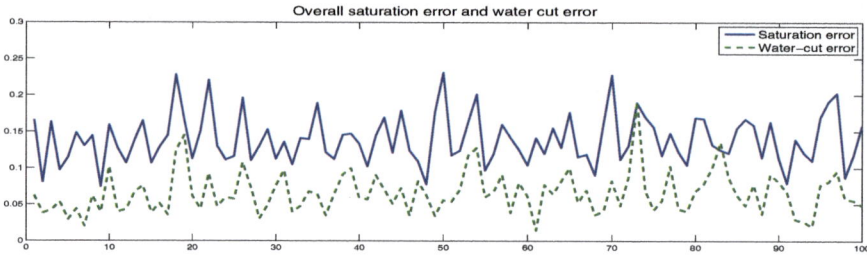

**Fig. 5.33.** $L^2$ errors of the saturation field and water-cut errors for 100 randomly chosen exponential variogram fields with $l_1 = 0.5$, $l_2 = 0.1$, and $\sigma^2 = 2$.

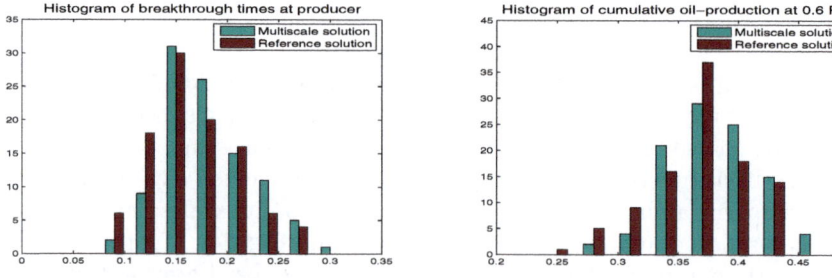

**Fig. 5.34.** Histograms of the breakthrough times and cumulative oil production values shown in Figure 5.32.

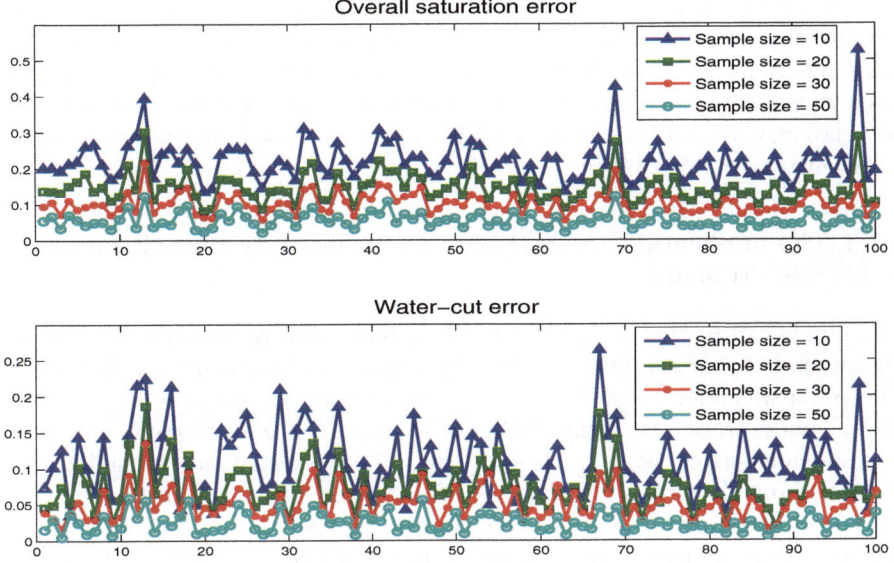

**Fig. 5.35.** Saturation and water-cut errors for solutions obtained using different number of permeability realizations to generate the multiscale basis functions.

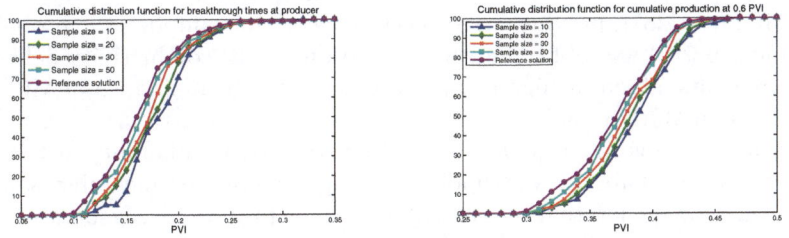

**Fig. 5.36.** Cumulative probability distribution for breakthrough time and cumulative oil production at 0.6 PVI.

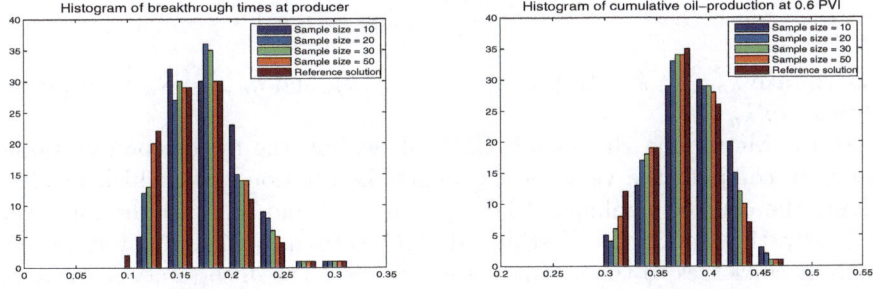

**Fig. 5.37.** Histograms of the breakthrough times and cumulative oil production.

solution on the coarse grid. We present numerical results for two-phase immiscible flow in stochastic porous media which show that one can use few basis functions in approximating the solutions of permeability fields with large uncertainty space. Finally, we would like to note that the proposed approaches can be easily combined with interpolation-based approaches in order to achieve greater flexibility.

### 5.7.2 The applications of MsFEMs to uncertainty quantification in inverse problems

In a number of papers [105, 89, 109, 88], applications of MsFEMs or upscaling methods to uncertainty quantification in inverse problems are discussed. The problem under consideration consists of finding stochastic realizations of the conductivity (or permeability) field given the measurement data and measurement errors (e.g., fractional flow (oil-cut) measurements defined by (2.43)). From the probabilistic point of view, this problem can be regarded as the conditioning of the permeability field to the measured data with associated measurement errors. Consequently, our goal is to sample from the conditional distribution $P(k|D)$, where $k$ is the fine-scale permeability field and $D$ is the measured data. Using the Bayes' theorem (e.g., [237]) we can write $\pi(k) = P(k|D) \propto P(D|k)P(k)$.

The techniques based on Metropolis–Hastings Markov chain Monte Carlo (MCMC) (see [237]) provide a rigorous framework for sampling the probability distribution $\pi(k)$ and obtaining the realizations of the conductivity field given measurements, albeit at high computational cost. The main idea of MCMC is to generate a Markov chain with $\pi(k)$ as its stationary distribution. A key step to this approach is to construct the desired transition probability distribution for the Markov chain. In Metropolis–Hastings MCMC, permeability samples, $k_1, ..., k_n, ...$ are generated. In particular, at $k_n$, a proposal $k$ is generated using instrumental probability distribution $q(k|k_n)$. Furthermore, $k$ is accepted as a sample with probability

$$p(k_n, k) = \min\left(1, \frac{q(k_n|k)\pi(k)}{q(k|k_n)\pi(k_n)}\right);$$

that is, take $k_{n+1} = k$ with probability $p(k_n, k)$, and $k_{n+1} = k_n$ with probability $1 - p(k_n, k)$.

In the Metropolis–Hastings MCMC algorithm, the major computational cost is to compute the value of the target distribution $\pi(k)$, which involves solving the coupled nonlinear PDE system (5.1) and (5.2) on the fine grid. Generally, the Metropolis–Hastings MCMC method requires many iterations before it converges to the steady state. To assess the uncertainty accurately, one needs to generate a large number of different samples. Thus, the direct (full) MCMC simulations are usually prohibitively expensive. Moreover, the acceptance rate of the direct MCMC method can be very low, due to the large dimensions of the permeability field. As a result, most of the CPU time is spent on rejected samples.

An important way to improve the direct MCMC method is to increase the acceptance rate by modifying the proposal distribution $q(k|k_n)$. Typically, some simplified models can be used to do so (e.g., [66, 179]). In [105, 89, 109], we discuss algorithms that use approximate and inexpensive coarse-scale simulations based on MsFEMs to speedup MCMC calculations. In particular, we consider an application to two-phase flow and transport simulations where the pressure equation is upscaled using the MsFVEM or mixed MsFEM or stochastic mixed MsFEM, and the saturation equation is upscaled using a simple volume averaging

$$\frac{\partial \overline{S}}{\partial t} + \overline{v} \cdot \nabla f(\overline{S}) = 0. \tag{5.57}$$

Although, this type of upscaling can introduce some errors (see Figure 2.11), it can be used in uncertainty quantification in inverse problems for the following reasons. First, this approach, which combines MsFEMs for the pressure equation and primitive upscaled model for the saturation equation, is very inexpensive. Second, we have observed that there is a strong correlation between the misfit corresponding to fine- and coarse-scale fractional flows.

Denote by $D_k^*$ the coarse-scale data computed using MsFEMs. In the applications to two-phase flow and transport, MsFEMs are used for the pressure equation with permeability $k$ and (5.57) for the saturation equation. Furthermore, we denote by $\pi^*(k) = P(k|D^*)$ the corresponding coarse-scale approximation of the target distribution $\pi(k)$. In general, one can perform offline simulations to estimate a statistical relation between the coarse-scale output $D_k^*$ and the fine-scale output $D_k$ via offline simulations for different $k$s sampled from the prior distribution. Based on this relation, $\pi^*$ can be estimated (see [109]). Our main emphasis is the use of physics-based coarse-scale models for uncertainty quantification in inverse problems.

Using the coarse-scale distribution $\pi^*(k)$ as a filter, the preconditioned MCMC was proposed in [105]. In this approach, the coarse-scale simulation is used in the second stage to screen the proposal before running a fine-scale simulation. More precisely, after making a proposal as in Metropolis–Hastings MCMC, the coarse-scale simulation is performed and the proposal is screened using $\pi^*$ distribution. If the proposal is accepted at this stage, only then a fine-scale simulation is performed for the proposed permeability field to decide whether to accept the proposal. Because the computation of the coarse-scale solution is very cheap, this step can be implemented very quickly to decide whether to run fine-scale simulations. The second step of the algorithm serves as a filter that avoids unnecessary fine-scale runs for the rejected samples. In [105], we show that the modified Markov chain is ergodic and converges to the correct distribution. We present numerical results for permeability fields generated using two-point correlation functions (see (5.50)) in [105]. Our results demonstrate that preconditioned MCMC has similar convergence properties, it has higher acceptance rates, and provides an order of magnitude of CPU saving. We refer to [105] for details.

An important type of proposal distribution can be derived from the Langevin diffusion, as proposed by Grenander and Miller [138], which uses the gradient of the posterior in the proposal. The use of the gradient information in inverse problems for subsurface characterization is not new (e.g., see [216]). The use of gradient information allows us to achieve high acceptance rates (e.g., [249]). In [89], we proposed the preconditioned coarse-gradient Langevin algorithm, where the gradient information based on $\pi^*$ was used for generating a proposal. This step is much cheaper than the corresponding step involving a fine-scale gradient of $\pi$ because the simulations are performed on the coarse grid. Furthermore, this proposal is screened using coarse-scale models as in preconditioned MCMC discussed above. The details of this algorithm can be found in [89], where we presented numerical results. Numerical results show that preconditioned coarse-gradient Langevin algorithms are efficient and can give similar performance as the fine-scale Langevin algorithms with much less computational cost. We refer to [89] for details.

The MCMC method used in these simulations employs either the mixed MsFEM or MsFVEM in the preconditioning step. If a proposal is accepted by the preconditioning step, the proposed algorithms compute the fine-scale solutions corresponding to the proposed permeability field. At this stage, we have already precomputed basis functions that can be further used to reconstruct the velocity field on the fine scale. Then the transport equation can be solved on the fine grid coupled with the coarse-grid pressure equation. This approach provides an accurate approximation to the production data on the fine grid as discussed earlier and can be used to replace the fine-scale computation in the last stage. In this procedure, the basis functions are not updated in time, or updated only in a few coarse blocks. Thus the fine-scale computation in the last stage of MCMC algorithms can also be implemented quickly. Because the basis functions from the first-stage is re-used for the fine-scale computation, this combined multiscale approach can be very efficient for our sampling problem.

For problems involving a very high dimensional uncertainty space, such as permeability fields described by the exponential variogram, it is often advantageous to use an approximate response surface in computing Langevin proposals. We proposed the use of sparse interpolation techniques based on coarse-scale models in obtaining the approximation of the response surface in [88]. In this case, the posterior distribution is interpolated using sparse interpolation techniques. We first compute the posterior distribution at sparse locations that correspond to some selected realizations of the permeability field. These computations are performed on the coarse grid as before with MsFVEM and thus they are inexpensive. Furthermore, the posterior distribution is approximated using polynomial interpolation. Based on the interpolated posterior distribution, Langevin samples are proposed using analytical gradients of the posterior distribution. The numerical simulations show that one can achieve further gains in CPU if interpolation is used. We refer to [88] for further details.

We note that there are other efficient approaches (e.g., [275, 185, 128, 129]) which are used in uncertainty quantification in inverse problems for porous media flows. Here, our goal was simply to discuss an application of MsFEM to porous media flows within MCMC methods.

## 5.8 Discussions

In this chapter, we discussed the applications of MsFEMs to porous media flow and transport in the context of two-phase immiscible flow and transport. In a number of recent findings, the latter has been extended to more complicated porous media equations involving compressibility, gravity, and three phases as demonstrated in Sections 5.5 and 5.6. In general, MsFEMs offer a great advantage when the heterogeneities do not change significantly or these changes can be localized. This allows us to solve the flow equations on a coarse grid. In a more complex situation, this may not be the case and one has to be careful applying multiscale methods.

Another interesting application of multiscale finite element methods is to inverse problems. In [246], the authors took advantage of the adaptivity of multiscale methods to speedup inverse problems associated with finding permeability fields given average flow rates at the well and some other prior information. During the inversion procedure, the permeability is updated only in local regions using time travel inversion. Because of local changes in the permeability heterogeneities, multiscale basis functions are constructed only in a few coarse blocks and the solution is rapidly computed. This leads to very fast inversion and the CPU time for finding appropriate permeability samples defined on a multi million grid block is very small (less than two minutes on a PC).

# 6

# Analysis

In this chapter, we present analysis only for some representative cases of Ms-FEMs from Chapters 2, 3, and 4. We consider simpler cases to convey the main difficulties that arise in the analysis of MsFEMs. Some of the technical details are avoided to keep the presentation simple and make it accessible to a broader audience.

In Section 6.1, the convergence analysis of MsFEMs for linear elliptic problems is presented. In this chapter, the MsFEM using local information is studied. First, we present a basic convergence analysis of the MsFEM which demonstrates the resonance errors. In Section 6.1.2, the analysis of MsFEMs with oversampling is studied. This analysis shows that an oversampling technique reduces the resonance errors. In Section 6.1.3, the analysis of mixed MsFEMs using local information is presented. The results obtained in Section 6.1 use homogenization theory.

In Section 6.2, the convergence analysis of MsFEM for nonlinear problems is considered. We show the convergence results only for nonlinear elliptic equations with periodic spatial heterogeneities. The proof relies on homogenization theory and uses a number of auxiliary results that can be found in [104].

In Section 6.3, the analysis of MsFEMs using limited global information is presented. We study the convergence of mixed MsFEM (Section 6.3.1) and a Galerkin MsFEM (Section 6.3.2). The convergence analysis is carried out under some suitable assumptions. We show that MsFEMs using global information converge independent of resonance errors.

Although only some representative cases of MsFEMs are analyzed here, we have attempted to illustrate basic mathematical tools and ideas used in the analysis of multiscale methods. We hope the analysis presented in this chapter will help the reader to understand essential error sources that arise in multiscale algorithms and guide them in estimating these errors. This will further help to design more efficient numerical methods for real-life multiscale processes.

Y. Efendiev, T.Y. Hou, *Multiscale Finite Element Methods: Theory and Applications*, 165
Surveys and Tutorials in the Applied Mathematical Sciences 4,
DOI 10.1007/978-0-387-09496-0_6, © Springer Science+Business Media LLC 2009

## 6.1 Analysis of MsFEMs for linear problems (from Chapter 2)

For the analysis here, we restrict ourselves to a periodic case $k(x) = (k_{ij}(x/\varepsilon))$. We assume $k_{ij}(y)$, $y = x/\epsilon$ are smooth periodic functions in $y$ in a unit cube $Y$. We assume that $f \in L^2(\Omega)$. The assumptions on $k_{ij}$ can be relaxed and one can extend the analysis to the locally periodic case, $k = k(x, x/\epsilon)$, random homogeneous case, and other cases. For simplicity, we consider the analysis in two dimensions. Denote $L$ introduced in (2.1) by $L_\epsilon$. Let $p_0$ be the solution of the homogenized equation (see Appendix B for the background material on homogenization)

$$L_0 p_0 := -\mathrm{div}(k^* \nabla p_0) = f \text{ in } \Omega, \quad p_0 = 0 \text{ on } \partial\Omega, \tag{6.1}$$

where

$$k_{ij}^* = \frac{1}{|Y|} \int_Y k_{il}(y)(\delta_{lj} + \frac{\partial \chi^j}{\partial y_l})\, dy,$$

and $\chi^j(y)$ is the periodic solution of the cell problem in the period $Y$

$$\mathrm{div}_y(k(y)\nabla_y\chi^j) = -\frac{\partial}{\partial y_i}k_{ij}(y) \text{ in } Y, \quad \int_Y \chi^j(y)\, dy = 0.$$

We note that $p_0 \in H^2(\Omega)$ because $\Omega$ is a convex polygon. Denote by $p_1(x, y) = \chi^j(y)(\partial p_0(x)/\partial x_j)$ and let $\theta_\varepsilon$ be the solution of the problem

$$L_\varepsilon \theta_\varepsilon = 0 \text{ in } \Omega, \quad \theta_\varepsilon(x) = -p_1(x, x/\epsilon) \text{ on } \partial\Omega. \tag{6.2}$$

For simplicity of presentation, we denote by $\| \cdot \|_{\alpha,\beta,\cdot}$ and $| \cdot |_{\alpha,\beta,\cdot}$, the norm and semi-norm in $W^{\alpha,\beta}(\cdot)$. If only one subscript is used, for example, $\| \cdot \|_{\alpha,\cdot}$, then the norm or semi-norm in $H^\alpha$ is assumed. Also, for simplicity, we consider when $\mathcal{T}_h$ is a triangular partition. Our analysis of the multiscale finite element method relies on the following homogenization result obtained by Moskow and Vogelius [204].

**Lemma 6.1.** *Let $p_0 \in H^2(\Omega)$ be the solution of (6.1), $\theta_\varepsilon \in H^1(\Omega)$ be the solution to (6.2) and $p_1(x) = \chi^j(x/\varepsilon)\partial p_0(x)/\partial x_j$. Then there exists a constant $C$ independent of $p_0, \varepsilon$ and $\Omega$ such that*

$$\| p - p_0 - \varepsilon(p_1 + \theta_\varepsilon) \|_{1,\Omega} \leq C\varepsilon(| p_0 |_{2,\Omega} + \| f \|_{0,\Omega}).$$

### 6.1.1 Analysis of conforming multiscale finite element methods

The analysis of conforming multiscale finite element methods uses Cea's lemma [55].

**Lemma 6.2.** *Let $p$ be the solution of (2.1) and $p_h$ be the solution of (2.3). Then we have*

$$\| p - p_h \|_{1,\Omega} \leq C \inf_{v_h \in \mathcal{P}_h} \| p - v_h \|_{1,\Omega}.$$

**Error Estimates ($h < \varepsilon$)**

Let $\Pi_h : C(\bar{\Omega}) \to W_h \subset H_0^1(\Omega)$ be the usual Lagrange interpolation operator:

$$\Pi_h p(x) = \sum_{j=1}^{J} p(x_j)\phi_i^0(x) \quad \forall p \in C(\bar{\Omega})$$

and $I_h : C(\bar{\Omega}) \to \mathcal{P}_h$ be the corresponding interpolation operator defined through the multiscale basis function $\phi_i$,

$$I_h p(x) = \sum_{j=1}^{J} p(x_j)\phi_j(x) \quad \forall p \in C(\bar{\Omega}).$$

From the definition of the basis function $\phi_i$ in (2.2) we have

$$L_\varepsilon(I_h p) = 0 \text{ in } K, \quad I_h p = \Pi_h p \text{ on } \partial K, \tag{6.3}$$

for any $K \in \mathcal{T}_h$.

**Lemma 6.3.** *Let $p \in H^2(\Omega)$ be the solution of (2.1). Then there exists a constant $C$ independent of $h, \varepsilon$ such that*

$$\| p - I_h p \|_{0,\Omega} + h\| p - I_h p \|_{1,\Omega} \leq Ch^2(|p|_{2,\Omega} + \| f \|_{0,\Omega}). \tag{6.4}$$

*Proof.* At first it is known from standard finite element interpolation theory that

$$\| p - \Pi_h p \|_{0,\Omega} + h\| p - \Pi_h p \|_{1,\Omega} \leq Ch^2(|p|_{2,\Omega} + \| f \|_{0,\Omega}). \tag{6.5}$$

On the other hand, because $\Pi_h p - I_h p = 0$ on $\partial K$, the standard scaling argument yields

$$\| \Pi_h p - I_h p \|_{0,K} \leq Ch|\Pi_h p - I_h p|_{1,K} \quad \forall K \in \mathcal{T}_h. \tag{6.6}$$

To estimate $|\Pi_h p - I_h p|_{1,K}$ we multiply the equation in (6.3) by $I_h p - \Pi_h p \in H_0^1(K)$ to get

$$\int_K k(\tfrac{x}{\varepsilon})\nabla I_h p \cdot \nabla(I_h p - \Pi_h p)dx = 0.$$

Thus, upon using the equation in (2.1), we get

$$\int_K k(\frac{x}{\varepsilon})\nabla(I_h p - \Pi_h p) \cdot \nabla(I_h p - \Pi_h p)dx$$
$$= \int_K k(\frac{x}{\varepsilon})\nabla(p - \Pi_h p) \cdot \nabla(I_h p - \Pi_h p)dx - \int_K k(\frac{x}{\varepsilon})\nabla p \cdot \nabla(I_h p - \Pi_h p)dx$$
$$= \int_K k(\frac{x}{\varepsilon})\nabla(p - \Pi_h p) \cdot \nabla(I_h p - \Pi_h p)dx - \int_K f(I_h p - \Pi_h p)dx.$$

This implies that

$$|I_h p - \Pi_h p|_{1,K} \leq Ch|p|_{2,K} + \| I_h p - \Pi_h p \|_{0,K} \| f \|_{0,K}.$$

Hence

$$|I_h p - \Pi_h p|_{1,K} \leq Ch(|p|_{2,K} + \| f \|_{0,K}), \qquad (6.7)$$

where we have used (6.6). Now the lemma follows from (6.5)–(6.7). $\square$

In conclusion, we have the following standard estimate by using Lemmas 6.2 and 6.3.

**Theorem 6.4.** *Let $p \in H^2(\Omega)$ be the solution of (2.1) and $p_h \in \mathcal{P}_h$ be the solution of (2.3). Then we have*

$$\| p - p_h \|_{1,\Omega} \leq Ch(|p|_{2,\Omega} + \| f \|_{0,\Omega}). \qquad (6.8)$$

Note that the estimate (6.8) blows up as does $h/\epsilon$ as $\varepsilon \to 0$ because $|p|_{2,\Omega} = O(1/\epsilon)$. This is insufficient for practical applications. In the next subsection, we derive an error estimate which is uniform as $\varepsilon \to 0$.

## Error Estimates ($h > \varepsilon$)

In this section, we show that the MsFEM gives a convergence result uniform in $\epsilon$ as $\epsilon$ tends to zero. This is the main feature of the MsFEM over the traditional finite element method. The main result in this subsection is the following theorem.

**Theorem 6.5.** *Let $p \in H^2(\Omega)$ be the solution of (2.1) and $p_h \in \mathcal{P}_h$ be the solution of (2.3). Then we have*

$$\| p - p_h \|_{1,\Omega} \leq C(h + \varepsilon)\| f \|_{0,\Omega} + C\left(\frac{\varepsilon}{h}\right)^{1/2} \| p_0 \|_{1,\infty,\Omega}, \qquad (6.9)$$

*where $p_0 \in H^2(\Omega) \cap W^{1,\infty}(\Omega)$ is the solution of the homogenized equation (6.1).*

To prove the theorem, we first denote

$$p_{\mathrm{I}}(x) = I_h p_0(x) = \sum_j p_0(x_j)\phi_j(x) \in \mathcal{P}_h.$$

From (6.3) we know that $L_\epsilon p_{\mathrm{I}} = 0$ in $K$ and $p_{\mathrm{I}} = \Pi_h p_0$ on $\partial K$ for any $K \in \mathcal{T}_h$. The homogenization theory implies that

$$\| p_{\mathrm{I}} - p_{\mathrm{I}0} - \varepsilon(p_{\mathrm{I}1} - \theta_{\mathrm{I}\varepsilon}) \|_{1,K} \leq C\varepsilon(\| f \|_{0,K} + |p_{\mathrm{I}0}|_{2,K}), \qquad (6.10)$$

where $p_{\mathrm{I}0}$ is the solution of the homogenized equation on $K$:

$$L_0 p_{I0} = 0 \ \text{in} \ K, \quad p_{I0} = \Pi_h p_0 \ \text{on} \ \partial K, \tag{6.11}$$

$p_{I1}$ is given by the relation

$$p_{I1}(x, y) = \chi^j(y) \frac{\partial p_{I0}}{\partial x_j} \ \text{in} \ K, \tag{6.12}$$

and $\theta_{I\varepsilon} \in H^1(K)$ is the solution of the problem:

$$L_\varepsilon \theta_{I\varepsilon} = 0 \ \text{in} \ K, \quad \theta_{I\varepsilon}(x) = -p_{I1}(x, x/\epsilon) \ \text{on} \ \partial K. \tag{6.13}$$

It is obvious from (6.11) that

$$p_{I0} = \Pi_h p_0 \ \text{in} \ K, \tag{6.14}$$

because $\Pi_h p_0$ is linear on $K$. From (6.10) and Lemma 6.1 we obtain that

$$\begin{aligned}
\| p - p_I \|_{1,\Omega} \leq \| p_0 - p_{I0} \|_{1,\Omega} + \| \varepsilon(p_1 - p_{I1}) \|_{1,\Omega} \\
+ \| \varepsilon(\theta_\varepsilon - \theta_{I\varepsilon}) \|_{1,\Omega} + C\varepsilon \| f \|_{0,\Omega},
\end{aligned} \tag{6.15}$$

where we have used the regularity estimate $\| p_0 \|_{2,\Omega} \leq C \| f \|_{0,\Omega}$. Now it remains to estimate the terms on the right-hand side of (6.15). We show that the dominating resonance error is due to $\theta_{I\varepsilon}$.

**Lemma 6.6.** *We have*

$$\| p_0 - p_{I0} \|_{1,\Omega} \leq Ch \| f \|_{0,\Omega}, \tag{6.16}$$
$$\| \varepsilon(p_1 - p_{I1}) \|_{1,\Omega} \leq C(h + \varepsilon) \| f \|_{0,\Omega}. \tag{6.17}$$

*Proof.* The estimate (6.16) is a direct consequence of standard finite element interpolation theory because $p_{I0} = \Pi_h p_0$ by (6.14). Next we note that $\chi^j(x/\varepsilon)$ satisfies

$$\| \chi^j \|_{0,\infty,\Omega} + \varepsilon \| \nabla \chi^j \|_{0,\infty,\Omega} \leq C \tag{6.18}$$

for some constant $C$ independent of $h$ and $\varepsilon$. Thus we have, for any $K \in \mathcal{T}_h$,

$$\| \varepsilon(p_1 - p_{I1}) \|_{0,K} \leq C\varepsilon \| \chi^j \frac{\partial}{\partial x_j}(p_0 - \Pi_h p_0) \|_{0,K} \leq Ch\varepsilon | p_0 |_{2,K},$$

$$\begin{aligned}
\| \varepsilon \nabla(p_1 - p_{I1}) \|_{0,K} = \varepsilon \| \nabla(\chi^j \frac{\partial(p_0 - \Pi_h p_0)}{\partial x_j}) \|_{0,K} \\
\leq C \| \nabla(p_0 - \Pi_h p_0) \|_{0,K} + C\varepsilon | p_0 |_{2,K} \\
\leq C(h + \varepsilon) | p_0 |_{2,K}.
\end{aligned}$$

This completes the proof. □

**Lemma 6.7.** *We have*

$$\| \varepsilon \theta_\varepsilon \|_{1,\Omega} \leq C\sqrt{\varepsilon} \| p_0 \|_{1,\infty,\Omega} + C\varepsilon | p_0 |_{2,\Omega}. \tag{6.19}$$

*Proof.* Let $\zeta \in C_0^\infty(\mathbb{R}^2)$ be the cut-off function that satisfies $\zeta \equiv 1$ in $\Omega \backslash \Omega_{\delta/2}$, $\zeta \equiv 0$ in $\Omega_\delta$, $0 \le \zeta \le 1$ in $\mathbb{R}^2$, and $|\nabla\zeta| \le C/\delta$ in $\Omega$, where for any $\delta > 0$ sufficiently small, we denote $\Omega_\delta$ as

$$\Omega_\delta = \{x \in \Omega : \text{dist}(x, \partial\Omega) \ge \delta\}.$$

With this definition, it is clear that $\theta_\varepsilon + \zeta p_1 = \theta_\varepsilon + \zeta(\chi^j \partial p_0/\partial x_j) \in H_0^1(\Omega)$. Multiplying the equation in (6.2) by $\theta_\varepsilon + \zeta p_1$, we get

$$\int_\Omega k(\frac{x}{\varepsilon})\nabla\theta_\varepsilon \cdot \nabla(\theta_\varepsilon + \zeta\chi^j\frac{\partial p_0}{\partial x_j})dx = 0,$$

which yields, by using (6.18),

$$\begin{aligned}
\|\nabla\theta_\varepsilon\|_{0,\Omega} &\le C\|\nabla(\zeta\chi^j\partial p_0/\partial x_j)\|_{0,\Omega} \\
&\le C\|\nabla\zeta \cdot \chi^j\partial p_0/\partial x_j\|_{0,\Omega} + C\|\zeta\nabla\chi^j\partial p_0/\partial x_j\|_{0,\Omega} \\
&\quad + C\|\zeta\chi^j\partial^2 p_0/\partial^2 x_j\|_{0,\Omega} \\
&\le C\sqrt{|\partial\Omega|\cdot\delta}\frac{D}{\delta} + C\sqrt{|\partial\Omega|\cdot\delta}\frac{D}{\varepsilon} + C|p_0|_{2,\Omega}, \quad (6.20)
\end{aligned}$$

where $D = \|p_0\|_{1,\infty,\Omega}$ and the constant $C$ is independent of the domain $\Omega$. From (6.20) we have

$$\begin{aligned}
\|\varepsilon\theta_\varepsilon\|_{0,\Omega} &\le C(\frac{\varepsilon}{\sqrt{\delta}} + \sqrt{\delta})\|p_0\|_{1,\infty,\Omega} + C\varepsilon|p_0|_{2,\Omega} \\
&\le C\sqrt{\varepsilon}\|p_0\|_{1,\infty,\Omega} + C\varepsilon|p_0|_{2,\Omega}, \quad (6.21)
\end{aligned}$$

where we have taken $\delta = \epsilon$. Moreover, by applying the maximum principle to (6.2), we get

$$\|\theta_\varepsilon\|_{0,\infty,\Omega} \le \|\chi^j\partial p_0/\partial x_j\|_{0,\infty,\partial\Omega} \le C\|p_0\|_{1,\infty,\Omega}. \quad (6.22)$$

Combining (6.21) and (6.22), we complete the proof. $\square$

**Lemma 6.8.** *We have*

$$\|\varepsilon\theta_{I\varepsilon}\|_{1,\Omega} \le C\left(\frac{\varepsilon}{h}\right)^{1/2}\|p_0\|_{1,\infty,\Omega}. \quad (6.23)$$

*Proof.* First we remember that for any $K \in \mathcal{T}_h$, $\theta_{I\varepsilon} \in H^1(K)$ satisfies

$$L_\varepsilon\theta_{I\varepsilon} = 0 \text{ in } K, \quad \theta_{I\varepsilon} = -\chi^j(\frac{x}{\varepsilon})\frac{\partial(\Pi_h p_0)}{\partial x_j} \text{ on } \partial K. \quad (6.24)$$

By applying the maximum principle and (6.18) we get

$$\|\theta_{I\varepsilon}\|_{0,\infty,K} \le \|\chi^j\partial(\Pi_h p_0)/\partial x_j\|_{0,\infty,\partial K} \le C\|p_0\|_{1,\infty,K}.$$

Thus we have

$$\| \, \varepsilon\theta_{I\varepsilon} \,\|_{0,\Omega} \le C\varepsilon\| \, p_0 \,\|_{1,\infty,\Omega}. \tag{6.25}$$

On the other hand, because the constant $C$ in (6.20) is independent of $\Omega$, we can apply the same argument leading to (6.20) to obtain

$$\| \, \varepsilon\nabla\theta_{I\varepsilon} \,\|_{0,K} \le C\varepsilon\| \, \Pi_h p_0 \,\|_{1,\infty,K}(\sqrt{|\partial K|}/\sqrt{\delta} + \sqrt{|\partial K|\delta}/\varepsilon) + C\varepsilon| \, \Pi_h p_0 \,|_{2,K}$$
$$\le C\sqrt{h}\| \, p_0 \,\|_{1,\infty,K}(\frac{\varepsilon}{\sqrt{\delta}} + \sqrt{\delta})$$
$$\le C\sqrt{h\varepsilon}\| \, p_0 \,\|_{1,\infty,K},$$

which implies that

$$\| \, \varepsilon\nabla\theta_{I\varepsilon} \,\|_{0,\Omega} \le C\Big(\frac{\varepsilon}{h}\Big)^{1/2}\| \, p_0 \,\|_{1,\infty,\Omega}.$$

This completes the proof.   □

*Proof.* Theorem 6.5 is now a direct consequence of (6.15) and Lemmas 6.6–6.8 and the regularity estimate $\| \, p_0 \,\|_{2,\Omega} \le C\| \, f \,\|_{0,\Omega}$.   □

*Remark 6.9.* As we pointed out earlier, the MsFEM indeed gives a correct homogenized result as $\epsilon$ tends to zero. This is in contrast to the traditional FEM which does not give the correct homogenized result as $\epsilon \to 0$. The $L_2$ error would grow as $O(h^2/\epsilon^2)$. On the other hand, we also observe that when $h \sim \epsilon$, the multiscale method attains a large error in both $H^1$ and $L^2$ norms. This is called the *resonance* effect between the coarse-grid scale ($h$) and the small scale ($\epsilon$) of the problem. This estimate reflects the intrinsic scale interaction between the two scales in the  discrete problem. Extensive numerical experiments confirm that this estimate is indeed generic and sharp. From the viewpoint of practical applications, it is important to reduce or completely remove the resonance error for problems with many scales because the chance of hitting a resonance sampling is high.

*Remark 6.10.* It can be shown that [147]

$$\|p - p_h\|_{0,\Omega} \le C\Big(h + \frac{\epsilon}{h}\Big).$$

### 6.1.2 Analysis of nonconforming multiscale finite element methods

Let $\phi_i$ be multiscale basis functions obtained using the oversampling technique on $K$ as introduced in Section 2.3.2 and $\phi_i^0$ (piecewise linear function if $\mathcal{T}_h$ is a triangulation) be its homogenized part. We keep the same notation for the space spanned by multiscale basis functions as in the conforming case; that is $\mathcal{P}_h = \text{span}\{\phi_i\}$. The analysis follows the proof presented in [143].

   The Petrov–Galerkin formulation of the original problem is to seek $p_h \in \mathcal{P}_h$ such that

$$k_h(p_h, v_h) = f(v_h), \quad \forall v_h \in W_h,$$

where

$$k_h(p, v) = \sum_{K \in \mathcal{T}_h} \int_K \nabla p \cdot k(\frac{x}{\epsilon}) \nabla v dx, \quad f(v) = \int_\Omega f v dx.$$

Define $\| \cdot \|_{h,\Omega}$ to be the discrete $H^1$ semi-norm as

$$\|v\|_{h,\Omega} = \left( \sum_{K \in \mathcal{T}_h} \int_K |\nabla v|^2 dx \right)^{1/2}.$$

We use the following result [107].

**Lemma 6.11.** *Assume that $K \subset K_E$ is at least a distance of $h$ away from $\partial K_E$. Then*

$$\|\nabla \eta^i\|_{L^\infty(K)} \leq C/h, \tag{6.26}$$

*where $C$ is a constant that is independent of $\epsilon$ and $h$. Here, $\eta^i$ is the solution of $L_\epsilon \eta^i = 0$ in $K$, $\eta^i = -\chi^i$ on $\partial K$.*

**Theorem 6.12.** *Let $p_h$ be the Petrov–Galerkin MsFEM solution. Assume Lemma 6.11 holds and $\epsilon/h$ is sufficiently small. If the homogenized part of $p$, $p_0$, is in $H^2(\Omega)$, we have*

$$\|p_h - p\|_{h,\Omega} \leq C_1 h + C_2 \frac{\epsilon}{h} + C_3 \sqrt{\epsilon}. \tag{6.27}$$

*Proof.* To estimate $\|p_h - p\|_{h,\Omega}$, we first show that the following inf-sup condition or coercivity condition of the bilinear form $k_h(\cdot, \cdot)$ holds for sufficiently small $\epsilon$. There exists $C > 0$, independent of $\epsilon$ and $h$ such that

$$\sup_{v \in W_h} \frac{|k_h(p_h, v)|}{\|v\|_{1,\Omega}} \geq C \|p_h\|_{h,\Omega}, \quad \forall p_h \in \mathcal{P}_h. \tag{6.28}$$

Define

$$\tilde{k}_{ij}(y) = k_{il}(y) \left( \delta_{lj} + \frac{\partial \chi^j(y)}{\partial y_l} \right)$$

and

$$\tilde{k}(u, v) = \sum_{K \in \mathcal{T}_h} \int_K \nabla v \cdot \tilde{k}(\frac{x}{\epsilon}) \nabla u dx, \quad v \in W_h.$$

Thus, by the expansion $p_h = p_h^0 + \epsilon \chi(x/\epsilon) \cdot \nabla p_h^0 + \epsilon \theta_\epsilon^h$, we have

$$k_h(p_h, v_h) = \tilde{k}(p_h^0, v_h) + \epsilon k_h(\theta_\epsilon^h, v_h) = f(v_h), \quad \forall v_h \in W_h. \tag{6.29}$$

Taking $v_h = p_h^0 \in W_h$ in (6.29), we get

$$k_h(p_h, p_h^0) = \tilde{k}(p_h^0, p_h^0) + \epsilon k_h(\theta_\epsilon^h, p_h^0). \tag{6.30}$$

Moreover, using $\|\theta_\epsilon^h\|_{h,\Omega} \leq (C/h)\|\nabla p_h^0\|_{0,\Omega}$ (which follows from Lemma 6.11), we obtain that

$$|k_h(\theta_\epsilon^h, p_h^0)| \leq C\|\theta_\epsilon^h\|_{h,\Omega}\|\nabla p_h^0\|_{0,\Omega} \leq \frac{C}{h}\|\nabla p_h^0\|_{0,\Omega}^2. \tag{6.31}$$

Next, we note that $\|p_h\|_{h,\Omega} \leq C(1 + \epsilon/h)\|\nabla p_h^0\|_{0,\Omega}$ (see [143]) and $\tilde{k}(p_h^0, p_h^0)$ is bounded below and bounded above uniformly when $\epsilon/h \leq C$ (see (3.5) in [143]). Consequently, (6.30) and (6.31) imply that when $\epsilon/h$ is sufficiently small

$$|k_h(p_h, p_h^0)| \geq |\tilde{k}(p_h^0, p_h^0)| - \epsilon|k_h(\theta_\epsilon^h, p_h^0)| \geq C(1 - \frac{\epsilon}{h})\|\nabla p_h^0\|_{0,\Omega}^2$$

$$\geq C\|\nabla p_h^0\|_{0,\Omega}\|p_h\|_{h,\Omega}.$$

Thus, (6.28) holds.

Let $p_I \in \mathcal{P}_h$ be the interpolation from $\mathcal{P}_h$. Using inf-sup condition (6.28) we have

$$\begin{aligned}
\|p_h - p\|_{h,\Omega} &\leq \|p_I - p\|_{h,\Omega} + \|p_h - p_I\|_{h,\Omega} \\
&\leq \|p_I - p\|_{h,\Omega} + C \sup_{v_h \in W_h} \frac{|k_h(p_h - p_I, v_h)|}{\|v_h\|_{1,\Omega}} \\
&= \|p_I - p\|_{h,\Omega} + C \sup_{v_h \in W_h} \frac{|k_h(p_I - p, v_h)|}{\|v_h\|_{1,\Omega}} \\
&\leq (1 + C)\|p_I - p\|_{h,\Omega}.
\end{aligned} \tag{6.32}$$

Here, we have used the fact

$$k_h(p_h - p, v_h) = 0, \quad \forall v_h \in W_h.$$

Following the derivation of the proof of Theorem 3.1 in [107] (where $p_I = \sum p_0(x_i)\phi_i(x)$ is chosen) and using Lemma 6.11, we can easily show that

$$\|p_I - p\|_{h,\Omega} \leq C_1 h + C_2 \frac{\epsilon}{h} + C_3\sqrt{\epsilon}.$$

Therefore, (6.27) follows from (6.32).

### 6.1.3 Analysis of mixed multiscale finite element methods

In this section, we present the analysis of mixed multiscale finite element methods. We slightly modify the problem and consider a more general case with varying smooth mobility $\lambda(x)$. We consider the elliptic equation

$$-\text{div}(\lambda(x)k_\epsilon(x)\nabla p) = f \quad \text{in} \quad \Omega$$

$$-\lambda(x)k_\epsilon(x)\nabla p \cdot n = g(x) \quad \text{on} \quad \partial\Omega, \quad \int_\Omega p\,dx = 0,$$

where $\lambda(x)$ is a positive smooth function and $k_\epsilon(x) = k(x/\epsilon)$ is a symmetric positive and definite periodic tensor with periodicity $\epsilon$. We note that $\lambda(x)$

appears in two-phase flows (see (2.40)). Under the assumption that $\lambda(x)$ is sufficiently smooth, one can analyze the convergence (dominant resonance error) of MsFEMs with basis functions constructed with $\lambda = 1$. The basis functions are constructed with $\lambda = 1$ and satisfy (2.16).

Let $\psi_i^K = k_\epsilon(x)\nabla\phi_i^K$ and the basis function space for the velocity field be defined by

$$\mathcal{V}_h = \bigoplus_K \{\psi_i^K\} \subset H(\text{div}, \Omega),$$

where $H(\text{div}, \Omega)$ is the space of functions such that $\|\cdot\|_{0,\Omega} + \|\text{div}(\cdot)\|_{0,\Omega}$ is bounded. The variational problem is to find $\{v, p\} \in H(\text{div}, \Omega) \times L^2(\Omega)/\mathbb{R}$ such that $v \cdot n = g$ on $\partial\Omega$ and they solve the following variational problem,

$$\int_\Omega (\lambda k_\epsilon)^{-1} v \cdot w dx - \int_\Omega \text{div}(w)\, p dx = 0 \quad \forall w \in H_0(\text{div}, \Omega)$$
$$\int_\Omega \text{div}(v)\, q dx = f \quad \forall q \in L^2(\Omega)/\mathbb{R}, \tag{6.33}$$

where $H_0(\text{div}, \Omega)$ is the subspace of $H(\text{div}, \Omega)$ which consists of functions with homogeneous Neumann boundary conditions.

Set $Q_h = \oplus_K P_0(K) \cap L^2(\Omega)/\mathbb{R}$, a set of piecewise constant functions. The approximation problem is to find $\{v_h, p_h\} \in \mathcal{V}_h \times Q_h$ such that $v_h \cdot n = g_h$ on $\partial\Omega$

$$\int_\Omega (\lambda k_\epsilon)^{-1} v_h \cdot w_h dx - \int_\Omega \text{div}(w_h)\, p_h dx = 0 \quad \forall w_h \in \mathcal{V}_h^0$$
$$\int_\Omega \text{div}(v_h)\, q_h dx = f \quad \forall q_h \in Q_h. \tag{6.34}$$

We state the convergence theorem as the following.

**Theorem 6.13.** *Let $\{v, p\} \in H(\text{div}, \Omega) \times L^2(\Omega)/\mathbb{R}$ solve variational problem (6.33) and $\{v_h, p_h\} \in \mathcal{V}_h \times Q_h$ solve the discrete variational problem (6.34). If the homogenized solution $p_0 \in H^2(\Omega) \cap W^{1,\infty}(\Omega)$, then*

$$\|v - v_h\|_{H(\text{div},\Omega)} + \|p - p_h\|_{0,\Omega} \leq C_1(p_0, \lambda)\epsilon$$
$$+ C_2(p_0, f, \lambda, g)h + C_3(p_0, \lambda)\sqrt{\epsilon h} + C_4(p_0, \lambda)\sqrt{\frac{\epsilon}{h}}, \tag{6.35}$$

*where the coefficients are defined in (6.38), (6.41), (6.39), and (6.40).*

First, we state a stability estimate [71].

**Lemma 6.14.** *If $\{v, p\}$ and $\{v_h, p_h\}$, respectively, solve the continuous variational problem (6.33) and the discrete variational problem (6.34), then*

$$\|v - v_h\|_{H(\text{div},\Omega)} + \|p - p_h\|_{0,\Omega}$$
$$\leq C(\inf_{\substack{u_h \in \mathcal{V}_h \\ u_h - g_{0,h} \in \mathcal{V}_h^0}} \|v - u_h\|_{H(\text{div},\Omega)} + \inf_{q_h \in Q_h} \|p - q_h\|_{0,\Omega}). \tag{6.36}$$

The well-posedness of the discrete problem is verified in [71]. To obtain the convergence rate, we need to estimate the right-hand side of (6.36). The following proposition is used in the proof.

**Proposition 6.15.** *Let $p$ and $p_h$ be the solutions of* (6.33) *and* (6.34), *respectively; then*

$$\inf_{q_h \in Q_h} \|p - q_h\|_{0,\Omega} \le Ch\|g\|_{H^{-1/2}(\partial\Omega)}.$$

*Proof.* Define $\bar{q}_h = (1/|K|) \int_K p\,dx$ in each coarse block $K$. Furthermore, we apply the Poincaré inequality and standard regularity estimate for elliptic equations to obtain

$$\inf_{q_h \in Q_h} \|p - q_h\|_{0,\Omega} \le \|p - \bar{q}_h\|_{0,\Omega} \le Ch\|\nabla p\|_{0,\Omega} \le Ch\|g\|_{H^{-1/2}(\partial\Omega)}.$$

Next, we define the interpolation operator $\Pi_h : H(\text{div}, \Omega) \bigcap H^1(\Omega) \longrightarrow \mathcal{V}_h$ by

$$\Pi_h v|_K = (\int_{e_i^K} v \cdot nds)\psi_i^K.$$

Let $RT_0 = \text{span}\{R_i^K, i = 1, 2, ..., n; \ K \in \mathcal{T}_h\}$ be the lowest–order Raviart–Thomas finite element space and define the interpolation operator $P_h : H(\text{div}, \Omega) \bigcap H^1(\Omega) \longrightarrow RT_0$ by

$$P_h v|_K = (\int_{e_i^K} v \cdot nds)R_i^K.$$

It is easy to check that $\text{div}\Pi_h v = \text{div}P_h v$ and $\Pi_h v \cdot n = P_h v \cdot n$.

Next, we need to estimate the first term on the right-hand side of (6.36). The basic idea is to choose a particular $u_h$ approximating $v$. Let the homogenized flux $v_0 = \lambda k^* \nabla p_0$ and choose $t_h|_K = \Pi_h v_0$. Then we have $t_h - g_{0,h} \in \mathcal{V}_h^0$, where $g_{0,h} = \sum_{e \in \partial\Omega}(\int_e gds)\psi_i^K$ . Consequently, it remains to estimate $\|v - t_h\|_{H(\text{div},\Omega)}$. From the definition of $t_h$, an easy calculation gives rise to $\text{div}(t_h|_K) = \langle f \rangle_K$ and $\text{div}(v) = f$, where $\langle f \rangle_K = (1/|K|) \int_K f\,dx$. Therefore, we have

$$\|\text{div}(v) - \text{div}(t_h)\|_{0,\Omega} \le C|f|_{1,\Omega}.$$

The next step is to estimate $\|v - t_h\|_{0,\Omega}$. We use the homogenization technique for this purpose. Set $\phi^K = \alpha_i^K \phi_i^K$, where $\alpha_i^K = \int_{e_i^K} v_0 \cdot nds$. Then $t_h = k_\epsilon \nabla \phi^K$ and $\text{div}(k_\epsilon \nabla \phi^K) = \text{div}(P_h v_0) = 0$ in $K$, where $\phi^K \in H^1(K)/\mathbb{R}$ satisfies the following equation

$$\text{div}(k_\epsilon \nabla \phi^K) = 0 \quad \text{in } K$$
$$k_\epsilon \nabla \phi^K \cdot n = P_h v_0 \cdot n \quad \text{on } e_i^K.$$

Let $\phi_0^K$ be the solution of the corresponding homogenization equation,

$$\operatorname{div}(k^*\nabla\phi_0^K) = 0 \qquad \text{in } K$$
$$k^*\nabla\phi_i^K \cdot n = P_h v_0 \cdot n \qquad \text{on } e_i^K.$$

To complete the estimation of $\|v - t_h\|_{0,\Omega}$, we need the following lemma.

**Lemma 6.16.** *Let* $p_1 = p_0 + \epsilon\chi\cdot\nabla p_0$ *and* $\phi_1^K = \phi_0^K + \epsilon\chi\cdot\nabla\phi_0^K$. *Then*

$$|\phi_0^K - p_0|_{1,K} \le Ch\|\lambda^{-1} - 1\|_{0,\infty,K}\|\lambda\|_{1,\infty,K}\|p_0\|_{2,K}$$
$$|p_1 - \phi_1^K|_{1,K} \le C(h\|\lambda^{-1} - 1\|_{0,\infty,K} + \epsilon)\|\lambda\|_{1,\infty,K}\|p_0\|_{2,K}$$
$$|\phi_0^K|_{1,\infty,K} \le Ch^{-\frac{d}{2}+1}\|\lambda\|_{1,\infty,K}\|p_0\|_{2,K} + C\|\lambda\|_{0,\infty,K}|p_0|_{1,\infty,K}. \tag{6.37}$$

*Proof.* It is easy to prove that $k^*\nabla\phi_0^K = P_h v_0 \in L^\infty(K)$. Then we have $\phi_0^K \in H^2(K) \cap W^{1,\infty}(K)$. Applying the interpolation estimate of Raviart–Thomas finite elements, we obtain

$$\begin{aligned}|\phi_0^K - p_0|_{1,K} &= \|(k^*)^{-1}P_h v_0 - (\lambda k^*)^{-1}v_0\|_{0,K}\\ &\le C\|\lambda^{-1} - 1\|_{0,\infty,K}\|P_h v_0 - v_0\|_{0,K}\\ &\le Ch\|\lambda^{-1} - 1\|_{0,\infty,K}|v_0|_{1,K}\\ &\le Ch\|\lambda^{-1} - 1\|_{0,\infty,K}\|\lambda\|_{1,\infty,K}\|p_0\|_{2,K}.\end{aligned}$$

Because $\nabla\phi_0^K = (k^*)^{-1}P_h v_0$ and $P_h$ is a bounded operator, it is easy to show that

$$|\phi_0^K|_{1,K} \le C|\lambda|_{0,\infty,K}|p_0|_{1,K}$$
$$|\phi_0^K|_{2,K} \le C\|\lambda\|_{1,\infty,K}|p_0|_{2,K}.$$

Applying the above estimates, we obtain

$$\begin{aligned}|p_1 - \phi_1^K|_{1,K} &\le |p_0 - \phi_0^K|_{1,K} + \|(\nabla_y\cdot\chi)\nabla(p_0 - \phi_0^K)\|_{0,K}\\ &\quad + \epsilon\|\chi(\nabla^2 p_0 - \nabla^2\phi_0^K)\|_{0,K} \le Ch\|\lambda^{-1} - 1\|_{0,\infty,K}\|\lambda\|_{1,\infty,K}\|p_0\|_{2,K}\\ &\quad + C\epsilon\|\lambda\|_{1,\infty,K}|p_0|_{2,K}.\end{aligned}$$

As for the estimation of (6.37), we invoke the inverse inequality of finite elements and get

$$\begin{aligned}|\phi_0^K|_{1,\infty,K} &\le C\|P_h v_0 - \langle v_0\rangle_K\|_{0,\infty,K} + C\|\langle v_0\rangle_K\|_{0,\infty,K}\\ &\le Ch^{-d/2}\|P_h v_0 - \langle v_0\rangle_K\|_{0,K} + C\|\langle v_0\rangle_K\|_{0,\infty,K}\\ &\le Ch^{-d/2+1}\|\lambda\|_{1,\infty,K}\|p_0\|_{2,K} + C\|\lambda\|_{0,\infty,K}|p_0|_{1,\infty,K},\end{aligned}$$

where $d = 2$. The proof of the lemma is complete.

Next, we return to estimate $\|v - t_h\|_{0,\Omega}$. Applying the definitions of $v$ and $t_h$ and the Lemma 6.16, we obtain that

$$\|v - t_h\|_{0,K} \leq C\|\lambda - 1\|_{0,\infty,K}\|\nabla p - \nabla\phi^K\|_{0,K}$$
$$\leq C\|\lambda - 1\|_{0,\infty,K}(\|\nabla p - \nabla p_1\|_{0,K} + \|\nabla p_1 - \nabla\phi_1^K\|_{0,K}$$
$$+ \|\nabla\phi_1^K - \nabla\phi^K\|_{0,K}) \leq C\|\lambda - 1\|_{0,\infty,K}[\epsilon(\|\lambda\|_{0,\infty,K}\|p_0\|_{2,K}$$
$$+ \|\phi_0^K\|_{2,K}) + \sqrt{\epsilon h^{d-1}}(\|\lambda\|_{0,\infty,K}|p_0|_{1,\infty,K} + |\phi_0^K|_{1,\infty,K})]$$
$$+ C\|\lambda - 1\|_{0,\infty,K}[(h\|\lambda^{-1} - 1\|_{0,\infty,K} + \epsilon)\|\lambda\|_{1,\infty,K}\|p_0\|_{2,K}]$$
$$\leq C_{K,1}(p_0,\lambda)\epsilon + C_{K,2}(p_0,\lambda)h + C_{K,3}(p_0,\lambda)\sqrt{\epsilon h}$$
$$+ C_{K,4}(p_0,\lambda)\sqrt{\epsilon h^{d-1}},$$

where $d$ refers to the dimension of the space $\mathbb{R}^d$ ($d = 2$ for simplicity). Here we have used the corrector estimates (see Appendix B for discussions on corrector estimates for the Dirichlet problem and [71] for the corrector results that are used in the Neumann problem). Note that the constants in the above inequality are given by

$$C_{K,1}(p_0,\lambda) = C\|\lambda - 1\|_{0,\infty,K}\|\lambda\|_{1,\infty,K}\|p_0\|_{2,K}$$
$$C_{K,2}(p_0,\lambda) = C\|\lambda - 1\|_{0,\infty,K}\|\lambda^{-1} - 1\|_{0,\infty,K}\|\lambda\|_{1,\infty,K}\|p_0\|_{2,K}$$
$$C_{K,3}(p_0,\lambda) = C\|\lambda - 1\|_{0,\infty,K}\|\lambda\|_{1,\infty,K}\|p_0\|_{2,K}$$
$$C_{K,4}(p_0,\lambda) = C\|\lambda - 1\|_{0,\infty,K}(1 + \|\lambda\|_{0,\infty,K})\|p_0\|_{1,\infty,K}.$$

Taking the summation all over $K$, we have

$$\|v - t_h\|_{0,\Omega} \leq C_1(p_0,\lambda)\epsilon + \tilde{C}_2(p_0,\lambda)h + C_3(p_0,\lambda)\sqrt{\epsilon h} + C_4(p_0,\lambda)\sqrt{\frac{\epsilon}{h}}.$$

Here we have used the assumption that the triangulation is quasi-uniform, and the notations of the above coefficients are

$$C_1(p_0,\lambda) = C\|\lambda - 1\|_{0,\infty,\Omega}\|\lambda\|_{1,\infty,\Omega}\|p_0\|_{2,\Omega} \tag{6.38}$$
$$\tilde{C}_2(p_0,\lambda) = C\|\lambda - 1\|_{0,\infty,\Omega}\|\lambda^{-1} - 1\|_{0,\infty,\Omega}\|\lambda\|_{1,\infty,\Omega}\|p_0\|_{2,\Omega}$$
$$C_3(p_0,\lambda) = C\|\lambda - 1\|_{0,\infty,\Omega}\|\lambda\|_{1,\infty,\Omega}\|p_0\|_{2,\Omega} \tag{6.39}$$
$$C_4(p_0,\lambda) = C\|\lambda - 1\|_{0,\infty,\Omega}(1 + \|\lambda\|_{0,\infty,\Omega})\|p_0\|_{1,\infty,\Omega}. \tag{6.40}$$

Finally, applying Proposition 6.15, we get

$$\|v - v_h\|_{H(\mathrm{div},\Omega)} + \|p_\epsilon - p_h\|_{0,\Omega} \leq C_1(p_0,\lambda)\epsilon + C_2(p_0,\lambda,g)h$$
$$+ C_3(p_0,\lambda)\sqrt{\epsilon h} + C_4(p_0,\lambda)\sqrt{\frac{\epsilon}{h}},$$

where

$$C_2(p_0,f,\lambda,g) = \tilde{C}_2(p_0,\lambda) + C\|g\|_{-1/2,\partial\Omega} + C|f|_{1,\Omega}. \tag{6.41}$$

*Remark 6.17.* From the proof, we see that the resonance term $O(\sqrt{\epsilon/h})$ comes from the terms estimated by $|p_0|_{1,\infty,K}$. If the $p_0$ can be exactly solved by some finite element method on the coarse grid, then we can use an inverse inequality to improve the convergence to $O(\epsilon + h + \sqrt{\epsilon h})$.

*Remark 6.18.* From the proof of the convergence theorem, one can see that it is sufficient to require $\lambda \in W^{1,\infty}(\Omega)$ and $\lambda^{-1} \in L^{\infty}(\Omega)$.

*Remark 6.19.* If the oversampling technique is used to approximate the flux $v$ (see [71]), the resonance error can be reduced to $O(\epsilon/h)$.

## 6.2 Analysis of MsFEMs for nonlinear problems (from Chapter 3)

For the analysis of MsFEMs, we assume the following conditions for $k(x, \eta, \xi)$ and $k_0(x, \eta, \xi)$, $\eta \in \mathbb{R}$ and $\xi \in \mathbb{R}^d$.

$$|k(x, \eta, \xi)| + |k_0(x, \eta, \xi)| \le C\,(1 + |\eta|^{\gamma-1} + |\xi|^{\gamma-1}), \qquad (6.42)$$

$$(k(x, \eta, \xi_1) - k(x, \eta, \xi_2)) \cdot (\xi_1 - \xi_2) \ge C\,|\xi_1 - \xi_2|^{\gamma}, \qquad (6.43)$$

$$k(x, \eta, \xi) \cdot \xi + k_0(x, \eta, \xi)\eta \ge C|\xi|^{\gamma}. \qquad (6.44)$$

Denote

$$H(\eta_1, \xi_1, \eta_2, \xi_2, r) = (1 + |\eta_1|^{r} + |\eta_2|^{r} + |\xi_1|^{r} + |\xi_2|^{r}), \qquad (6.45)$$

for arbitrary $\eta_1,\ \eta_2 \in \mathbb{R}$, $\xi_1,\ \xi_2 \in \mathbb{R}^d$, and $r > 0$. We further assume that

$$
\begin{aligned}
&|k(x, \eta_1, \xi_1) - k(x, \eta_2, \xi_2)| + |k_0(x, \eta_1, \xi_1) - k_0(x, \eta_2, \xi_2)| \\
&\le C\,H(\eta_1, \xi_1, \eta_2, \xi_2, \gamma - 1)\,\nu(|\eta_1 - \eta_2|) \\
&+ C\,H(\eta_1, \xi_1, \eta_2, \xi_2, \gamma - 1 - s)\,|\xi_1 - \xi_2|^{s},
\end{aligned}
\qquad (6.46)
$$

where $s > 0$, $\gamma > 1$, $s \in (0, \min(\gamma - 1, 1))$ and $\nu$ is the modulus of continuity, a bounded, concave, and continuous function in $\mathbb{R}_+$ such that $\nu(0) = 0$, $\nu(t) = 1$ for $t \ge 1$, and $\nu(t) > 0$ for $t > 0$. Throughout, $\gamma'$ is defined by $1/\gamma + 1/\gamma' = 1$, $y = x/\epsilon$. In further analysis $K \in \mathcal{T}_h$ is referred to simply by $K$. Inequalities (6.42)-(6.46) are the general conditions that guarantee the existence of a solution and are used in homogenization of nonlinear operators [220]. Here $\gamma$ represents the rate of the polynomial growth of the fluxes with respect to the gradient and, consequently, it controls the summability of the solution. We do not assume any differentiability with respect to $\eta$ and $\xi$ in the coefficients. Our objective is to present a MsFEM and study its convergence for general nonlinear equations, where the fluxes can be discontinuous functions in space. These kinds of equations arise in many applications such as nonlinear heat conduction, flow in porous media, and so on. (see, e.g., [207, 244, 245]).

We present the main part of the analysis. For additional proofs of some auxiliary lemma, we refer to [104]. The analysis is presented for problems with scale separation. For this reason, we assume that the smallest scale is $\epsilon$ and denote the coefficients by $k(x, \cdot, \cdot) = k_\epsilon(x, \cdot, \cdot)$ and $k_0(x, \cdot, \cdot) = k_{0,\epsilon}(x, \cdot, \cdot)$.

In [111] we have shown using $G$-convergence theory that

$$\lim_{h \to 0} \lim_{\epsilon \to 0} \|p_h - p_0\|_{W_0^{1,\gamma}(\Omega)} = 0, \tag{6.47}$$

(up to a subsequence) where $p_0$ is a solution of (3.26) and $p_h$ is a MsFEM solution given by (3.6). This result can be obtained without any assumption on the nature of the heterogeneities and cannot be improved because there could be infinitely many scales $\alpha(\epsilon)$ present such that $\alpha(\epsilon) \to 0$ as $\epsilon \to 0$.

Next we present the convergence results for MsFEM solutions. In the proof of this theorem we show the form of the truncation error (in a weak sense) in terms of the resonance errors between the mesh size and small-scale $\epsilon$. The resonance errors are derived explicitly. To obtain the convergence rate from the truncation error, one needs some lower bounds. Under the general conditions, such as (6.42)–(6.46), one can prove strong convergence of MsFEM solutions without an explicit convergence rate (cf. [245]). To convert the obtained convergence rates for the truncation errors into the convergence rate of MsFEM solutions, additional assumptions, such as monotonicity, are needed.

**Theorem 6.20.** *Assume $k_\epsilon(x, \eta, \xi)$ and $k_{0,\epsilon}(x, \eta, \xi)$ are periodic functions with respect to $x$, let $p_0$ be a solution of (3.26), and $p_h$ is a MsFEM solution given by (3.6). Moreover, we assume that $\nabla p_h$ is uniformly bounded in $L^{\gamma+\alpha}(\Omega)$ for some $\alpha > 0$[1]. Then*

$$\lim_{\epsilon \to 0} \|p_h - p_0\|_{W_0^{1,\gamma}(\Omega)} = 0, \tag{6.48}$$

*where $h = h(\epsilon) \gg \epsilon$ and $h \to 0$ as $\epsilon \to 0$ (up to a subsequence).*

**Theorem 6.21.** *Let $p_0$ and $p_h$ be the solutions of the homogenized problem (3.26) and MsFEM (3.6), respectively, with the coefficient $k_\epsilon(x, \eta, \xi) = k(x/\epsilon, \xi)$ and $k_{0,\epsilon} = 0$. Then*

$$\|p_h - p_0\|_{W_0^{1,\gamma}(\Omega)}^\gamma \leq C \left( \left(\frac{\epsilon}{h}\right)^{s/((\gamma-1)(\gamma-s))} + \left(\frac{\epsilon}{h}\right)^{\gamma/(\gamma-1)} + h^{\gamma/(\gamma-1)} \right). \tag{6.49}$$

We first prove Theorem 6.20. Then, using the estimates obtained in the proof of this theorem, we show (6.49). The main idea of the proof of Theorem 6.20 is the following. First, the boundedness of the discrete solutions independent of $\epsilon$ and $h$ is shown. This allows us to extract a weakly converging subsequence. The next task is to prove that a limit is a solution of the homogenized equation. For this reason correctors for $v_{r,h}$ (see (3.2)) are used and their convergence is demonstrated. We would like to note that the known convergence results for the correctors assume a fixed (given) homogenized solution, whereas the correctors for $v_{r,h}$ are defined for only a uniformly bounded sequence $v_h$, that is, the homogenization limits of $v_{r,h}$ (with respect to $\epsilon$) depend on $h$, and are only uniformly bounded. Because of this, more

---

[1] Please see Remark 6.28 at the end of the proof of Theorem 6.20 for more discussions and partial results regarding this assumption.

precise corrector results need to be obtained where the homogenized limit of the solution is tracked carefully in the analysis. Note that to prove (6.47) (see [112]), one does not need correctors and can use the fact of the convergence of fluxes, and, thus, the proof of the periodic case differs from the one in [112]. Some results (Lemmas 6.22, 6.23, and their proofs) do not require periodicity assumptions. For these results we use the notations $k_\epsilon(x, \eta, \xi)$ and $k_{0,\epsilon}(x, \eta, \xi)$ to distinguish the two cases. The rest of the proofs require periodicity, and we use $k(x/\epsilon, \eta, \xi)$ and $k_0(x/\epsilon, \eta, \xi)$ notations.

**Lemma 6.22.** *There exists a constant $C > 0$ such that for any $v_h \in W_h$*

$$\langle k_{r,h} v_h, v_h \rangle \geq C \|\nabla v_h\|_{L^\gamma(\Omega)}^\gamma,$$

*for sufficiently small $h$.*

The proof of this lemma is provided in [104]. The following lemma is used in the proof of Lemma 6.24.

**Lemma 6.23.** *Let $v_\epsilon - v_0 \in W_0^{1,\gamma}(K)$ and $w_\epsilon - w_0 \in W_0^{1,\gamma}(K)$ satisfy the following problems, respectively,*

$$- \operatorname{div} k_\epsilon(x, \eta, \nabla v_\epsilon) = 0 \ \ in \ K \tag{6.50}$$

$$- \operatorname{div} k_\epsilon(x, \eta, \nabla w_\epsilon) = 0 \ \ in \ K, \tag{6.51}$$

*where $\eta$ is constant in $K$. Then the following estimate holds:*

$$\|\nabla(v_\epsilon - w_\epsilon)\|_{L^\gamma(K)} \leq C H_0 \|\nabla(v_0 - w_0)\|_{L^\gamma(K)}^{\gamma/(\gamma-s)}, \tag{6.52}$$

*where*

$$H_0 = \left( |K| + \|\eta\|_{L^\gamma(K)}^\gamma + \|\nabla v_0\|_{L^\gamma(K)}^\gamma + \|\nabla w_0\|_{L^\gamma(K)}^\gamma \right)^{(\gamma-s-1)/(\gamma-s)},$$

*where $s \in (0, \min(1, \gamma - 1))$, $\gamma > 1$.*

For the proof of this lemma, we refer to [104].

Next, we introduce, as before, the fast variable $y = x/\epsilon$. Regarding $\eta^{v_h}$, where $\eta^{v_h} = (1/|K|) \int_K v_h dx$ in each $K$, we note that Jensen's inequality implies

$$\|\eta^{v_h}\|_{L^\gamma(\Omega)} \leq C \|v_h\|_{L^\gamma(\Omega)}. \tag{6.53}$$

In addition, the following estimates hold for $\eta^{v_h}$,

$$\|v_h - \eta^{v_h}\|_{L^\gamma(K)} \leq C h \|\nabla v_h\|_{L^\gamma(K)}. \tag{6.54}$$

At this stage we define a numerical corrector associated with $v_{r,h} = E^{MsFEM} v_h$, $v_h \in W_h$. First, let

$$P_{\eta,\xi}(y) = \xi + \nabla_y N_{\eta,\xi}(y), \tag{6.55}$$

for $\eta \in \mathbb{R}$ and $\xi \in \mathbb{R}^d$, where $N_{\eta,\xi} \in W^{1,\gamma}_{per}(Y)$ is the periodic solution (with average zero) of

$$- \operatorname{div}(k(y, \eta, \xi + \nabla_y N_{\eta,\xi}(y))) = 0 \text{ in } Y, \tag{6.56}$$

where $Y$ is a unit period. The homogenized fluxes are defined as follows:

$$k^*(\eta, \xi) = \int_Y k(y, \eta, \xi + \nabla_y N_{\eta,\xi}(y)) \, dy, \tag{6.57}$$

$$k_0^*(\eta, \xi) = \int_Y k_0(y, \eta, \xi + \nabla_y N_{\eta,\xi}(y)) \, dy, \tag{6.58}$$

where $k^*$ and $k_0^*$ satisfy the conditions similar to (6.42)–(6.46). We refer to [220] for further details. Using (6.55), we denote our numerical corrector by $P_{\eta^{v_h}, \nabla v_h}$ which is defined as

$$P_{\eta^{v_h}, \nabla v_h} = \nabla v_h + \nabla_y N_{\eta^{v_h}, \nabla v_h}(y). \tag{6.59}$$

Here $\eta^{v_h}$ is a piecewise constant function defined in each $K \in \mathcal{T}_h$ by $\eta^{v_h} = (1/|K|) \int_K v_h dx$. Consequently, $P_{\eta^{v_h}, \nabla v_h}$ is defined in $\Omega$ by using (6.59) in each $K \in \mathcal{T}_h$. For the linear problem $P_{\eta^{v_h}, \nabla v_h} = \nabla v_h + N(y) \cdot \nabla v_h$. Our goal is to show the convergence of these correctors for the uniformly bounded family of $v_h$ in $W^{1,\gamma}(\Omega)$. We note that the corrector results known in the literature are for a fixed homogenized solution.

**Lemma 6.24.** *Let $v_{r,h}$ satisfy (3.2), where $k_\epsilon(x, \eta, \xi)$ is a periodic function with respect to $x$, and assume that $v_h$ is uniformly bounded in $W^{1,\gamma}_0(\Omega)$. Then*

$$\|\nabla v_{r,h} - P_{\eta^{v_h}, \nabla v_h}\|_{L^\gamma(\Omega)}$$
$$\leq C \left(\frac{\epsilon}{h}\right)^{1/(\gamma(\gamma - s))} \left(|\Omega| + \|v_h\|^\gamma_{L^\gamma(\Omega)} + \|\nabla v_h\|^\gamma_{L^\gamma(\Omega)}\right)^{1/\gamma}. \tag{6.60}$$

We note that here $s \in (0, \min(\gamma - 1, 1))$, $\gamma > 1$. For the proof of this lemma, we need the following proposition.

**Proposition 6.25.** *For every $\eta \in \mathbb{R}$ and $\xi \in \mathbb{R}^d$ we have*

$$\|P_{\eta,\xi}\|^\gamma_{L^\gamma(Y_\epsilon)} \leq c (1 + |\eta|^\gamma + |\xi|^\gamma) |Y_\epsilon|, \tag{6.61}$$

*where $Y_\epsilon$ is a period of size $\epsilon$.*

An easy consequence of this proposition is the following estimate for $N_{\eta,\xi}$ (see (6.56)).

**Corollary 6.26.** *For every $\eta \in \mathbb{R}$ and $\xi \in \mathbb{R}^d$ we have*

$$\|\nabla_y N_{\eta,\xi}\|^\gamma_{L^\gamma(Y_\epsilon)} \leq c (1 + |\eta|^\gamma + |\xi|^\gamma) |Y_\epsilon|. \tag{6.62}$$

The proof of Proposition 6.25 is presented in [104].

*Proof.* (Lemma 6.24) Recall that by definition

$$P_{\eta^{v_h},\nabla v_h} = \nabla v_h + \nabla_y N_{\eta^{v_h},\nabla v_h}(y) = \nabla v_h + \epsilon \nabla N_{\eta^{v_h},\nabla v_h}(x/\epsilon),$$

where by using (6.56) $N_{\eta^{v_h},\nabla v_h}(y)$ is a zero-mean periodic function satisfying the following,

$$-\operatorname{div}(k(x/\epsilon,\eta^{v_h},\nabla v_h + \epsilon \nabla N_{\eta^{v_h},\nabla v_h})) = 0 \text{ in } K. \tag{6.63}$$

We expand $v_{r,h}$ as

$$v_{r,h} = v_h(x) + \epsilon N_{\eta^{v_h},\nabla v_h}(x/\epsilon) + \theta(x,x/\epsilon). \tag{6.64}$$

We note that here $\theta(x,x/\epsilon)$ is similar to the correction terms that arise in linear problems because of the mismatch between linear boundary conditions and the oscillatory corrector, $N_{\eta^{v_h},\nabla v_h}(x/\epsilon) = N(x/\epsilon)\cdot\nabla v_h$. Next we denote by $w_{r,h} = v_h(x) + \epsilon N_{\eta^{v_h},\nabla v_h}(x/\epsilon)$. Clearly $w_{r,h}$ satisfies (6.63). Taking all these into account, the claim in the lemma is the same as proving

$$\|\nabla\theta\|_{L^\gamma(\Omega)} = \|\nabla(v_{r,h} - w_{r,h})\|_{L^\gamma(\Omega)}$$
$$\leq C\left(\frac{\epsilon}{h}\right)^{1/(\gamma(\gamma-s))}\left(|\Omega| + \|v_h\|^\gamma_{L^\gamma(\Omega)} + \|\nabla v_h\|^\gamma_{L^\gamma(\Omega)}\right)^{1/\gamma}. \tag{6.65}$$

Here we may write $w_{r,h}$ as a solution of the following boundary value problem:

$$-\operatorname{div}(k(x/\epsilon,\eta^{v_h},\nabla w_{r,h})) = 0 \text{ in } K \text{ and } w_{r,h} = v_h + \epsilon \tilde{N}_{\eta^{v_h},\nabla v_h} \text{ on } \partial K,$$

with $\tilde{N}_{\eta^{v_h},\nabla v_h} = \zeta N_{\eta^{v_h},\nabla v_h}$, where $\zeta$ is a sufficiently smooth function whose value is 1 on a strip of width $\epsilon$ adjacent to $\partial K$ and 0 elsewhere. We denote this strip by $S_\epsilon$. This idea has been used in [164]. By Lemma 6.23 we have the following estimate:

$$\|\nabla\theta\|^\gamma_{L^\gamma(K)} = \|\nabla(v_{r,h} - w_{r,h})\|^\gamma_{L^\gamma(K)}$$
$$\leq C\,H_0\,\|\nabla(v_h - v_h - \epsilon\tilde{N}_{\eta^{v_h},\nabla v_h})\|^{\gamma/(\gamma-s)}_{L^\gamma(K)} \tag{6.66}$$
$$\leq C\,H_0\,\|\epsilon\nabla\tilde{N}_{\eta^{v_h},\nabla v_h}\|^{\gamma/(\gamma-s)}_{L^\gamma(K)},$$

where

$$H_0 =$$
$$\left(|K| + \|\eta^{v_h}\|^\gamma_{L^\gamma(K)} + \|\nabla v_h\|^\gamma_{L^\gamma(K)} + \|\nabla(v_h + \epsilon\tilde{N}_{\eta^{v_h},\nabla v_h})\|^\gamma_{L^\gamma(K)}\right)^{\frac{(\gamma-s-1)}{(\gamma-s)}}. \tag{6.67}$$

We need to show that $H_0$ is bounded and $\|\epsilon\nabla\tilde{N}_{\eta^{v_h},\nabla v_h}\|^\gamma_{L^\gamma(\Omega)}$ uniformly vanishes as $\epsilon \to 0$. For this purpose, we use the following notations. Let $J_\epsilon^K = \{i \in \mathbb{Z}^d : Y^i \bigcap K \neq 0, K\backslash Y^i \neq 0\}$ and $F_\epsilon^K = \cup_{i\in J_\epsilon^K} Y^i$. In other words,

$F_\epsilon^K$ is the union of all periods $Y^i$ that cover the strip $S_\epsilon$. Using these notations and because $\zeta$ is zero everywhere in $K$, except in the strip $S_\epsilon$, we may write the following

$$
\begin{aligned}
\|\epsilon \nabla \tilde{N}_{\eta^{v_h},\nabla v_h}\|_{L^\gamma(K)}^\gamma &= \epsilon^\gamma \int_K |\nabla(\zeta\, N_{\eta^{v_h},\nabla v_h})|^\gamma \, dx \\
&= \epsilon^\gamma \int_{S_\epsilon} |\nabla(\zeta\, N_{\eta^{v_h},\nabla v_h})|^\gamma \, dx \\
&\le \epsilon^\gamma \int_{F_\epsilon^K} |\nabla(\zeta\, N_{\eta^{v_h},\nabla v_h})|^\gamma \, dx \\
&= \epsilon^\gamma \sum_{i\in J_\epsilon^K} \int_{Y_\epsilon^i} |\nabla(\zeta\, N_{\eta^{v_h},\nabla v_h})|^\gamma \, dx \\
&\le \epsilon^\gamma \sum_{i\in J_\epsilon^K} \int_{Y_\epsilon^i} \left( |\nabla N_{\eta^{v_h},\nabla v_h}|^\gamma |\zeta|^\gamma + |N_{\eta^{v_h},\nabla v_h}|^\gamma |\nabla\zeta|^\gamma \right) \, dx,
\end{aligned}
$$
$$(6.68)$$

where we have used the product rule on the partial derivative in the last line of (6.68). Our aim now is to show that the sum of integrals in the last line of (6.68) is uniformly bounded. We note that (see Corollary 6.26)

$$\|\nabla_y N_{\eta^{v_h},\nabla v_h}\|_{L^\gamma(Y_\epsilon^i)}^\gamma \le C(1 + |\eta^{v_h}|^\gamma + |\nabla v_h|^\gamma)\,|Y_\epsilon^i|, \qquad (6.69)$$

from which, using the Poincaré-Friedrich inequality we have

$$\|N_{\eta^{v_h},\nabla v_h}\|_{L^\gamma(Y_\epsilon^i)}^\gamma \le C(1 + |\eta^{v_h}|^\gamma + |\nabla v_h|^\gamma)\,|Y_\epsilon^i|. \qquad (6.70)$$

We note also that $\eta^{v_h}$ and $\nabla v_h$ are constant in $K$. Because $\zeta$ is sufficiently smooth, and whose value is one on the strip $S_\epsilon$ and zero elsewhere, we know that $|\nabla\zeta| \le C/\epsilon$ (cf. [164]). Applying all these facts to (6.68) we have

$$
\begin{aligned}
\|\epsilon \nabla \tilde{N}_{\eta^{v_h},\nabla v_h}\|_{L^\gamma(K)}^\gamma &\le C\,\epsilon^\gamma \, (1 + |\eta^{v_h}|^\gamma + |\nabla v_h|^\gamma) \sum_{i\in J_\epsilon^K} (1 + \epsilon^{-\gamma})\,|Y_\epsilon^i| \\
&= C\,(\epsilon^\gamma + 1)\,(1 + |\eta^{v_h}|^\gamma + |\nabla v_h|^\gamma) \sum_{i\in J_\epsilon^K} |Y_\epsilon^i| \\
&\le C\,(1 + |\eta^{v_h}|^\gamma + |\nabla v_h|^\gamma) \sum_{i\in J_\epsilon^K} |Y_\epsilon^i|.
\end{aligned}
$$

Moreover, because all $Y_\epsilon^i$, $i \in J_\epsilon^K$, cover the strip $S_\epsilon$, we know that $\sum_{i\in J_\epsilon^K} |Y_\epsilon^i| \le C\, h^{d-1}\, \epsilon$. Hence, we have

$$
\begin{aligned}
\|\epsilon \nabla \tilde{N}_{\eta^{v_h},\nabla v_h}\|_{L^\gamma(K)}^\gamma &\le C\,\frac{h^d}{h^d}\,(1 + |\eta^{v_h}|^\gamma + |\nabla v_h|^\gamma)\, h^{d-1}\,\epsilon \\
&\le C\,\frac{\epsilon}{h}\,\left( |K| + \|\eta^{v_h}\|_{L^\gamma(K)}^\gamma + \|\nabla v_h\|_{L^\gamma(K)}^\gamma \right).
\end{aligned}
$$
$$(6.71)$$

Furthermore, using this estimate and noting that $\epsilon/h < 1$, we obtain from (6.67) that

$$H_0 \le C \left( |K| + \|\eta^{v_h}\|^\gamma_{L^\gamma(K)} + \|v_h\|^\gamma_{L^\gamma(K)} + \|\nabla v_h\|^\gamma_{L^\gamma(K)} \right)^{(\gamma-s-1)/(\gamma-s)}.$$

(6.72)

Summarizing the results from (6.66) combined with (6.72) and (6.71), we get

$$\|\nabla \theta\|^\gamma_{L^\gamma(K)} \le C\, H_0 \, \|\epsilon \, \nabla \tilde{N}_{\eta^{v_h}, \nabla v_h}\|^{\gamma/(\gamma-s)}_{L^\gamma(K)}$$

$$\le C \left( \frac{\epsilon}{h} \right)^{1/(\gamma-s)} \left( |K| + \|\eta^{v_h}\|^\gamma_{L^\gamma(K)} + \|v_h\|^\gamma_{L^\gamma(K)} + \|\nabla v_h\|^\gamma_{L^\gamma(K)} \right).$$

Finally summing over all $K \in \mathcal{T}_h$ and applying (6.53) to $\sum_{K \in \mathcal{T}_h} \|\eta^{v_h}\|^\gamma_{L^\gamma(K)}$, we obtain

$$\|\nabla\theta\|^\gamma_{L^\gamma(\Omega)} = \sum_K \|\nabla\theta\|^\gamma_{L^\gamma(K)}$$

$$\le C \left( \frac{\epsilon}{h} \right)^{1/(\gamma-s)} \sum_K \left( |K| + \|v_h\|^\gamma_{L^\gamma(K)} + \|\nabla v_h\|^\gamma_{L^\gamma(K)} \right) \quad (6.73)$$

$$= C \left( \frac{\epsilon}{h} \right)^{1/(\gamma-s)} \left( |\Omega| + \|v_h\|^\gamma_{L^\gamma(\Omega)} + \|\nabla v_h\|^\gamma_{L^\gamma(\Omega)} \right).$$

The last inequality uniformly vanishes as $\epsilon$ approaches zero, thus we have completed the proof of Lemma 6.24.

The next lemma is crucial for the proof of Theorem 6.20 and it guarantees the convergence of MsFEM solutions to a solution of the homogenized equation. This lemma also provides us with the estimate for the truncation error (in a weak sense).

**Lemma 6.27.** *Suppose* $v_h, w_h \in W_h$ *where* $\nabla v_h$ *and* $\nabla w_h$ *are uniformly bounded in* $L^{\gamma+\alpha}(\Omega)$ *and* $L^\gamma(\Omega)$, *respectively, for some* $\alpha > 0$. *Let* $\kappa^*$ *be the operator associated with the homogenized problem (3.26), such that*

$$\langle \kappa^* v_h, w_h \rangle = \sum_{K \in \mathcal{T}_h} \int_K \left( k^*(v_h, \nabla v_h) \cdot \nabla w_h + k_0^*(v_h, \nabla v_h) w_h \right) dx, \quad \forall v_h, w_h \in W_h.$$

(6.74)

*Then we have*

$$\lim_{\epsilon \to 0} \langle \kappa_{r,h} v_h - \kappa^* v_h, w_h \rangle = 0.$$

The proof of this lemma is presented in [104]. Now we are ready to prove Theorem 6.20.

*Proof.* (Theorem 6.20) Because $\kappa_{r,h}$ is coercive, it follows that $p_h$ is bounded, which implies that it has a subsequence (which we also denote by $p_h$) such that $p_h \rightharpoonup \tilde{p}$ in $W^{1,\gamma}(\Omega)$ as $\epsilon \to 0$. Because the operator $\kappa^*$ is of type $S_+$ (see, e.g., [245], page 3, for the definition), then by its definition, the strong convergence would be true if we can show that $\limsup_{\epsilon \to 0} \langle \kappa^* p_h, p_h - \tilde{p} \rangle \to 0$. Moreover, by adding and subtracting the term, we have the following equality

$$\langle \kappa^* p_h, p_h - \tilde{p} \rangle = \langle \kappa^* p_h - \kappa_{r,h} p_h, p_h - \tilde{p} \rangle + \langle \kappa_{r,h} p_h, p_h - \tilde{p} \rangle$$

$$= \langle \kappa^* p_h - \kappa_{r,h} p_h, p_h \rangle - \langle \kappa^* p_h - \kappa_{r,h} p_h, \tilde{p} \rangle + \int_{\Omega} f(p_h - \tilde{p}) dx.$$

$$(6.75)$$

Lemma 6.27 implies that the first and second term vanish as $\epsilon \to 0$ provided $\nabla p_h$ is uniformly bounded in $L^{\gamma+\alpha}$ for $\alpha > 0$, and the last term vanishes as $\epsilon \to 0$ (up to a subsequence) by the weak convergence of $p_h$. One can assume additional mild regularity assumptions [201] for input data and obtain Meyers type estimates, $\|\nabla p_0\|_{L^{\gamma+\alpha}(\Omega)} \le C$, for the homogenized solutions. In this case it is reasonable to assume that the discrete solutions are uniformly bounded in $L^{\gamma+\alpha}(\Omega)$. We have obtained results on Meyers type estimates for our approximate solutions in the case $\gamma = 2$ [114]. Finally, because $\kappa^*$ is also of type M (see, e.g., [244], page 38, for the definition), all these conditions imply that $\kappa^* \tilde{p} = f$, which means that $\tilde{p} = p_0$.

*Remark 6.28.* We would like to point out that for the proof of Theorem 6.20 it is assumed that $\nabla p_h$ is uniformly bounded in $L^{\gamma+\alpha}(\Omega)$ for some $\alpha > 0$ (see discussions after (6.75)). This has been shown for $\gamma = 2$ in [114]. To avoid this assumption, one can impose additional restrictions on $k^*(\eta, \xi)$ (see, [112], pages 254, 255). We note that the assumption, $\nabla p_h$ is uniformly bounded in $L^{\gamma+\alpha}(\Omega)$, is not used for the estimation of the resonance errors.

Next we present some explicit estimates for the convergence rates of Ms-FEM. First, we note that from the proof of the Lemma 6.27 it follows that the truncation error of MsFEM (in a weak sense) is given by

$$\langle k_{r,h} p_h - \kappa^* p_h, w_h \rangle = \langle f - A^* p_h, w_h \rangle$$

$$\le C \left( \frac{\epsilon}{h} \right)^{s/(\gamma(\gamma-s))} \left( |\Omega| + \|p_h\|_{L^{\gamma}(\Omega)}^{\gamma} + \|\nabla p_h\|_{L^{\gamma}(\Omega)}^{\gamma} \right)^{(1/\gamma')} \|\nabla w_h\|_{L^{\gamma}(\Omega)}$$

$$+ C \frac{\epsilon}{h} \left( |\Omega| + \|p_h\|_{L^{\gamma}(\Omega)}^{\gamma} + \|\nabla p_h\|_{L^{\gamma}(\Omega)}^{\gamma} \right)^{1/\gamma'} \|\nabla w_h\|_{L^{\gamma}(\Omega)} + e(h) \|\nabla w_h\|_{L^{\gamma}(\Omega)}$$

$$= C \left( \left( \frac{\epsilon}{h} \right)^{s/(\gamma(\gamma-s))} + \frac{\epsilon}{h} \right) \left( |\Omega| + \|p_h\|_{L^{\gamma}(\Omega)}^{\gamma} + \|\nabla p_h\|_{L^{\gamma}(\Omega)}^{\gamma} \right)^{1/\gamma'} \|\nabla w_h\|_{L^{\gamma}(\Omega)}$$

$$+ e(h) \|\nabla w_h\|_{L^{\gamma}(\Omega)},$$

$$(6.76)$$

where $e(h)$ is a generic sequence independent of small-scale $\epsilon$, such that $e(h) \to 0$ as $h \to 0$. We note that the first term on the right side of (6.76) is the leading order resonance error caused by the linear boundary conditions imposed on $\partial K$, and the second term is due to the mismatch between the mesh size and the small scale of the problem. These resonance errors are also present in the linear case as we discussed in Section 6.1. If one uses the periodic solution of the auxiliary problem for constructing the solutions of the local problems, then the resonance error can be removed. To obtain explicit convergence rates, we first derive upper bounds for $\langle \kappa^* p_h - \kappa^* P_h p_0, p_h - P_h p_0 \rangle$, where $P_h u$ denotes a finite element projection of $u$ onto $W_h$; that is,

$$\langle \kappa^* P_h p_0, v_h \rangle = \int_\Omega f v_h dx, \quad \forall v_h \in W_h,$$

and $\langle \kappa^* p_h, v_h \rangle$ is defined by (6.74). Then using estimate (6.76), we have

$$
\begin{aligned}
&\langle \kappa^* p_h - \kappa^* P_h p_0, p_h - P_h u \rangle = \langle \kappa^* p_h - k_{r,h} p_h, p_h - P_h p_0 \rangle \\
&+ \langle k_{r,h} p_h - \kappa^* P_h p_0, p_h - P_h p_0 \rangle = \langle \kappa^* p_h - k_{r,h} p_h, p_h - P_h p_0 \rangle \\
&+ \langle f - \kappa^* P_h p_0, p_h - P_h p_0 \rangle = \langle \kappa^* p_h - k_{r,h} p_h, p_h - P_h p_0 \rangle \\
&\leq C \left( \left( \frac{\epsilon}{h} \right)^{s/(\gamma(\gamma-s))} + \frac{\epsilon}{h} \right) \left( |\Omega| + \|p_h\|_{L^\gamma(\Omega)}^\gamma + \|\nabla p_h\|_{L^\gamma(\Omega)}^\gamma \right)^{1/\gamma'} \times
\end{aligned}
\tag{6.77}
$$

$$\|\nabla(p_h - P_h p_0)\|_{L^\gamma(\Omega)} + e(h)\|\nabla(p_h - P_h p_0)\|_{L^\gamma(\Omega)}.$$

The estimate (6.77) does not allow us to obtain an explicit convergence rate without some lower bound for the left side of the expression. In the proof of Theorem 6.20, we only use the fact that $\kappa^*$ is the operator of type $S_+$, which guarantees that the convergence of the left side of (6.77) to zero implies the convergence of the discrete solutions to a solution of the homogenized equation. Explicit convergence rates can be obtained by assuming some kind of an inverse stability condition, $\|\kappa^* u - \kappa^* v\| \geq C\|u - v\|$. In particular, we may assume that $\kappa^*$ is a monotone operator; that is,

$$\langle \kappa^* u - \kappa^* v, u - v \rangle \geq C\|\nabla(u - v)\|_{L^\gamma(\Omega)}^\gamma. \tag{6.78}$$

A simple way to achieve monotonicity is to assume $k_\epsilon(x, \eta, \xi) = k_\epsilon(x, \xi)$ and $k_{0,\epsilon}(x, \eta, \xi) = 0$, although one can impose additional conditions on $k_\epsilon(x, \eta, \xi)$ and $k_{0,\epsilon}(x, \eta, \xi)$, such that monotonicity condition (6.78) is satisfied. For our further calculations, we only assume (6.78). Then from (6.77) and (6.78), and using the Young inequality, we have

$$\|\nabla(p_h - P_h p_0)\|_{L^\gamma(\Omega)}^\gamma \leq C \left( \left( \frac{\epsilon}{h} \right)^{s/((\gamma-1)(\gamma-s))} + \left( \frac{\epsilon}{h} \right)^{\gamma/(\gamma-1)} \right) + e(h).$$

Next taking into account the convergence of standard finite element solutions of the homogenized equation we write

$$\|\nabla P_h p - \nabla p_0\|_{L^\gamma(\Omega)} \leq e_1(h),$$

where $e_1(h) \to 0$ (as $h \to 0$) is independent of $\epsilon$. Consequently, using the triangle inequality we have

$$\|\nabla(p_h - p_0)\|_{L^\gamma(\Omega)}^\gamma \leq C \left( \left( \frac{\epsilon}{h} \right)^{s/((\gamma-1)(\gamma-s))} + \left( \frac{\epsilon}{h} \right)^{\gamma/(\gamma-1)} \right) + e(h) + e_1(h).$$

*Proof.* (Theorem 6.21).

For monotone operators, $k_\epsilon(x, \eta, \xi) = k_\epsilon(x, \xi)$ and $k_{0,\epsilon}(x, \eta, \xi) = 0$, $\eta \in \mathbb{R}$ and $\xi \in \mathbb{R}^d$, the estimates for $e(h)$ and $e_1(h)$ can be easily derived. In particular, because of the absence of $\eta$ in $k_\epsilon$, $e(h) = 0$, and $e_1(h) \leq Ch^{1/(\gamma-1)}$ (see, e.g., [75]). Combining these estimates we have

$$\|\nabla(p_h - p_0)\|^\gamma_{L^\gamma(\Omega)} \leq C\left(\left(\frac{\epsilon}{h}\right)^{s/((\gamma-1)(\gamma-s))} + \left(\frac{\epsilon}{h}\right)^{\gamma/(\gamma-1)} + h^{\gamma/(\gamma-1)}\right).$$

From here one obtains (6.49).

*Remark 6.29.* One can impose various conditions on the operators to obtain different kinds of convergence rates. For example, under the additional assumptions (cf. [207])

$$\left|\frac{\partial k^*(\eta, \xi)}{\partial \eta}\right| + \left|\frac{\partial k^*(\eta, \xi)}{\partial \xi}\right| \leq C, \quad \frac{\partial k_i^*(\eta, \xi)}{\partial \xi_j}\beta_i\beta_j \geq C|\beta|^2,$$

where $\beta \in \mathbb{R}^d$ is an arbitrary vector, and $\gamma = 2$, following the analysis presented in [207] (pages 51, 52), the convergence rate in terms of the $L^\gamma$-norm of $p_h - P_h p$ can be obtained,

$$\|\nabla(p_h - P_h p_0)\|^\gamma_{L^\gamma(\Omega)} \leq C\left(\left(\frac{\epsilon}{h}\right)^{s/((\gamma-1)(\gamma-s))} + \left(\frac{\epsilon}{h}\right)^{\gamma/(\gamma-1)}\right) \tag{6.79}$$
$$+e(h) + C\|p_h - P_h p_0\|^\gamma_{L^\gamma(\Omega)},$$

where $s \in (0, 1)$, $\gamma = 2$.

*Remark 6.30.* For the linear operators ($\gamma = 2$, $s = 1$), we recover the convergence rate $Ch + C_1\sqrt{\epsilon/h}$.

*Remark 6.31.* We have shown that the MsFEM for nonlinear problems has the same error structure as for linear problems. In particular, our studies revealed two kinds of resonance errors for nonlinear problems with the same nature as those that arise in linear problems.

## 6.3 Analysis for MsFEMs with limited global information (from Chapter 4)

### 6.3.1 Mixed finite element methods with limited global information

**Elliptic case**

We begin by restating the main assumption in a rigorous way.

*Assumption A1. There exist functions $v_1, ..., v_N$ and sufficiently smooth $A_1(x), ..., A_N(x)$ such that*

$$v(x) = \sum_{i=1}^{N} A_i(x) v_i, \qquad (6.80)$$

*where $v_i = k \nabla p_i$ and $p_i$ solves $\text{div}(k(x) \nabla p_i) = 0$ in $\Omega$ with appropriate boundary conditions.*

For our analysis, we assume $A_i(x) \in W^{1,\xi}(\Omega)$, and $v_i = k(x) \nabla p_i \in L^\eta(\Omega)$ for some $\xi$ and $\eta$, $i = 1, ..., N$. *Throughout this section, we do not use the Einstein summation convention.*

*Remark 6.32.* As an example of two global fields in $\mathbb{R}^2$ (similar results hold in $\mathbb{R}^d$; see [218] for details), we use the results of Owhadi and Zhang [218]. Let $v_i = k(x) \nabla p_i$ ($i = 1, 2$) be defined by the elliptic equation

$$\begin{aligned} \text{div}(k(x) \nabla p_i) &= 0 \quad \text{in} \ \ \Omega \\ p_i &= x_i \quad \text{on} \ \ \partial\Omega, \end{aligned} \qquad (6.81)$$

where $x = (x_1, x_2)$. In the harmonic coordinate $(p_1, p_2)$, $p = p(p_1, p_2) \in W^{2,s}$ ($s \geq 2$). Consequently, $v = \lambda(x) k(x) \nabla p = \sum_i \lambda(\partial p / \partial p_i) k \nabla p_i := \sum_i A_i(x) v_i$, where $A_i(x) = \lambda(\partial p / \partial p_i) \in W^{1,s}$.

To avoid the possibility that $\int_{e_l} v_i \cdot n ds$ is zero or unbounded, we make the following assumption for our analysis.

*Assumption A2. There exist positive constants $C$ such that*

$$\int_{e_l} |v_i \cdot n| ds \leq C h^{\beta_1} \quad \text{and} \quad \left\| \frac{v_i \cdot n}{\int_{e_l} v_i \cdot n ds} \right\|_{L^r(e_l)} \leq C h^{-\beta_2 + 1/r - 1} \qquad (6.82)$$

*uniformly for all edges $e_l$, where $\beta_1 \leq 1$, $\beta_2 \geq 0$, and $r \geq 1$.*

*Remark 6.33.* The second part of Assumption A2 is to assure $|\int_{e_l} v_i \cdot n ds|$ remains positive. It can be also written as

$$\left\| \frac{v_i \cdot n}{\int_{e_l} v_i \cdot n ds} - \langle \frac{v_i \cdot n}{\int_{e_l} v_i \cdot n ds} \rangle_{e_l} \right\|_{L^r(e_l)} \leq C h^{-\beta_2 + 1/r - 1},$$

where $\langle \cdot \rangle = (1/|e_l|) \int_{e_l} (\cdot) ds$, which is used to estimate the velocity basis function. If $v_i$ are bounded, then $\beta_2 = 0$. Note that

$$\left\| \frac{v_i \cdot n}{\int_{e_l} v_i \cdot n ds} - \langle \frac{v_i \cdot n}{\int_{e_l} v_i \cdot n ds} \rangle_{e_l} \right\|_{L^r(e_l)} = 0$$

if $v_i|_K$ is an $RT_0$ basis function or standard mixed MsFEM basis functions introduced in [71]. Finally, we note that if $r = 1$ and $|\int_{e_l} v_i \cdot n ds| \geq C h^{\beta_1}$, then $\beta_2 = 0$.

We recall the definition of basis functions $\psi_{ij}^K = k(x)\nabla\phi_{ij}^K$ and

$$\mathcal{V}_h = \bigoplus_K \{\psi_{ij}^K\} \bigcap H(\mathrm{div}, \Omega), \quad \mathcal{V}_h^0 = \bigoplus_K \{\psi_{ij}^K\} \bigcap H_0(\mathrm{div}, \Omega).$$

Let $Q_h = \oplus_K P_0(K) \subset L^2(\Omega)/\mathbb{R}$ (i.e., piecewise constants), be the basis function for the pressure. We define

$$g_{0,h} = \sum_{e \in \{\partial K \cap \partial\Omega, K \in \mathcal{T}_h\}} (\int_e g\, ds)\psi_{i,e}$$

for some fixed $i \in \{1, 2, ..., N\}$, where $\psi_{i,e}$ is the corresponding multiscale basis function to the edge $e$. Let $g_h = g_{0,h} \cdot n$ on $\partial\Omega$. The numerical mixed formulation is to find $\{v_h, p_h\} \in \mathcal{V}_h \times Q_h$ which satisfies (4.7) and $v_h \cdot n = g_h$ on $\partial\Omega$.

First, we note the following result.

**Lemma 6.34.**

$$v_i|_K \in \mathrm{span}\{\psi_{ij}^K\}, \quad i = 1, .., N; \quad j = 1, 2, 3.$$

*Proof.* First, we prove the lemma for $v_1$. For this proof, we would like to find constants $\beta_{ij}^K$s such that $\sum_{i,j} \beta_{ij}^K \psi_{ij}^K = v_1$. That is,

$$\begin{aligned}
\sum_{i,j} \beta_{ij}^K \mathrm{div}(k(x)\nabla\phi_{ij}^K) &= \frac{1}{|K|} \sum_{i,j} \beta_{ij}^K = 0 \\
\sum_{i,j} \beta_{ij}^K k(x)\nabla\phi_{ij}^K \cdot n_{e_l} &= \sum_{i,j} \beta_{ij}^K \delta_{jl} \frac{v_i \cdot n_{e_l}}{\int_{e_l} v_i \cdot n ds} = v_1 \cdot n_{e_l}.
\end{aligned} \tag{6.83}$$

Noticing that $v_i = k(x)\nabla p_i$ and $\mathrm{div}(k(x)\nabla p_i) = 0$, we have $p_i = \sum_{i,j} \beta_{ij}^K \phi_{ij}^K + C$ for some constant $C$ because $p_i$ and $\sum_{i,j} \beta_{ij}^K \phi_{ij}^K$ satisfy the same elliptic equation with Neumann boundary condition as $p_i$, and then we have $v_i = \sum_{i,j} \beta_{ij}^K \psi_{ij}^K$. The second equation in (6.83) implies that we can take $\beta_{1j}^K = \int_{e_j} v_1 \cdot n ds$ and $\beta_{ij}^K = 0$ for $i \neq 1$. Consequently,

$$\sum_{i,j} \beta_{ij}^K = \sum_j \int_{e_j} v_1 \cdot n ds = \int_K \mathrm{div}(v_1)dx = 0,$$

which is the first equation in (6.83). One can obtain similar results for other $v_i$ ($i = 2, ..., N$).

Following our assumption, let

$$X = \{u | u = \sum_{i=1}^N a_i(x)v_i\}$$

be a subspace of $H(\mathrm{div}, \Omega)$. For our analysis, we require that the integrals $\int_{e_j} a_i(x)v_i \cdot nds$ are well defined. This is also needed in our computations be-cause $\int_{e_j} a_i(x)v_i \cdot nds$ determines the fluxes along the edges in two-phase flow simulations. One way to achieve this is to assume, as we did earlier, that $a_i(x) \in W^{1,\xi}(\Omega)$, $v_i \in L^{\eta}(\Omega)$, $\frac{1}{2} = 1/\xi + 1/\eta$. Because $a_i(x) \in W^{1,\xi}(\Omega)$ and $v_i \in L^{\eta}(\Omega)$ ($\frac{1}{2} = \frac{1}{\xi} + \frac{1}{\eta}$), Hölder inequality implies that $(\nabla a_i)v_i \in L^2(\Omega)$. Noticing that $\mathrm{div}(v_i) = 0$, we have $\mathrm{div}(a_i(x)v_i) \in L^2(\Omega)$ immediately. Invok-ing the Sobolev embedding theorem (see [18]), we get $a_iv_i \in L^{\eta}(\Omega)$ because $W^{1,\xi}(\Omega) \hookrightarrow L^{\infty}(\Omega)$. The integrals $\int_{e_j} a_i(x)v_i \cdot nds$ are well defined by the fact that $a_iv_i \in L^{\rho}(\Omega)$ ($\rho > 2$) and $\mathrm{div}(a_i(x)v_i) \in L^2(\Omega)$ (see page 125 of [57]). We define an interpolation operator $\Pi_h : X \longrightarrow \mathcal{V}_h$ such that in each element $K$, for any $v = \sum_i a_i(x)v_i \in X$

$$\Pi_h|_K(\sum_i a_i(x)v_i) = \sum_{i,j} a_{ij}^K \psi_{ij}^K,$$

where $a_{ij}^K = \int_{e_j} a_i(x)v_i \cdot nds$.

The proof of the following lemma can be found in [8].

**Lemma 6.35.** *Let $\Pi_h$ be defined as above. Then $\forall v = \sum_{i=1}^N a_iv_i \in X$,   $q_h \in Q_h$,*
*(1) $\int_{\Omega} \mathrm{div}(v - \Pi_h v)q_h dx = 0$;*
*(2) $\|\Pi_h v\|_{H(\mathrm{div},\Omega)} \leq C\|v\|_{X,\Omega}$, if $\beta_1 \geq 2\beta_2$,*
*where $\|v\|_{X,\Omega} := \|\mathrm{div}(v)\|_{0,\Omega} + \sum_{i=1}^N \|a_i\|_{1,\Omega}$ and $C$ only depends on $N$, the constants in Assumption A2 (see (6.82)) and the pre-computed global fields $v_i$.*

*Remark 6.36.* If $v_i \in L^{\infty}(\Omega)$, then $\beta_1 = 1$, $\beta_2 = 0$, and the proof of Lemma 6.35 implies that $\|\Pi_h v\|_{H(\mathrm{div},\Omega)} \leq C(\max_i \|v_i\|_{L^{\infty}(\Omega)}) \sum_i \|a_i\|_{1,\Omega}$.

*Remark 6.37.* For $v = \sum_{i=1}^N a_iv_i$, where $a_i \in W^{1,\xi}(\Omega)$ and $v_i \in L^{\eta}(\Omega)$ ($1/2 = 1/\xi + 1/\eta$), one can also show that

$$\|\Pi_h v\|_{H(\mathrm{div},\Omega)} \leq C\sum_i \|a_i\|_{1,\xi,\Omega},$$

if $\alpha + \beta_1 - \beta_2 - 1 \geq 0$, where $C$ only depends on $N$, the constants in Assumption A2 (see (6.82)), and the pre-computed global fields $v_i$.

*Remark 6.38.* We note that $\|v\|_{X,\Omega}$ may not be a norm in general because $v = \sum_i a_iv_i = 0$ may not imply that $a_i$ are zero (this does not affect the derivation of the discrete inf-sup condition). In the problem setting considered here, one can assume that $\|v\|_{X,\Omega}$ is a norm. Indeed, $a_i$ are coarse-scale functions, and $v_i$ are fine-scale functions. Thus, in each coarse-grid block, the linear combination $\sum_i a_i v_i$ zero will imply that $a_i$ are zero unless $v_i$ are also coarse-scale functions. In the latter case, one can use standard mixed finite element basis functions. If $N = d$ ($d$ being the dimension of the space), $\|v\|_{X,\Omega}$ is a norm when $v_i$ are linearly independent. In the discrete setting, $a_i$ are vectors defined on the coarse grid, whereas $v_i$ are defined on the fine grid. If $\sum_i a_i v_i$ is zero, this implies that the vectors $v_i$ are linearly dependent, and thus, the basis functions are linearly dependent.

Lemma 6.35 and the continuous inf-sup condition imply the discrete inf-sup condition (see page 58 of [57]). We assume that the continuous inf-sup condition holds (see [8] for more details). Assuming a continuous inf-sup condition, we have that for any $q_h \in Q_h$, there exists a constant $C$ such that

$$\sup_{v_h \in \mathcal{V}_h} \frac{\int_\Omega \operatorname{div}(v_h)q_h\,dx}{\|v_h\|_{H(\operatorname{div},\Omega)}} \geq C\|q_h\|_{0,\Omega}. \tag{6.84}$$

Because of the inf-sup condition (6.84), we have the following optimal approximation (see [57, 71]).

**Lemma 6.39.** *Let $\{v,p\}$ and $\{v_h,p_h\}$ be the solution of (4.4) and (4.7) respectively. Then*

$$\|v - v_h\|_{H(\operatorname{div},\Omega)} + \|p - p_h\|_{0,\Omega} \leq C \inf_{w_h \in \mathcal{V}_h, w_h - g_{0,h} \in \mathcal{V}_h^0} \|v - w_h\|_{H(\operatorname{div},\Omega)}$$

$$+ C \inf_{q_h \in Q_h} \|p - q_h\|_{0,\Omega}. \tag{6.85}$$

Next, we formulate our main result.

**Theorem 6.40.** *Let $\{v,p\}$ and $\{v_h,p_h\}$ be the solution of (4.4) and (4.7), respectively. If $\alpha + \beta_1 - \beta_2 - 1 > 0$, we have*

$$\|v - v_h\|_{H(\operatorname{div},\Omega)} + \|p - p_h\|_{0,\Omega} \leq Ch^{\alpha+\beta_1-\beta_2-1},$$

*where $\alpha = 1 - 2/\xi$, $\xi$ and $A_i$ are defined in Assumption A1, and $\beta_i$ ($i = 1, 2$) are defined in Assumption A2. Here $C$ is independent of $h$ and depends on $N$, the constants in Assumption A2, $\|A_i\|_{1,\xi,\Omega}$ ($i = 1,..,N$) and $\|f\|_{1,\Omega}$.*

*Proof.* For the proof, we need to choose a proper $u_h$ and a proper $q_h$ such that the right-hand side of (6.85) is small.

The second term on the right hand in (6.85) can be easily estimated. In fact, with the choice $q_h|_K = \langle p \rangle_K$ (i.e., the average of $p$ in $K$), we have

$$\inf_{q_h \in Q_h} \|p - p_h\|_{0,\Omega} \leq Ch|p|_{1,\Omega}.$$

Next, we try to find a $u_h \in V_h$, say $u_h|_K = \sum_{i,j} c_{ij}^K \psi_{ij}^K$, and estimate the first term on the right-hand side in (6.85). Invoking Lemma 6.34 and its proof, it follows that in each $K$,

$$
\begin{aligned}
v - u_h &= \sum_i A_i(x)v_i - \sum_{i,j} c_{ij}^K \psi_{ij} \\
&= \sum_i (A_i(x) \sum_j \beta_{ij}^K \psi_{ij}^K) - \sum_{i,j} c_{ij}^K \psi_{ij}^K \\
&= \sum_{i,j} (A_i(x)\beta_{ij}^K - c_{ij}^K)\psi_{ij}^K,
\end{aligned}
\tag{6.86}
$$

where $\beta_{ij}^K = \int_{e_j} v_i \cdot nds$ . Set $c_{ij}^K = A_{ij}^K = \int_{e_j} A_i(x)v_i \cdot nds$.

Because $\int_K \sum_i \operatorname{div}(A_i(x)v_i)dx = f$, we get by the divergence theorem

$$
\int_{\partial K} \sum_i A_i(x)v_i \cdot nds = f.
$$

This gives rise to

$$
\begin{aligned}
\|\operatorname{div}(v - \sum_{i,j} c_{ij}^K \psi_{ij}^K)\|_{0,K} &= \|f - \sum_{i,j} c_{ij}^K \frac{1}{|K|}\|_{0,K} \\
&= \|f - \sum_{i,j} \int_{e_j} A_i(x)v_i \cdot nds \frac{1}{|K|}\|_{0,K} = \|f - \langle f \rangle_K\|_{0,K} \le Ch|f|_{1,K}.
\end{aligned}
\tag{6.87}
$$

After summation over all $K$ for (6.87), we have

$$
\|\operatorname{div}(v - u_h)\|_{0,\Omega} \le Ch|f|_{1,\Omega}.
\tag{6.88}
$$

Next we estimate $\|v - \sum_{i,j} c_{ij}^K \psi_{ij}^K\|_{0,K}$. Because $A_i(x) \in W^{1,\xi}(\Omega)$, by using the Sobolev embedding theorem and Taylor expansion (or definition of $C^\alpha$) we have

$$
|A_i(x)|_{e_j} - \bar{A}_i^j| \le Ch^\alpha \|A_i\|_{C^\alpha(\Omega)},
$$

where $\bar{A}_i^j$ is the average $A_i(x)$ along $e_j$ and $\alpha = 1 - 2/\xi$. So

$$
\begin{aligned}
|A_{ij}^K - \bar{A}_i^j \beta_{ij}^K| &= |\int_{e_j} A_i v_i \cdot nds - \bar{A}_i^j \int_{e_j} v_i \cdot nds| \\
&= |\int_{e_j} (A_i - \bar{A}_i^j)(v_i \cdot n)ds| \le Ch^{\alpha+\beta_1} \|A_i\|_{C^\alpha(\Omega)},
\end{aligned}
\tag{6.89}
$$

where we have used the  Assumption A2 (see (6.82)).

Next, we present an estimate for $\|\psi_{ij}^K\|_{0,K}$. For this reason, we introduce the lowest Raviart–Thomas basis functions $R_j^K$ for velocity. We know that $\operatorname{div}(R_j^K) = 1/|K|$ and $R_j^K \cdot n = \delta_{jl}/|e_j|$ (e.g., [57]). We multiply (4.6) by a test function $w$; we have

$$\int_K k\nabla\phi_{ij}^K \nabla w\, dx = -\int_K w\, \mathrm{div}(k\nabla\phi_{ij}^K)dx + \int_{\partial K}(k\nabla\phi_{ij}^K \cdot n)w\, ds$$

$$= -\int_K w\, \mathrm{div}R_j^K\, dx + \int_{\partial K}(k\nabla\phi_{ij}^K \cdot n)w\, ds$$

$$= \int_K (\nabla w)R_j^K\, dx + \int_{\partial K}(k\nabla\phi_{ij}^K \cdot n - R_j^K \cdot n)w\, ds$$

$$= \int_K (\nabla w)R_j^K\, dx + \int_{\partial K}\delta_{jl}\left(\frac{v_i \cdot n}{\int_{e_l} v_i \cdot n\, ds} - \langle\frac{v_i \cdot n}{\int_{e_l} v_i \cdot n\, ds}\rangle_{e_l}\right)w\, ds,$$

(6.90)

where we have used that $\langle\frac{v_i \cdot n}{\int_{e_j} v_i \cdot n\, ds}\rangle_{e_j} = R_j^K \cdot n_{e_j} = \frac{1}{|e_j|}$.

If we set $w = \phi_{ij}^K$, then it follows that

$$C\|\nabla\phi_{ij}^K\|_{0,K}^2 \leq \|\nabla\phi_{ij}^K\|_{0,K}\|R_j^K\|_{0,K}$$

$$+ \|\frac{v_i \cdot n}{\int_{e_j} v_i \cdot n\, ds} - \langle\frac{v_i \cdot n}{\int_{e_j} v_i \cdot n\, ds}\rangle_{e_j}\|_{L^r(e_j)}\|\phi_{ij}^K\|_{L^{r'}(\partial K)}$$

$$\leq C\|\nabla\phi_{ij}^K\|_{0,K} + Ch^{-\beta_2+1/r-1}\|\phi_{ij}^K\|_{L^{r'}(\partial K)}$$

$$\leq C\|\nabla\phi_{ij}^K\|_{0,K} + Ch^{-\beta_2+1/r-1}(h^{-1+1/r'}\|\phi_{ij}^K\|_{0,K} + h^{\frac{1}{r'}}\|\nabla\phi_{ij}^K\|_{0,K})$$

$$\leq C\|\nabla\phi_{ij}^K\|_{0,K} + Ch^{-\beta_2+1/r-1}h^{1/r'}\|\nabla\phi_{ij}^K\|_{0,K}$$

$$\leq C\|\nabla\phi_{ij}^K\|_{0,K} + Ch^{-\beta_2}\|\nabla\phi_{ij}^K\|_{0,K},$$

where $r'$ satisfies $1/r + 1/r' = 1$ ($r$ is defined in Assumption A2), and we have used Assumption A2 (see (6.82)) and $\|R_j^K\|_{0,K} \leq C$ (e.g., [57]) in the second step, the trace inequality (by rescaling) in the third step, and $\langle\phi_{ij}^K\rangle_K = 0$ along with the Poincaré–Friedrichs inequality (by rescaling) in the fourth step. Consequently, we have

$$\|\psi_{ij}^K\|_{0,K} \leq C(1 + h^{-\beta_2}),$$

(6.91)

where $C$ only depends on *Assumption A2* and the constants in trace inequality and Poincaré inequality in a fixed reference domain. Combining (6.89) and (6.91), it follows immediately

$$\|v - u_h\|_{0,K} = \|\sum_{i,j}(A_i(x)\beta_{ij}^K - A_{ij}^K)\psi_{ij}^K\|_{0,K}$$

$$\leq \|\sum_{i,j}(A_i(x) - \bar{A}_i^j)\beta_{ij}^K\psi_{ij}^K\|_{0,K} + \|\sum_{i,j}(\bar{A}_i^j\beta_{ij}^K - A_{ij}^K)\psi_{ij}^K\|_{0,K}$$

$$\leq \|\sum_{i,j}|A_i(x) - \bar{A}_i^j|\beta_{ij}^K\psi_{ij}^K\|_{0,K} + \|\sum_{i,j}|\bar{A}_i^j\beta_{ij}^K - A_{ij}^K|\psi_{ij}^K\|_{0,K}$$

(6.92)

$$\leq Ch^{\alpha+\beta_1}(\sum_i \|A_i\|_{C^\alpha(\Omega)})\sum_{i,j}\|\psi_{ij}^K\|_{0,K}$$

$$\leq Ch^{\alpha+\beta_1-\beta_2}(\sum_i \|A_i\|_{C^\alpha(\Omega)}),$$

where we have used Assumption A2 (see (6.82)) and $C$ depends on $N$ and the constants in Assumption A2. After summation over all $K$ for (6.92) we have

$$\|v - u_h\|_{0,\Omega}^2 = \sum_K \|u - u_h\|_{0,K}^2$$

$$\leq C(\sum_i \|A_i\|_{C^\alpha(\Omega)})^2 \sum_K h^{2(\alpha+\beta_1-\beta_2)}$$

$$\leq C(\sum_i \|A_i\|_{C^\alpha(\Omega)})^2 \frac{1}{h^2} h^{2(\alpha+\beta_1-\beta_2)}$$

$$= C(\sum_i \|A_i\|_{C^\alpha(\Omega)})^2 h^{2(\alpha+\beta_1-\beta_2-1)}.$$

Consequently,

$$\|v - v_h\|_{0,\Omega} \leq C(\sum_i \|A_i\|_{C^\alpha(\Omega)}) h^{\alpha+\beta_1-\beta_2-1}. \tag{6.93}$$

According to (6.85), for those $K$, $\partial K \cap \partial \Omega$, we adjust proper $c_{ij}^K$ such that $\sum_{i,j} c_{ij}^K \psi_{i,j}^K - g_{0,h} \in V_h^0$, but this does not affect our convergence rate. Therefore, invoking Lemma 6.39, (6.88), (6.93), and the Sobolev embedding theorem from $W^{1,\xi}$ into $C^\alpha$, Theorem 6.40 follows.

From the proof of Theorem 6.40, one can easily get the following result. Let $v$ and $v_h$ be the velocity in (4.4) and (4.7), respectively; then we have

$$\|v - v_h\|_{0,\Omega} \leq C(\sum_i \|A_i\|_{C^\alpha(\Omega)}) h^{\alpha+\beta_1-\beta_2-1}.$$

Remark 6.41. If $A_i(x) \in C^1(\Omega)$ in Assumption A1 and $v_i$ are defined such that $\beta_1 = 1$ and $\beta_2 = 0$ (e.g., $v_i$ are bounded), then Theorem 6.40 implies that

$$\|v - v_h\|_{H(\text{div},\Omega)} + \|p - p_h\|_{0,\Omega} \leq Ch.$$

Remark 6.42. We note that the local mixed MsFEMs suffer from a resonance error and a typical convergence rate for periodic coefficients is

$$\|v_\epsilon - v_h\|_{H(\text{div},\Omega)} + \|p_\epsilon - p_h\|_{0,\Omega} \leq C(h + \left(\frac{\epsilon}{h}\right)^\gamma),$$

where $\gamma = 1/2$ for the mixed multiscale method introduced in [71]. In our global mixed MsFEM, the boundary condition for the velocity basis is heterogeneous and Theorem 6.40 implies that stability is independent of the small scale and the resonance error is removed.

Remark 6.43. One can relax the main assumption used here and assume that

$$\|v(x) - \sum_i A_i(x)v_i(x)\|_{H(\text{div},\Omega)} \leq C\delta.$$

In this case, we can expect the convergence as

$$\|v - v_h\|_{H(\text{div},\Omega)} + \|p - p_h\|_{0,\Omega} \leq C(h^{\alpha+\beta_1-\beta_2-1} + \delta).$$

## Parabolic equations

Next, we extend the analysis to parabolic equations. We use the following assumption for the parabolic equation.

   *Assumption A1p. There exist functions $v_1, ..., v_N$ and sufficiently smooth $A_1(t,x), ..., A_N(t,x)$ such that*

$$v(t,x) = \sum_{i=1}^{N} A_i(t,x)v_i,$$

*where $v_i = k\nabla p_i$ and $p_i$ solves $\mathrm{div}(k(x)\nabla p_i) = 0$ in $\Omega$ with appropriate boundary conditions.*

   For our analysis, we assume, as before, $A_i(t,x) \in L^2(0,T;W^{1,\xi}(\Omega))(\xi > 2)$ and $v_i = k(x)\nabla p_i \in L^\eta(\Omega)$ $(1/2 = 1/\xi + 1/\eta)$, $i = 1, ..., N$.

*Remark 6.44.* Let $v_i = k(x)\nabla p_i$ $(i = 1, 2)$ be defined in (6.81), then Owhadi and Zhang in [217] show that $p(t,x) = p(t,p_1,p_2) \in L^2(0,T;W^{2,s})$ $(s > 2)$. Consequently, $v(t,x) = k(x)\nabla p = \sum_i(\partial p/\partial p_i)k\nabla p_i := \sum_i A_i(t,x)v_i$, where $A_i(t,x) = \partial p/\partial p_i \in L^2(0,T;W^{1,s})$.

   We define

$$\|u\|^2_{L^2_k(\Omega)} = \int_\Omega u \cdot k^{-1}(x)u dx$$

and

$$\|u\|^2_{L^2(0,T;L^2_k(\Omega))} = \int_0^T \int_\Omega u \cdot k^{-1}(x)u dx ds.$$

Let $\Pi_h : H(\mathrm{div}) \longrightarrow V_h$ be the interpolation operator defined as in Section 6.3.1 and $P_{Q_h} : L^2(\Omega) \longrightarrow Q_h$ be the $L^2$ projection onto $Q_h$.
   From (4.9) and (4.10), we have

$$\int_\Omega \frac{\partial}{\partial t}(p - p_h)q_h dx + \int_\Omega \mathrm{div}(v - v_h)q_h dx = 0, \quad \forall q_h \in Q_h$$
$$\int_\Omega k^{-1}(v - v_h) \cdot w_h dx - \int_\Omega \mathrm{div}(w_h)(p - p_h)dx = 0, \quad \forall w_h \in V_h. \tag{6.94}$$

Taking $w_h = \Pi_h v - v_h$ and $q_h = P_{Q_h}p - p_h$, we have

$$\int_\Omega \frac{\partial}{\partial t}(p - p_h)(P_{Q_h}p - p_h)dx + \int_\Omega \mathrm{div}(v - v_h)(P_{Q_h}p - p_h)dx = 0$$
$$\int_\Omega k^{-1}(v - v_h) \cdot (\Pi_h v - v_h)dx - \int_\Omega \mathrm{div}(\Pi_h v - v_h)(p - p_h dx) = 0. \tag{6.95}$$

Rewriting $p - p_h = p - P_{Q_h}p + P_{Q_h}p - p_h$ and $v - v_h = v - \Pi_h v + \Pi_h v - v_h$ in (6.95) and summation of the two equalities, we obtain

$$\int_\Omega \frac{\partial}{\partial t}(P_{Q_h}p - p_h)(P_{Q_h}p - p_h)dx + \int_\Omega k^{-1}(\Pi_h v - v_h) \cdot (\Pi_h v - v_h)dx$$

$$= -\int_\Omega \frac{\partial}{\partial t}(p - P_{Q_h}p)(P_{Q_h}p - p_h)dx - \int_\Omega k^{-1}(v - \Pi_h v) \cdot (\Pi_h v - v_h)dx$$

$$+ \int_\Omega [\mathrm{div}(\Pi_h v - v_h)(p - P_{Q_h}p) - \mathrm{div}(v - \Pi_h v)(P_{Q_h}p - p_h)]dx.$$

$$(6.96)$$

Because $P_{Q_h}$ is the $L^2(\Omega)$ projection onto $Q_h$, $P_{Q_h}$ commutes with the time derivative operator $\partial/\partial t$. Consequently, the first and third terms of the right-hand side in (6.96) vanish. By Lemma 6.35, the fourth term of the right-hand side in (6.96) also vanishes. Consequently, (6.96) becomes

$$\int_\Omega \frac{\partial}{\partial t}(P_{Q_h}p - p_h)(P_{Q_h}p - p_h)dx + \int_\Omega k^{-1}(\Pi_h v - v_h) \cdot (\Pi_h v - v_h)dx$$

$$= -\int_\Omega k^{-1}(v - \Pi_h v) \cdot (\Pi_h v - v_h)dx.$$

The Schwarz inequality and Young's inequality give rise to

$$\frac{1}{2}\frac{\partial}{\partial t}\|P_{Q_h}p - p_h\|_{0,\Omega}^2 + 2\|\Pi_h v - v_h\|_{L_k^2(\Omega)}^2$$

$$\leq \lambda\|\Pi_h v - v_h\|_{L_k^2(\Omega)}^2 + \frac{1}{4\lambda}\|v - \Pi_h v\|_{L_k^2(\Omega)}^2.$$

Integrating with respect to time and applying Gronwall's inequality and after choosing the proper value for $\lambda$, we have

$$\|P_{Q_h}p - p_h\|_{C^0(0,T;L^2(\Omega))}^2 + \|\Pi_h v - v_h\|_{L^2(0,T;L_k^2(\Omega))}^2$$

$$\leq C(\|P_{Q_h}p(0) - p_{0,h}\|_{0,\Omega}^2 + \|v - \Pi_h v\|_{L^2(0,T;L_k^2(\Omega))}^2).$$

Invoking the triangle inequality, we have

$$\|p - p_h\|_{C^0(0,T;L^2(\Omega))}^2 + \|v - v_h\|_{L^2(0,T;L_k^2(\Omega))}^2$$

$$\leq C(\|P_{Q_h}p(0) - p_{0,h}\|_{0,\Omega}^2 + \|v - \Pi_h v\|_{L^2(0,T;L_k^2(\Omega))}^2) \qquad (6.97)$$

$$+ \|p - P_{Q_h}p\|_{C^0(0,T;L^2(\Omega))}^2.$$

Hence, we obtain the following lemma.

**Lemma 6.45.** *Let $\{v, p\}$ and $\{v_h, p_h\}$ be the solution of (4.9) and (4.10), respectively. Under Assumption A1p and the definition of $\mathcal{V}_h$ in Section 6.3.1, the estimate (6.97) holds.*

Utilizing Lemma 6.45 and the proof of Theorem 6.40, we can derive the convergence result.

**Theorem 6.46.** *Let $\{v, p\}$ and $\{v_h, p_h\}$ be the solution of (4.9) and (4.10), respectively. If $\alpha + \beta_1 - \beta_2 - 1 > 0$ then*

$$\|p - p_h\|_{C^0(0,T;L^2(\Omega))} + \|v - v_h\|_{L^2(0,T;L^2_k(\Omega))} \leq Ch^{\alpha+\beta_1-\beta_2-1},$$

*where $\alpha = 1 - 2/\xi$ and $\xi$ is from Assumption A1p, and $\beta_i$ $(i = 1, 2)$ are defined in Assumption A2.*

*Proof.* Owing to the fact that $P_{Q_h}$ is the $L^2(\Omega)$ projection onto $Q_h$,

$$\|p - P_{Q_h}p\|_{C^0(0,T;L^2(\Omega))} \leq Ch|p|_{C^0(0,T;H^1(\Omega))}, \tag{6.98}$$

we estimate the first and the third term of right-hand side in (6.97). Next we estimate the term $\|v - \Pi_h v\|^2_{L^2(0,T;L^2_k(\Omega))}$. Define

$$A^K_{ij}(t) = \int_{e_j} A_i(t, s)(v_i \cdot n) ds$$

in each element $K$. Because $k^{-1}(x)$ is bounded, we have in each element $K$,

$$\|v - \Pi_h v\|^2_{L^2(0,T;L^2_k(K))}$$

$$= \int_0^T \int_K \sum_{i,j}(A_i(t,x)\beta^K_{ij} - A^K_{ij}(t))\psi^K_{ij} \cdot k^{-1} \sum_{i,j}(A_i(t,x)\beta^K_{ij} - A^K_{ij}(t))\psi^K_{ij} dx dt$$

$$\leq C \int_0^T \int_K (\sum_{i,j}(A_i(t,x)\beta^K_{ij} - A^K_{ij}(t))\psi^K_{ij})^2 dx dt$$

$$= C\|\sum_{i,j}(A_i(t,x)\beta^K_{ij} - A^K_{ij}(t))\psi^K_{ij}\|^2_{L^2(0,T;L^2(K))} \tag{6.99}$$

$$\leq C\|\sum_{i,j}(A_i(t,x) - \bar{A}^j_i(t))\beta^K_{ij}\psi^K_{ij}\|^2_{L^2(0,T;L^2(K))}$$

$$+ C\|\sum_{i,j}(\bar{A}^j_i(t)\beta^K_{ij} - A^K_{ij}(t))\psi^K_{ij}\|^2_{L^2(0,T;L^2(K))}$$

$$\leq Ch^{2(\alpha+\beta_1)} \sum \|\psi^K_{ij}\|^2_{0,K}.$$

In the last step, we used that facts that $A_i \in L^2(0,T;W^{1,\xi})$, Assumption A2 (see (6.82)) and proof of Theorem 6.40 (see (6.92)). After summation over all $K$ for (6.99), we have

$$\|v - \Pi_h v\|_{L^2(0,T;L^2_k(\Omega))} \leq Ch^{(\alpha+\beta_1-\beta_2-1)}. \tag{6.100}$$

Now, the proof can be completed taking into account (6.98) and (6.100).

## 6.3.2 Galerkin finite element methods with limited global information

We have proposed some analysis for modified MsFEMs in [103] and [3]. The main idea is to show that the pressure evolution in two-phase flow simulations is strongly influenced by the initial pressure. To demonstrate this, we consider a channelized permeability field, where the value of the permeability in the channel is large. We assume the permeability has the form $kI$, where $I$ is an identity matrix. In a channelized medium, the dominant flow is within the channels. Our analysis assumes a single channel and is restricted to 2D. Here, we briefly mention the main findings. Denote the initial stream function and pressure by $\eta = \psi(x, t = 0)$ and $\zeta = p(x, t = 0)$ ($\zeta$ is also denoted by $p^{sp}$ previously). The stream function is defined as

$$\partial\psi/\partial x_1 = -v_2, \quad \partial\psi/\partial x_2 = v_1. \tag{6.101}$$

Then the equation for the pressure can be written as

$$\frac{\partial}{\partial\eta}\left(|k|^2\lambda(S)\frac{\partial p}{\partial\eta}\right) + \frac{\partial}{\partial\zeta}\left(\lambda(S)\frac{\partial p}{\partial\zeta}\right) = 0. \tag{6.102}$$

For simplicity, $S = 0$ at time zero is assumed. We consider a typical boundary condition that gives high flow within the channel, such that the high flow channel will be mapped into a large slab in $(\eta, \zeta)$ coordinate system. If the heterogeneities within the channel in the $\eta$ direction are not strong (e.g., a narrow channel in Cartesian coordinates), the saturation within the channel will depend on $\zeta$. In this case, the leading-order pressure will depend only on $\zeta$, and it can be shown that

$$p(\eta, \zeta, t) = p_0(\zeta, t) + \text{high-order terms}, \tag{6.103}$$

where $p_0(\zeta, t)$ is the dominant pressure. Note that this result is shown when $\lambda$ is smooth. This asymptotic expansion shows that the time-varying pressure strongly depends on the initial pressure (i.e., the leading-order term in the asymptotic expansion is a function of initial pressure and time only). We note that (6.103) does not hold when $\lambda$ has discontinuities. In this case, our results hold away from the sharp interfaces and one can localize the interface by updating some basis functions. Our numerical results show that this update does not improve the results substantially. We believe this is because the discontinuities in $\lambda$ are small compared to heterogeneities in porous media, the effects of which we capture using limited global information. In our analysis, we assume that $|p(x, t) - \hat{p}(p^{sp}, t)|_{H^1}$ is small.

Because the analysis of the multiscale finite element methods is carried out only for the pressure equation, we assume $t$ (time) is fixed. We recall the assumption.

*Assumption G. There exists a sufficiently smooth scalar-valued function* $G(\eta)$ *($G \in W^{3,2s/(s-4)}$, $s > 4$), such that*

$$|p - G(p^{sp})|_{1,\Omega} \leq C\delta, \tag{6.104}$$

*where $p^{sp}$ is single-phase flow pressure and $\delta$ is sufficiently small.*

We note $G$ is $p_0(\zeta, t)$ at fixed $t$ in (6.103). Moreover, one does not need to know the function $G$ for computing the multiscale approximation of the solution. It is only necessary that $G$ have certain smoothness properties, however, it is important that the basis functions span $p^{sp}$ in each coarse block.

**Theorem 6.47.** *Under Assumption $G$ and $p^{sp} \in W^{1,s}(\Omega)$ ($s > 4$), the Ms-FEM converges with the rate given by*

$$|p - p_h|_{1,\Omega} \leq C\delta + Ch^{1-2/s}. \tag{6.105}$$

The proof of this theorem is given in [3]. Note that Theorem 6.47 shows that MsFEM converges for problems without any scale separation and the proof of this theorem does not use homogenization techniques. Next, we present the proof.

*Proof.* Following standard practice of finite element estimation, we seek $p_I = c_i\phi_i$, where $\phi_i$ are single-phase flow-based multiscale finite element basis functions. In the proof, we assume that $|\phi_i^K|_{1,K} \leq C$. Then from Cea's lemma, we have

$$|p - p_h|_{1,\Omega} \leq |p - G(p^{sp})|_{1,\Omega} + |G(p^{sp}) - c_i\phi_i|_{1,\Omega}. \tag{6.106}$$

Next, we present an estimate for the second term. We choose $c_i = G(p^{sp}(x_i))$, where $x_i$ are vertices of $K$. Furthermore, using a Taylor expansion of $G$ around $\overline{p}_K$, which is the average of $p^{sp}$ over $K$,

$$\begin{aligned}
G(p^{sp}(x_i)) =&G(\overline{p}_K) + G'(\overline{p}_K)(p^{sp}(x_i) - \overline{p}_K) \\
&+ (p^{sp}(x_i) - \overline{p}_K)^2 \int_0^1 sG''(p^{sp}(x_i) + s(\overline{p}_K - p^{sp}(x_i)))ds.
\end{aligned} \tag{6.107}$$

We have in each $K$

$$\begin{aligned}
c_i\phi_i =&G(\overline{p}_K) \sum_i \phi_i + G'(\overline{p}_K)(p^{sp}(x_i) - \overline{p}_K)\phi_i \\
&+ (p^{sp}(x_i) - \overline{p}_K)^2\phi_i \int_0^1 sG''(p^{sp}(x_i) + s(\overline{p}_K - p^{sp}(x_i)))ds \\
=&G(\overline{p}_K) + G'(\overline{p}_K)(p^{sp}(x_i)\phi_i - \overline{p}_K) \\
&+ (p^{sp}(x_i) - \overline{p}_K)^2\phi_i \int_0^1 sG''(p^{sp}(x_i) + s(\overline{p}_K - p^{sp}(x_i)))ds.
\end{aligned} \tag{6.108}$$

In the last step, we have used $\sum_i \phi_i = 1$. Similarly, in each $K$,

$$\begin{aligned}
G(p^{sp}(x)) =&G(\overline{p}_K) + G'(\overline{p}_K)(p^{sp}(x) - \overline{p}_K) \\
&+ (p^{sp}(x) - \overline{p}_K)^2 \int_0^1 sG''(p^{sp}(x) + s(\overline{p}_K - p^{sp}(x)))ds.
\end{aligned} \tag{6.109}$$

Using (6.108) and (6.109), we get

$$
|G(p^{sp}) - c_i\phi_i|_{1,K} \le |G'(\overline{p}_K)(p^{sp}(x) - p^{sp}(x_i)\phi_i)|_{1,K}
$$

$$
+ |(p^{sp}(x_i) - \overline{p}_K)^2\phi_i \int_0^1 sG''(p^{sp}(x_i) + s(\overline{p}_K - p^{sp}(x_i)))ds|_{1,K} \quad (6.110)
$$

$$
+ |(p^{sp}(x) - \overline{p}_K)^2 \int_0^1 sG''(p^{sp}(x) + s(\overline{p}_K - p^{sp}(x)))ds|_{1,K}.
$$

Because $|p^{sp}(x) - p^{sp}(x_i)\phi_i|_{1,K} \le Ch\|f\|_{0,K}$, the estimate of the first term is the following,

$$
|G'(\overline{p}_K)(p^{sp}(x) - p^{sp}(x_i)\phi_i)|_{1,K} \le Ch\|f\|_{0,K}.
$$

For the second term on the right-hand side of (6.110), assuming $p^{sp}(x) \in W^{1,s}(\Omega)$ and $s > 4$, we have

$$
|(p^{sp}(x_i) - \overline{p}_K)^2\phi_i^K \int_0^1 sG''(p^{sp}(x_i) + s(\overline{p}_K - p^{sp}(x_i)))ds|_{1,K}
$$

$$
\le Ch|p^{sp}|_{1,4,K}^2 |\phi_i^K|_{1,K}
$$

$$
\le Ch|p^{sp}|_{1,4,K}^2,
$$

where we have used the assumption $|\phi_i^K|_{1,K} \le C$ and $W^{1,s} \subset W^{1,4}$ $(s \ge 4)$. Here, we have used the inequality (e.g., [18])

$$
|u(x) - u(y)| \le C|x - y|^{1-2/s}|u|_{1,s,K}.
$$

For the third term, a straightforward calculation gives

$$
|(p^{sp}(x) - \overline{p}_K)^2 \int_0^1 sG''(p^{sp}(x) + s(\overline{p}_K - p^{sp}(x)))ds|_{1,K}
$$

$$
\le \|(p^{sp}(x) - \overline{p}_K)^2\nabla p^{sp}(x) \int_0^1 (1 - s)sG'''(p^{sp}(x) + s(\overline{p}_K - p^{sp}(x)))ds\|_{0,K}
$$

$$
+ \|2(p^{sp}(x) - \overline{p}_K)\nabla p^{sp}(x) \int_0^1 sG''(p^{sp}(x) + s(\overline{p}_K - p^{sp}(x)))ds\|_{0,K}
$$

$$
\le Ch^{2-2/s}\|\nabla p^{sp}\|_{L^s(K)}^3\|G'''\|_{L^{2s/(s-4)}(K)} + Ch^{1-2/s}|p^{sp}|_{1,s,K}|p^{sp}|_{1,K}
$$

$$
\le Ch^{2-2/s}\|\nabla p^{sp}\|_{L^s(K)}^3 + Ch^{1-2/s}|p^{sp}|_{1,K}
$$

where we used the Hölder inequality in the second step.

Combining the above estimates, we have for $s > 4$

$$
|G(p^{sp}) - c_i\phi_i^K|_{1,K} \le Ch|p^{sp}|_{1,4,K}^2
$$
$$
+ Ch^{2-2/s} + Ch^{1-2/s}|p^{sp}|_{1,K} + Ch\|f\|_{0,K}. \quad (6.111)
$$

Summing (6.111) over all $K$ and taking into account Assumption G, we have

$$|p - p_h|_{1,\Omega} \leq C(\delta + h^{1-2/s}) + Ch|p^{sp}|^2_{1,4,\Omega} + Ch^{1-2/s}|p^{sp}|_{1,\Omega} + Ch\|f\|_{0,\Omega}$$
$$\leq C(\delta + h^{1-2/s}) + Ch|p^{sp}|^2_{1,s,\Omega} + Ch^{1-2/s}|p^{sp}|_{1,s,\Omega} + Ch\|f\|_{0,\Omega}.$$

Consequently, if $s > 4$ (see e.g., [28]), the single-phase flow-based MsFEM converges.

*Remark 6.48.* We can relax the assumption on $G$. In particular, it is sufficient to assume $G \in W^{2,m}$ ($m \geq 1$). In this case, the proof can be carried out using Taylor polynomials in Sobolev spaces. Also, if $\nabla p^{sp} \in L^\infty(\Omega)$, then the convergence rate in (6.105) is $C\delta + Ch$.

*Remark 6.49.* One can similarly analyze Galerkin MsFEMs using multiple global fields (see [3]). This analysis can be extended to parabolic equations (see [163]).

# A

# Basic notations

$k(x)$ - coefficients (heterogeneous)
$p$ - solution
$v$ - flux (velocity)
$x$ - space variable
$t$ - time variable
$\mathbb{R}^d$ - $d$-dimensional vector space
$\mathcal{T}_h$ - coarse-scale partition
$W_h$ - standard finite element spaces (e.g., piecewise linear functions)
$\mathcal{P}_h$ - multiscale finite element "space" for the solution[1]
$\mathcal{V}_h$ - multiscale finite element space for the flux
$p_h$ - approximate solution obtained using MsFEM[2]
$p_{r,h}$ - fine-scale approximation of the solution (for nonlinear MsFEM only)
$\Omega$ - global domain
$K$ - coarse grid block
$h$ - coarse mesh size
$\epsilon$ - small physical (characteristic) scale
$\phi_j$ - multiscale basis functions from $\mathcal{P}_h$
$\phi_j^0$ - standard (e.g., linear) basis functions from $W_h$
$\chi$ - solution of auxiliary periodic problem (linear case)
$N$ - solution of auxiliary periodic problem (nonlinear case)
$q_t$, $q_w$ - source terms
$f$ - source term; also the flux function
$n$ - outward normal
*fractional flow* - fraction of the displaced fluid (see (2.43))
*water-cut* - fraction of water in the produced fluid
*PVI* - pore volume injected (see (2.44))

---

[1] for nonlinear problems, it is not a linear space
[2] fine-scale approximation for linear problems and coarse-scale approximation for nonlinear problems

# B

## Review of homogenization

### B.1 Linear problems

In this appendix, we use the notations commonly used in the homogenization literature and these notations can be different from those used in the main text of the book. Consider the second-order elliptic equation

$$-\frac{\partial}{\partial x_i}\left(a_{ij}\left(x/\epsilon\right)\frac{\partial}{\partial x_j}\right)u_\epsilon + a_0(x/\epsilon)u_\epsilon = f, \quad u_\epsilon|_{\partial\Omega} = 0, \qquad \text{(B.1)}$$

where $a_{ij}(y)$ and $a_0(y)$ are 1-periodic in both variables of $y$, and satisfy $a_{ij}(y)\xi_i\xi_j \geq \alpha\xi_i\xi_i$, with $\alpha > 0$, $a_0 > \alpha_0 > 0$, and bounded. Here we have used the Einstein summation notation; that is a repeated index means summation with respect to that index.

This model equation represents a common difficulty shared by several physical problems. For porous media, it is the pressure equation described by Darcy's law with the coefficient $a_\epsilon = (a_{ij}(x/\epsilon))$ representing the permeability tensor. For composite materials, it is the steady heat conduction equation and the coefficient $a_\epsilon$ represents the thermal conductivity. For steady transport problems, it is a symmetrized form of the governing equation. In this case, the coefficient $a_\epsilon$ is a combination of transport velocity and viscosity tensor.

Homogenization theory studies the limiting behavior $u_\epsilon \to u_0$ as $\epsilon \to 0$. The main task is to find the homogenized coefficients, $a_{ij}^*$ and $a_0^*$, and the homogenized equation for the limiting solution $u$

$$-\frac{\partial}{\partial x_i}\left(a_{ij}^*\frac{\partial}{\partial x_j}\right)u_0 + a_0^*u_0 = f, \quad u_0|_{\partial\Omega} = 0.$$

We define the bilinear form

$$a^\epsilon(u,v) = \int_\Omega a_{ij}^\epsilon(x)\frac{\partial u}{\partial x_j}\frac{\partial v}{\partial x_i}\ dx + \int_\Omega a_0^\epsilon uv\ dx.$$

The elliptic problem can also be formulated as a variational problem: find $u_\epsilon \in H_0^1$,

$$a^\epsilon(u_\epsilon, v) = (f, v), \quad \text{for all} \quad v \in H_0^1(\Omega),$$

where $(f, v)$ is the usual $L^2$ inner product, $\int_\Omega f v \, dx$.

### B.1.1 Special case: One-dimensional problem

Let $\Omega = (x_0, x_1)$ and take $a_0 = 0$. We have

$$-\frac{d}{dx}\left(a(x/\epsilon)\frac{du_\epsilon}{dx}\right) = f, \quad \text{in} \quad \Omega,$$

where $u_\epsilon(x_0) = u_\epsilon(x_1) = 0$, and $a(y) > \alpha_0 > 0$ is $y$-periodic with period $y_0$.
By taking $v = u_\epsilon$ in the variational problem, we have

$$\|u_\epsilon\|_{1,\Omega} \leq C.$$

Therefore one can extract a subsequence, still denoted by $u_\epsilon$, such that

$$u_\epsilon \rightharpoonup u \quad \text{in} \quad H_0^1(\Omega) \quad \text{weakly}.$$

Next, we introduce

$$\xi^\epsilon = a^\epsilon \frac{du^\epsilon}{dx}.$$

Because $a^\epsilon$ is bounded, and $du^\epsilon/dx$ is bounded in $L^2(\Omega)$, so $\xi^\epsilon$ is bounded in $L^2(\Omega)$. Moreover, because $-d\xi^\epsilon/dx = f$, we have $\xi^\epsilon \in H^1(\Omega)$. Thus we get

$$\xi^\epsilon \to \xi \quad \text{in} \quad L^2(\Omega) \quad \text{strongly},$$

so that

$$\frac{1}{a^\epsilon}\xi^\epsilon \to m(1/a)\xi \text{ in } L^2(\Omega) \quad \text{weakly}.$$

Furthermore, we note that $\xi^\epsilon/a^\epsilon = du^\epsilon/dx$. Therefore, we arrive at

$$\frac{du_0}{dx} = m(1/a)\xi.$$

On the other hand, $-d\xi^\epsilon/dx = f$ implies $-d\xi/dx = f$. This gives

$$-\frac{d}{dx}\left(\frac{1}{m(1/a)}\frac{du_0}{dx}\right) = f.$$

This is the correct homogenized equation for $u$. Note that $a^* = 1/m(1/a)$ is the harmonic average of $a^\epsilon$. It is in general not equal to the arithmetic average $\overline{a^\epsilon} = m(a)$.

## B.1.2 Multiscale asymptotic expansions.

The above analysis does not generalize to multidimensions. In this subsection, we introduce the multiscale expansion technique in deriving homogenized equations.

We look for $u_\epsilon(x)$ in the form of asymptotic expansion

$$u_\epsilon(x) = u_0(x, x/\epsilon) + \epsilon u_1(x, x/\epsilon) + \epsilon^2 u_2(x, x/\epsilon) + \cdots,$$

where the functions $u_j(x, y)$ are periodic in $y$ with period 1.

Denote by $A^\epsilon$ the second-order elliptic operator

$$A^\epsilon = -\frac{\partial}{\partial x_i}\left(a_{ij}(x/\epsilon)\frac{\partial}{\partial x_j}\right).$$

When differentiating a function $\phi(x, x/\epsilon)$ with respect to $x$, we have

$$\frac{\partial}{\partial x_j} = \frac{\partial}{\partial x_j} + \frac{1}{\epsilon}\frac{\partial}{\partial y_j},$$

where $y$ is evaluated at $y = x/\epsilon$. With this notation, we can expand $A^\epsilon$ as follows,

$$A^\epsilon = \epsilon^{-2}A_1 + \epsilon^{-1}A_2 + \epsilon^0 A_3, \tag{B.2}$$

where

$$A_1 = -\frac{\partial}{\partial y_i}\left(a_{ij}(y)\frac{\partial}{\partial y_j}\right),$$

$$A_2 = -\frac{\partial}{\partial y_i}\left(a_{ij}(y)\frac{\partial}{\partial x_j}\right) - \frac{\partial}{\partial x_i}\left(a_{ij}(y)\frac{\partial}{\partial y_j}\right),$$

$$A_3 = -\frac{\partial}{\partial x_i}\left(a_{ij}(y)\frac{\partial}{\partial x_j}\right) + a_0. \tag{B.3}$$

Substituting the expansions for $u_\epsilon$ and $A^\epsilon$ into $A^\epsilon u_\epsilon = f$, and equating the terms of the same power, we get

$$A_1 u_0 = 0, \tag{B.4}$$

$$A_1 u_1 + A_2 u_0 = 0, \tag{B.5}$$

$$A_1 u_2 + A_2 u_1 + A_3 u_0 = f. \tag{B.6}$$

Equation (B.4) can be written as

$$-\frac{\partial}{\partial y_i}\left(a_{ij}(y)\frac{\partial}{\partial y_j}\right)u_0(x, y) = 0,$$

where $u_0$ is periodic in $y$. The theory of second-order elliptic PDEs [132] implies that $u_0(x, y)$ is independent of $y$; that is $u_0(x, y) = u_0(x)$. This simplifies (B.5) for $u_1$,

$$-\frac{\partial}{\partial y_i}\left(a_{ij}(y)\frac{\partial}{\partial y_j}\right)u_1 = \left(\frac{\partial}{\partial y_i}a_{ij}(y)\right)\frac{\partial u}{\partial x_j}(x).$$

Define $\chi^j = \chi^j(y)$ as the solution to the following cell problem

$$\frac{\partial}{\partial y_i}\left(a_{ij}(y)\frac{\partial}{\partial y_j}\right)\chi^j = -\frac{\partial}{\partial y_i}a_{ij}(y) ,$$

where $\chi^j$ is periodic in $y$. The general solution of (B.5) for $u_1$ is then given by

$$u_1(x,y) = \chi^j(y)\frac{\partial u}{\partial x_j}(x) + \tilde{u}_1(x) .$$

Finally, we note that the equation for $u_2$ is given by

$$\frac{\partial}{\partial y_i}\left(a_{ij}(y)\frac{\partial}{\partial y_j}\right)u_2 = A_2 u_1 + A_3 u_0 - f . \tag{B.7}$$

The solvability condition implies that the right-hand side of (B.7) must have mean zero in $y$ over one periodic cell $Y = [0,1] \times [0,1]$; that is

$$\int_Y (A_2 u_1 + A_3 u_0 - f)\, dy = 0.$$

This solvability condition for second-order elliptic PDEs with periodic boundary condition [132] requires that the right-hand side of (B.7) have mean zero with respect to the fast variable $y$. This solvability condition gives rise to the homogenized equation for $u$:

$$-\frac{\partial}{\partial x_i}\left(a_{ij}^*\frac{\partial}{\partial x_j}\right)u + m(a_0)u = f , \tag{B.8}$$

where $m(a_0) = (1/|Y|)\int_Y a_0(y)\, dy$ and

$$a_{ij}^* = \frac{1}{|Y|}\left(\int_Y (a_{ij} - a_{ik}\frac{\partial \chi^j}{\partial y_k})\, dy\right) . \tag{B.9}$$

It is often difficult to compute the homogenized coefficients when the periodic cell problem requires very fine discretization. In this case, the bounds for the homogenized coefficients can be very useful. Finding accurate bounds depending on heterogeneities is a difficult task. There have been many works in the literature where bounds are computed and the corresponding optimal microstructures are determined. In the presence of tight bounds, one can avoid solving the cell problems for the computation of the homogenized solutions. We refer to [202, 74] for descriptions of various bounds and the literature reviews.

## B.1.3 Justification of formal expansions

The above multiscale expansion is based on a formal asymptotic analysis. However, we can justify its convergence rigorously.

Let $z_\epsilon = u_\epsilon - (u_0 + \epsilon u_1 + \epsilon^2 u_2)$. Applying $A^\epsilon$ to $z_\epsilon$, we get

$$A^\epsilon z_\epsilon = -\epsilon r_\epsilon ,$$

where $r_\epsilon = A_2 u_2 + A_3 u_1 + \epsilon A_3 u_2$. Thus we have $\|r_\epsilon\|_{\infty,\Omega} \leq C$.

On the other hand, we have

$$z_\epsilon|_{\partial\Omega} = -(\epsilon u_1 + \epsilon^2 u_2)|_{\partial\Omega}.$$

Thus, we obtain

$$\|z_\epsilon\|_{\infty,\partial\Omega} \leq c\epsilon.$$

It follows from the maximum principle [132] that

$$\|z_\epsilon\|_{\infty,\Omega} \leq C\epsilon$$

and therefore we conclude that

$$\|u_\epsilon - u_0\|_{\infty,\Omega} \leq C\epsilon.$$

## B.1.4 Boundary corrections

The above asymptotic expansion does not take into account the boundary condition of the original elliptic PDEs. If we add a boundary correction, we can obtain higher-order approximations.

Let $\theta_\epsilon \in H^1(\Omega)$ denote the solution to

$$\mathrm{div}_x(a^\epsilon \nabla_x \theta_\epsilon) = 0 \text{ in } \Omega, \quad \theta_\epsilon = u_1(x, x/\epsilon) \text{ on } \partial\Omega.$$

Then we have

$$(u_\epsilon - (u_0 + \epsilon u_1(x, x/\epsilon) - \epsilon\theta_\epsilon))|_{\partial\Omega} = 0.$$

Moskow and Vogelius [204] have shown that

$$\begin{aligned}
\|u_\epsilon - u_0 - \epsilon u_1(x, x/\epsilon) + \epsilon\theta_\epsilon\|_{0,\Omega} &\leq C_\omega \epsilon^{1+\omega} \|u_0\|_{2+\omega,\Omega}, \\
\|u_\epsilon - u_0 - \epsilon u_1(x, x/\epsilon) + \epsilon\theta_\epsilon\|_{1,\Omega} &\leq C\epsilon \|u_0\|_{2,\Omega},
\end{aligned} \tag{B.10}$$

where we assume $u \in H^{2+\omega}(\Omega)$ with $0 \leq \omega \leq 1$, and $\Omega$ is assumed to be a bounded, convex curvilinear polygon of class $C^\infty$.

## B.1.5 Nonlocal memory effect of homogenization

It is interesting to note that for certain degenerate problems, the homogenized equation may have a nonlocal memory effect.

Consider the simple 2D linear convection equation:

$$\frac{\partial u_\epsilon(x, y, t)}{\partial t} + a_\epsilon(y) \frac{\partial u_\epsilon(x, y, t)}{\partial x} = 0,$$

with initial condition $u_\epsilon(x, y, 0) = u_0(x, y)$. Note that $y = x_2$ is not a fast variable here.

We assume that $a_\epsilon$ is bounded and $u_0$ has compact support. It is easy to write down the solution explicitly,

$$u_\epsilon(x, y, t) = u_0(x - a_\epsilon(y)t, y),$$

however, it is not an easy task to derive the homogenized equation for the weak limit of $u_\epsilon$.

Using the Laplace transform and measure theory, Luc Tartar [255] showed that the weak limit $u$ of $u_\epsilon$ satisfies

$$\frac{\partial}{\partial t} u(x, y, t) + A_1(y) \frac{\partial}{\partial x} u(x, y, t) = \int_0^t \int \frac{\partial^2}{\partial x^2} u(x - \lambda(t - s), y, s) d\mu_y(\lambda) \, ds,$$

with $u(x, y, 0) = u_0(x, y)$, where $A_1(y)$ is the weak limit of $a_\epsilon(y)$, and $\mu_y$ is a probability measure of $y$ and has support in $[\min(a_\epsilon), \max(a_\epsilon)]$.

As we can see, the convection induces a nonlocal history-dependent diffusion term in the propagating direction $(x)$. The homogenized equation is not amenable to coarse-scale computation in general because the measure $\mu_y$ cannot be expressed explicitly in terms of $a_\epsilon$.

## B.1.6 Convection of microstructure

It is most interesting to see if one can apply the homogenization technique to obtain an averaged equation for the large-scale quantity for incompressible Euler or Navier–Stokes equations. In 1985, McLaughlin, Papanicolaou, and Pironneau [200] attempted to obtain a homogenized equation for the 3D incompressible Euler equations with a highly oscillatory velocity field. More specifically, they considered the following initial value problem,

$$\frac{\partial u}{\partial t} + (u \cdot \nabla) u = -\nabla p,$$

with $\nabla \cdot u = 0$ and highly oscillatory initial data

$$u(x, 0) = U(x) + W(x, x/\epsilon).$$

They then constructed multiscale expansions for both the velocity field and the pressure. In doing so, they made an important assumption that the microstructure is convected by the mean flow. Under this assumption, they constructed a multiscale expansion for the velocity field as follows:

$$u^\epsilon(x,t) = u_0(x,t) + w\left(\frac{\theta(x,t)}{\epsilon}, \frac{t}{\epsilon}, x, t\right) + \epsilon u_1\left(\frac{\theta(x,t)}{\epsilon}, \frac{t}{\epsilon}, x, t\right) + O(\epsilon^2).$$

The pressure field $p^\epsilon$ is expanded similarly. From this ansatz, one can show that $\theta$ is convected by the mean velocity:

$$\frac{\partial\theta}{\partial t} + u_0 \cdot \nabla\theta = 0, \quad \theta(x,0) = x .$$

It is a very challenging problem to develop a systematic approach to study the large-scale solution in three-dimensional Euler and Navier–Stokes equations. The work of McLaughlin, Papanicolaou, and Pironneau provided some insightful understanding into how small scales interact with large scales and how to deal with the closure problem. However, the problem is still not completely resolved because the cell problem obtained this way does not have a unique solution. Additional constraints need to be enforced in order to derive a large-scale averaged equation. With additional assumptions, they managed to derive a variant of the $k - \epsilon$ model in turbulence modeling.

*Remark B.1.* One possible way to improve the work of [200] is to take into account the oscillation in the Lagrangian characteristics $\theta_\epsilon$. The oscillatory part of $\theta_\epsilon$ in general could have an order-one contribution to the mean velocity of the incompressible Euler equation. In [148, 149, 150], Hou, Yang and co-workers have studied convection of the microstructure of the 2D and 3D incompressible Euler equations using a new approach. They do not assume that the oscillation is propagated by the mean flow. In fact, they found that it is crucial to include the effect of oscillations in the characteristics on the mean flow. Using this new approach, they can derive a well-posed cell problem that can be used to obtain an effective large-scale average equation.

More can be said for a passive scalar convection equation.

$$\frac{\partial v}{\partial t} + \frac{1}{\epsilon}\text{div}\left(u(x/\epsilon)v\right) = \alpha\Delta v,$$

with $v(x,0) = v_0(x)$. Here $u(y)$ is a known incompressible periodic (or stationary random) velocity field with zero mean. Assume that the initial condition is smooth.

Expand the solution $v^\epsilon$ in powers of $\epsilon$

$$v^\epsilon = v(t,x) + \epsilon v_1(t,x,x/\epsilon) + \epsilon^2 v_2(t,x,x/\epsilon) + \cdots .$$

The coefficients of $\epsilon^{-1}$ lead to

$$\alpha\Delta_y v_1 - u \cdot \nabla_y v_1 - u \cdot \nabla_x v = 0.$$

Let $e_k$, $k = 1, 2, 3$ be the unit vectors in the coordinate directions and let $\chi^k(y)$ satisfy the cell problem:

$$\alpha \Delta_y \chi^k - u \cdot \nabla_y \chi^k - u \cdot e_k = 0.$$

Then we have

$$v_1(t, x, y) = \sum_{k=1}^{3} \chi^k(y) \frac{v(t, x)}{\partial x_k}.$$

The coefficients of $\epsilon^0$ give

$$\alpha \Delta_y v_2 - u \cdot \nabla_y v_2 = u \cdot \nabla_x v_1 - 2\alpha \nabla_x \cdot \nabla_y v_1 - \alpha \Delta_x v + \frac{\partial v}{\partial t}.$$

The solvability condition for $v_2$ requires that the right-hand side have zero mean with respect to $y$. This gives rise to the equation for homogenized solution $v$,

$$\frac{\partial v}{\partial t} = \alpha \Delta_x v - \overline{u \cdot \nabla_x v_1}.$$

Using the cell problem, McLaughlin, Papanicolaou, and Pironneau obtained [200]

$$\frac{\partial v}{\partial t} = \sum_{i,j=1}^{3} (\alpha \delta_{ij} + \alpha_{T_{ij}}) \frac{\partial^2 v}{\partial x_i \partial x_j},$$

where $\alpha_{T_{ij}} = -\overline{u_i \chi^j}$.

## B.2 Nonlinear problems

We briefly discuss homogenization for general nonlinear elliptic equations, $u_\epsilon \in W_0^{1,p}(\Omega)$,

$$- \operatorname{div} a_\epsilon(x, u_\epsilon, \nabla u_\epsilon) + a_{0,\epsilon}(x, u_\epsilon, \nabla u_\epsilon) = f, \tag{B.11}$$

where $a_\epsilon(x, \eta, \xi)$ and $a_{0,\epsilon}(x, \eta, \xi)$, $\eta \in \mathbb{R}$, $\xi \in \mathbb{R}^d$ satisfy assumptions given by (6.42)–(6.46), which guarantee the well-posedness of the nonlinear elliptic problem (B.11). Here $\Omega \subset \mathbb{R}^d$ is a Lipschitz domain and $\epsilon$ denotes the small scale of the problem. The homogenization of nonlinear partial differential equations has been studied previously (see, e.g., [220]). It can be shown that a solution $u_\epsilon$ converges (up to a subsequence) to $u_0$ in an appropriate norm, where $u_0 \in W_0^{1,p}(\Omega)$ is a solution of a homogenized equation

$$- \operatorname{div} a^*(x, u_0, \nabla u_0) + a_0^*(x, u, \nabla u_0) = f. \tag{B.12}$$

The homogenized coefficients can be computed if we make an additional assumption on the heterogeneities, such as periodicity, almost periodicity, or when the fluxes are strictly stationary fields with respect to spatial variables.

In these cases, an auxiliary problem is formulated and used in the calculations of the homogenized fluxes $a^*$ and $a_0^*$. Next, we discuss this.

We assume that $a$ and $a_0$ are periodic functions with respect to the spatial variable. Then, the homogenized fluxes are defined as follows,

$$a^*(\eta, \xi) = \int_Y a(y, \eta, \xi + \nabla_y N_{\eta,\xi}(y)) \, dy, \tag{B.13}$$

$$a_0^*(\eta, \xi) = \int_Y a_0(y, \eta, \xi + \nabla_y N_{\eta,\xi}(y)) \, dy, \tag{B.14}$$

where $a^*$ and $a_0^*$ satisfy the conditions similar to (6.42)–(6.46). Here $N_{\eta,\xi} \in W_{per}^{1,p}(Y)$ is the periodic solution (with average zero) of

$$- \operatorname{div}(a(y, \eta, \xi + \nabla_y N_{\eta,\xi}(y))) = 0 \text{ in } Y, \tag{B.15}$$

where $Y$ is a unit period. We do not present the proof of the homogenization here and refer to [220], for example.

Next, we also present the homogenization results for the random homogeneous case. Homogenization in random homogeneous media for linear problems ([43, 164]) has been a pioneering work in this direction. We start with a description of random homogeneous fields on $\mathbb{R}^d$ which is shown to be useful in homogenization problems (e.g., [164]). Let $(U, \Sigma, \mu)$ be a probability space. A random homogeneous field is a measurable function on $U$ and $f(T(x)\omega)$ are realizations of the random field. The realizations are well-defined measurable functions on $\mathbb{R}^d$ for almost all $\omega \in U$. Consider a $d$-dimensional dynamical system on $U$, $T(x) : U \to U$, $x \in \mathbb{R}^d$, that satisfies the following conditions: (1) $T(0) = I$, and $T(x + y) = T(x)T(y)$; (2) $T(x) : U \to U$ preserve the measure $\mu$ on $U$; and (3) for any measurable function $f(\omega)$ on $U$, the function $f(T(x)\omega)$ defined on $\mathbb{R}^d \times U$ is also measurable (see [164]). Let $L^p(U)$ denote the space of all $p$-integrable functions on $U$. Then $U(x)f(\omega) = f(T(x)\omega)$ defines a $d$-parameter group of isometries in the space $L^p(U)$, and $U(x)$ is strongly continuous [164, 220]. We further assume that the dynamical system $T$ is ergodic; that is, any measurable $T$-invariant function on $U$ is constant. Denote by $\langle \cdot \rangle_\mu$ the mean value over $U$,

$$\langle f \rangle_\mu = \int_U f(\omega) d\mu(\omega) = E(f).$$

Denote by $D_\omega^i$ the generator of $U(x)$ along the $i$th coordinate direction; that is,

$$D_\omega^i = \lim_{\delta \to 0} \frac{f(T(x_i)\omega) - f(\omega)}{\delta}.$$

The domains $\partial_i$ of $D_\omega^i$ are dense in $L^2(U)$, and the intersection of all $D_\omega^i$ is also dense.

Next following [220] we define potential and solenoidal fields. A vector field $f \in L^p(U)$ is said to be potential (or solenoidal, respectively) if its generic realization $f(T_x\omega)$ is a potential (or solenoidal respectively) vector field in $\mathbb{R}^d$. Denote by $L_{pot}^p(U)$ (respectively, $L_{sol}^p(U)$) the subspace of $L^p(U)$ that consists of all potential (respectively, solenoidal) vector fields. Introduce the following notations,

$$V_{\text{pot}}^p = \{f \in L_{\text{pot}}^p(U), \langle f \rangle_\mu = 0\}, \quad V_{\text{sol}}^p = \{f \in L_{\text{sol}}^p(U), \langle f \rangle_\mu = 0\}.$$

The following properties are known (see [220], page 138)

$$L_{\text{pot}}^p(U) = V_{\text{pot}}^p \oplus \mathbb{R}^d, \quad L_{\text{sol}}^p(U) = V_{\text{sol}}^p \oplus \mathbb{R}^d,$$

$$L_{\text{sol}}^q(U) = (V_{\text{pot}}^p)^\perp, \quad L_{\text{pot}}^q(U) = (V_{\text{sol}}^p)^\perp.$$

Next, we consider (B.11) with the assumptions given by (6.42)–(6.46), which guarantee the well-posedness of the nonlinear elliptic problem.

It is known (e.g., [220]) that as $\epsilon \to 0$ $\nabla u_\epsilon$ converges to $\nabla u_0$ weakly in $L^p(\Omega)$ for a.e. $\omega$, and $u_0$ is the solution of

$$- \operatorname{div}(a^*(u_0, \nabla u_0)) + a_0^*(u_0, \nabla u_0) = f, \ u_0 \in W_0^{1,p}(\Omega). \tag{B.16}$$

Furthermore, $a^*$ and $a_0^*$ can be constructed using the solution of the following auxiliary problem. Given $\eta \in R$ and $\xi \in \mathbb{R}^d$ define $w_{\eta,\xi} \in V_{\text{pot}}^p$ such that

$$a(\omega, \eta, \xi + w_{\eta,\xi}(\omega)) \in L_{\text{sol}}^q(U)^d. \tag{B.17}$$

Then $a^*(\eta, \xi)$ and $a_0(\eta, \xi)$ are defined as

$$\begin{aligned} a^*(\eta, \xi) &= \langle a(\omega, \eta, \xi + w_{\eta,\xi}(\omega)) \rangle_\mu, \\ a_0^*(\eta, \xi) &= \langle a_0(\omega, \eta, \xi + w_{\eta,\xi}(\omega)) \rangle_\mu. \end{aligned} \tag{B.18}$$

Moreover, $a^*(\eta, \xi)$ and $a_0^*(\eta, \xi)$ satisfy similar estimates as $a$ and $a_0$ with different constants [220].

For parabolic problems, the homogenization also yields the macroscopic equations of the same class. If we consider

$$\frac{\partial u_\epsilon}{\partial t} - \operatorname{div}(a_\epsilon(x, t, u_\epsilon, \nabla u_\epsilon)) + a_{0,\epsilon}(x, t, u_\epsilon, \nabla u_\epsilon) = f, \tag{B.19}$$

where $a_\epsilon(x, t, \eta, \xi) = a(x/\epsilon^\beta, t/\epsilon^\alpha, \eta, \xi)$ $a_{0,\epsilon}(x, t, \eta, \xi) = a_0(x/\epsilon^\beta, t/\epsilon^\alpha, \eta, \xi)$.

The homogenization of nonlinear parabolic equations depends on the ratio between $\alpha$ and $\beta$ and is presented in [111]. The following cases are distinguished: (1) self-similar case ($\alpha = 2\beta$); (2) nonself-similar case ($\alpha < 2\beta$); (3) nonself-similar case ($\alpha > 2\beta$); (4) spatial case ($\alpha = 0$); and (5) temporal case ($\beta = 0$). To introduce the homogenized operator, we introduce fast variables $y = x/\epsilon^\beta$ and $\tau = t/\epsilon^\alpha$. Moreover, denote by $\langle \cdot \rangle_{y,\tau}$ the average over $y$ and $\tau$. If a single variable $y$ or $\tau$ is used as a subscript, then the average is taken with respect to that variables. Similarly, we denote by $\Pi_{y,\tau}$ the periodic box in space and time, and correspondingly $\Pi_y$ and $\Pi_\tau$ are periods in space and temporal variable. The homogenized operator is given by

$$\frac{\partial u_0}{\partial t} - \text{div}(a^*(x, t, u, \nabla u_0)) + a_0^*(x, t, u, \nabla u_0) = f,$$

where the homogenized coefficients are defined below.

- For self-similar case ($\alpha = 2\beta$),

$$a^*(\eta, \xi) = \langle a(y, \tau, \eta, \xi + \nabla N_{\eta,\xi}) \rangle_{y,\tau},$$
$$a_0^*(\eta, \xi) = \langle a_0(y, \tau, \eta, \xi + \nabla N_{\eta,\xi}) \rangle_{y,\tau},$$

where $N_{\eta,\xi}$ is the unique solution of

$$\frac{\partial N_{\eta,\xi}}{\partial \tau} - \text{div}_y \, a(\omega, \eta, \xi + \nabla_y N_{\eta,\xi}) = 0 \qquad (B.20)$$

in $\Pi_{y,\tau}$.

- For nonself-similar case ($\alpha < 2\beta$),

$$a^*(\eta, \xi) = \langle a(y, \tau, \eta, \xi + \nabla N_{\eta,\xi}) \rangle_{y,\tau},$$
$$a_0^*(\eta, \xi) = \langle a_0(y, \tau, \eta, \xi + \nabla N_{\eta,\xi}) \rangle_{y,\tau},$$

where $N_{\eta,\xi}$ is the unique solution of

$$- \text{div}_y \, a(y, \tau, \eta, \xi + \nabla_y N_{\eta,\xi}) = 0 \qquad (B.21)$$

in $\Pi_{y,\tau}$.

- For nonself-similar case ($\alpha > 2\beta$),

$$a^*(\eta, \xi) = \langle a(y, \tau, \eta, \xi + \nabla N_{\eta,\xi}) \rangle_{y,\tau},$$
$$a_0^*(\eta, \xi) = \langle a_0(y, \tau, \eta, \xi + \nabla N_{\eta,\xi}) \rangle_{y,\tau},$$

where $N_{\eta,\xi}$ is the unique solution of

$$- \text{div}_y \, \overline{a}(y, \eta, \xi + \nabla_y N_{\eta,\xi}) = 0. \qquad (B.22)$$

$\overline{a}(y, \eta, \xi) = \langle a(y, \tau, \eta, \xi) \rangle_\tau.$

- For spatial case ($\alpha = 0$),

$$a^*(t, \eta, \xi) = \langle a(y, t, \eta, \xi + \nabla N_{\eta,\xi}) \rangle_y$$
$$a_0^*(t, \eta, \xi) = \langle a_0(y, t, \eta, \xi + \nabla N_{\eta,\xi}) \rangle_y,$$

where $N_{\eta,\xi}$ satisfies

$$- \text{div}_y \, a(y, t, \eta, \xi + \nabla_y N_{\eta,\xi}) = 0 \qquad (B.23)$$

in $\Pi_y$ for each $t$ (assuming the coefficients smoothly depend on $t$).

- For temporal case ($\beta = 0$), the homogenized fluxes are defined by

$$a^*(x, \eta, \xi) = \langle a(x, \tau, \eta, \xi) \rangle_\tau,$$
$$a_0^*(x, \eta, \xi) = \langle a_0(\omega, \eta, \xi) \rangle_\tau. \qquad (B.24)$$

For the results concerning the random homogenization of nonlinear parabolic equations we refer to [111].

# References

1. J.E. Aarnes. *On the use of a mixed multiscale finite element method for greater flexibility and increased speed or improved accuracy in reservoir simulation*. SIAM MMS, 2:421–439, 2004.
2. J.E. Aarnes, P. Dostert, and Y. Efendiev. *Uncertainty quantification in subsurface applications using stochastic multiscale finite element methods*. In preparation.
3. J.E. Aarnes, Y. Efendiev, and L. Jiang. *Analysis of multiscale finite element methods using global information for two-phase flow simulations*. Submitted.
4. J.E. Aarnes, Y. Efendiev, T.Y. Hou, and L. Jiang. *Mixed multiscale finite element methods on adaptive unstructured grids using limited global information*. Submitted.
5. J.E. Aarnes and Y. Efendiev. *An adaptive multiscale method for simulation of fluid flow in heterogeneous porous media*. SIAM MMS, 5(30):918–939, 2006.
6. J.E. Aarnes and Y. Efendiev. *A multiscale method for modeling transport in porous media on unstructured corner-point grids*. J. A. & Comput. Technol., 2(2):299–318, 2008.
7. J.E. Aarnes and Y. Efendiev. *Mixed multiscale finite element for stochastic porous media flows*. SIAM Sci. Comp., 30(5):2319–2339, 2007.
8. J.E. Aarnes, Y. Efendiev, and L. Jiang. *Analysis of multiscale finite element methods using global information for two-phase flow simulations*. SIAM MMS, 7(2):655–676, 2007.
9. J.E. Aarnes, V. Hauge, and Y. Efendiev. *Coarsening of three-dimensional structured and unstructured grids for subsurface flow*. Adv. Water Resour., 30(11):2177–2193, 2007.
10. J.E. Aarnes and B.O. Heimsund. *Multiscale discontinuous Galerkin methods for elliptic problems with multiple scales*. LNCSE, Volume 44, Multiscale Methods in Science and Engineering, Springer Berlin, pp.1–20, 2005.
11. J.E. Aarnes, S. Krogstad, and K.-A. Lie. *A hierarchical multiscale method for two-phase flow based upon mixed finite elements and nonuniform grids*. SIAM MMS, 5(2):337–363, 2007.
12. J.E. Aarnes, S. Krogstad, and K.-A. Lie. *A multiscale framework for three-phase black-oil reservoir simulation*. Preprint.

13. J.E. Aarnes, S. Krogstad, and K.-A. Lie. *Multiscale mixed/mimetic methods on corner-point grids.* Comput. Geosci., DOI: 10.1007/s10596-007-9072-8

14. J.E. Aarnes and T.Y. Hou. *An efficient domain decomposition preconditioner for multiscale elliptic problems with high aspect ratios.* Acta Math. Applicat. Sinica, 18:63–76, 2002.

15. A. Abdulle and B. Engquist. *Finite element heterogeneous multiscale methods with near optimal computational complexity.* SIAM MMS, 6(4):1059–1084, 2007.

16. A. Abdulle. *Multiscale method based on discontinuous Galerkin methods for homogenization problems.* C.R. Math. Acad. Sci. Paris 346(1–2):97–102, 2008.

17. A. Abdulle. *On a priori error analysis of fully discrete heterogeneous multiscale FEM.* Multiscale Model. Simul. 4(2):447–459, 2005.

18. R.A. Adams. *Sobolev spaces.* Academic Press, New York-London, Pure and Applied Mathematics, Vol. 65, 1975.

19. R. Ahmadov. *Petrophysics and permeability of fault zones in sandstone with a focus on fault slip surfaces and slip bands.* MS Thesis, Stanford University, 2006.

20. R. Ahmadov, A. Aydin, M. Karimi-Fard, and L. Durlofsky. *Permeability upscaling of fault zones in the Aztec Sandstone, Valley of Fire State Park, Nevada, with a focus on slip surfaces and slip bands.* Hydrogeol. J., 15:1239–1250, 2007.

21. G. Allaire. *Homogenization and two-scale convergence.* SIAM Math. Anal., 23(6):1482–1518, 1992.

22. G. Allaire. *Shape optimization by the homogenization method.* Appl. Math. Sci., 146:1482–1518, Springer-Verlag, New York, 2002.

23. G. Allaire and R. Brizzi. *A multiscale finite element method for numerical homogenization.* SIAM MMS, 4(3):790–812, 2005.

24. T. Arbogast. *Numerical subgrid upscaling of two-phase flow in porous media.* in Numerical Treatment of Multiphase Flows in Porous Media, Z. Chen et al., Eds., Lecture Notes in Physics 552:35–49, Springer, Berlin, 2000.

25. T. Arbogast. *Implementation of a locally conservative numerical subgrid upscaling scheme for two-phase Darcy flow.* Comput. Geosci., 6:453–481, 2002.

26. T. Arbogast and K. Boyd. *Subgrid upscaling and mixed multiscale finite elements.* SIAM Num. Anal. vol. 44:1150–1171, 2006.

27. T. Arbogast, G. Pencheva, M.F. Wheeler, and I. Yotov. *A multiscale mortar mixed finite element method.* SIAM J. Multiscale Model. Simul., 6:319–346, 2007.

28. M. Avellaneda and F.-H. Lin. *Compactness methods in the theory of homogenization.* Comm. Pure Appl. Math., 40:803-847, 1987.

29. K. Aziz and A. Settari. *Petroleum Reservoir Simulation.* Elsevier Applied Scientific Pub., New York, 1979.

30. I. Babuška, U. Banerjee, and J.E. Osborn. *Survey of meshless and generalized finite element methods: A unified approach.* Acta Numerica, pp. 1–125, 2003,

31. I. Babuška, G. Caloz, and E. Osborn. *Special finite element methods for a class of second order elliptic problems with rough coefficients*. SIAM J. Numer. Anal., 31:945–981, 1994.

32. I. Babuška and J. M. Melenk. *The partition of unity method*. Internat. J. Numer. Meth. Eng., 40:727–758, 1997.

33. I. Babuška and E. Osborn. *Generalized finite element methods: Their performance and their relation to mixed methods*. SIAM J. Numer. Anal., 20:510–536, 1983.

34. I. Babuška and W.G. Szymczak. *An error analysis for the finite element method applied to convection-diffusion problems*. Comput. Meth. Appl. Math. Eng., 31:19–42, 1982.

35. N.S. Bakhvalov and G. Panasenko. *Homogenization of processes in periodic media*. "Nauka", Moscow, 1984.

36. V. Barthelmann, E. Novak, and K. Ritter. *High dimensional polynomial interpolation on sparse grids*. Adv. in Comput. Math., 12:273–288, 2000.

37. J.W. Barker and S. Thibeau, *A critical review of the use of pseudorelative permeabilities for upscaling*. SPE Reservoir Eng., 12:138–143, 1997

38. P.T. Bauman, J.T. Oden, and S. Prudhomme. *Adaptive multiscale modeling of polymeric materials with Arlequin coupling and goals algorithms*. Submitted to Comput. Meth. Appl. Math. Eng.

39. J.H. Bramble, J. Pasciak, J. Wang, and J. Xu. *Convergence Estimates for Product Iterative Methods with Applications to Domain Decomposition*. Math. Comp., 57(195):1–21, 1991.

40. T. Breitzman, R. Lipton, and E. Iarve. *Local field assessment inside multiscale composite architectures*. SIAM MMS, 6(3):937–962, 2007.

41. F. Brezzi, K. Lipnikov, M. Shashkov, and V. Simoncini. *A new discretization methodology for diffusion problems on generalized polyhedral meshes*. Comput. Meth. Appl. Mech. Eng., 196(37–40):3682–3692, 2007.

42. A. Beliaev. *The homogenization of Stokes flows in random porous domains of general type*. Asymptot. Anal., 19(2):81–94, 1999.

43. A. Bensoussan, J.L. Lions, and G. Papanicolaou. *Asymptotic analysis for periodic structures,* Volume 5 of Studies in Mathematics and Its Applications, North-Holland Publ., New York, 1978.

44. L. Berlyand, L. Borcea, and A. Panchenko. *Network approximation for effective viscosity of concentrated suspensions with complex geometry*. SIAM J. Math. Anal., 36(5):1580–1628, 2005.

45. L. Berlyand, Y. Gorb, and A. Novikov. *Discrete network approximation for highly-packed composites with irregular geometry in three dimensions*. Multiscale methods in science and engineering, 21-57, LNCSE, 44, Springer, Berlin, 2005.

46. L. Berlyand and A. Novikov. *Error of the network approximation for densely packed composites with irregular geometry*. SIAM J. Math. Anal., 34(2):385–408, 2002.

47. A. Bourgeat and A. Piatnitski. *Approximations of effective coefficients in stochastic homogenization*. Ann. Inst. H. Poincare Probab. Statist., 40(2):153–165, 2004.

48. L. Borcea and G. Papanicolaou. *Network approximation for transport properties of high contrast materials*. SIAM J. Appl. Math., 58(2):501–539, 1998.

49. A. Bourgeat. *Homogenized behavior of two-phase flows in naturally fractured reservoirs with uniform fractures distribution.* Comp. Meth. Appl. Mech. Eng., 47:205–215, 1984.

50. A. Bourgeat and A. Mikelić. *Homogenization of two-phase immiscible flows in a one-dimensional porous medium.* Asymptotic Anal., 9:359–380, 1994.

51. A. Brandt. *Multiscale computational methods: research activities.* In Proc. 1991 Hang Zhou International Conf. on Scientific Computation (Chan, T. and Shi, Z.-C., Eds), World Scientific, Singapore, pp. 1–7, 1992.

52. A. Brandt. *Multiscale scientic computation: Six year summary.* Gauss Center Report WI/GC-12, May 1999, Also in MGNET.

53. A. Brandt. *Multiscale computation from fast solvers to systematic upscaling.* In: Computational Fluid and Solid Mechanics (K.J. Bathe, ed.), Elsevier, The Netherlands, pp. 1871–1873, 2003.

54. A. Brandt. *Multiscale solvers and systematic upscaling in computational physics.* Comput. Phys. Commun., 169:438–441, 2005.

55. S. Brenner and L. Ridgway Scott. *The mathematical theory of finite element methods. Second edition.* Texts in Applied Mathematics, 15. Springer-Verlag, New York, 2002.

56. M. Brewster and G. Beylkin. *A multiresolution strategy for numerical homogenization.* ACHA, 2:327–349, 1995.

57. F. Brezzi and M. Fortin. *Mixed and hybrid finite element methods.* Springer-Verlag, Berlin, 1991.

58. F. Brezzi and A. Russo. *Choosing bubbles for advection-diffusion problems.* Math. Models Meth. Appl. Sci, 4:571–587, 1994.

59. F. Brezzi, L.P. Franca, T.J.R. Hughes and A. Russo. $b = \int g$. Comput. Meth. in Appl. Mech. Eng., 145:329–339, 1997.

60. J.E. Broadwell. *Shock structure in a simple discrete velocity gas,* Phys. Fluids, 7:1243–1246, 1964.

61. R. Caflisch. *Multiscale modeling for epitaxial growth.* International Congress of Mathematicians. III,:1419-1432, Eur. Math. Soc., Zurich, 2006.

62. L.Q. Cao. *Multiscale asymptotic expansion and finite element methods for the mixed boundary value problems of second order elliptic equation in perforated domains.* Numer. Math. 103(1):11–45, 2006.

63. L.Q. Cao, J.Z. Cui, D.C. Zhu, and J.L. Luo, *Multiscale finite element method for subdivided periodic elastic structures of composite materials.* J. Comput. Math. 19(2):205–212, 2001.

64. M.A. Celia, E. Bouloutas, and R. Zarba. *A general mass-conservative numerical solution for the unsaturated flow equation.* Water Resour. Res., 26(7):1483–1496, 1990.

65. L. Chamoin, J.T. Oden, and S. Prudhommee. *A stochastic coupling method for atomic-to-continuum Monte-Carlo simulations.* To appear in CMAME.

66. A. Christen and C. Fox. *MCMC using an approximation.* Technical report, Department of Mathematics, The University of Auckland, New Zealand.

67. E. Chung and Y. Efendiev. *Multiscale methods for elliptic equations with high contrast coefficients.* In preparation.

68. G. Chechkin, A. Piatnitski, and A. Shamaev. *Homogenization. Methods and applications.* Translations of Mathematical Monographs, 234. American Mathematical Society, Providence, RI, 2007.
69. Y. Chen and L.J. Durlofsky. *Adaptive local-global upscaling for general flow scenarios in heterogeneous formations.* Transport Porous Media, 62:157–185, 2006.
70. Y. Chen and L.J. Durlofsky. *An ensemble level upscaling approach for efficient estimation of fine scale production statistics using coarse scale simulations.* SPE paper 106086 presented in SPE Reservoir Simulation Sympoisum, Houston, February, 2007.
71. Z. Chen and T.Y. Hou. *A mixed multiscale finite element method for elliptic problems with oscillating coefficients.* Math. Comp., 72:541–576, 2002.
72. Z. Chen and T. Savchuk. *Analysis of the multiscale finite element method for nonlinear and random homogenization problems.* SIAM J. Numer. Anal., 46(1):260–279, 2007.
73. Z. Chen, M. Cui, T. Savchuk, and X. Yu. *The multiscale finite element method with nonconforming elements for elliptic homogenization problems.* SIAM MMS, to appear.
74. A. Cherkaev. *Variational methods for structural optimization.* Springer-Verlag, New York, 2000.
75. S.-S. Chow. *Finite element error estimates for nonlinear elliptic equations of monotone type.* Numer. Math., 54:373–393, 1989.
76. C.C. Chu, I. Graham, and T.Y. Hou. *The accuracy of multiscale finite element methods for high-contrast elliptic interface problems.* In preparation.
77. C.C. Chu, Y. Efendiev, and T.Y. Hou. *Localization of global multiscale finite element methods in the presence of sharp interfaces.* In preparation.
78. M. Christie and M. Blunt. *Tenth SPE comparative solution project: A comparison of upscaling techniques.* SPE Reser. Eval. Eng., 4:308–317, 2001.
79. B.C. Craft and M.F. Hawkins. *Petroleum reservoir engineering.* Prentice-Hall, Englewood Cliffs, NJ, 1959.
80. M. Cruz and A. Petera. *A Parallel Monte-Carlo finite element procedure for the analysis of multicomponent random media.* Int. J. Numer. Meth. Eng., 38:1087–1121, 1995.
81. J.H. Cushman, L.S. Bennethum, and P.P. Singh. *Toward rational design of drug delivery substrates: I. Mixture theory for two-scale biocompatible polymers.* SIAM MMS, 2(2):302–334, 2004.
82. G. Dal Maso and A. Defranceschi. *Correctors for the homogenization of monotone operators.* Differential Integral Eq., 3:1151–1166, 1990.
83. A. Datta-Gupta and M.J. King. *Streamline simulation: Theory and practice.* Society of Petroleum Engineers, 2007.
84. J. Dendy, J. Hyman, and J. Moulton. *The black box multigrid numerical homogenization algorithm.* J. Comput. Phys., 142:80–108, 1998.
85. C. Deutsch and A. Journel. *GSLIB: Geostatistical software library and user's guide, 2nd edition.* Oxford University Press, New York, 1998.
86. M. Dobson and M. Luskin. *Analysis of a force-based quasicontinuum approximation.* M2AN Math. Model. Numer. Anal., 42(1):113–139, 2008.

87. M. Dorobantu and B. Engquist. *Wavelet-based numerical homogenization.* SIAM J.Numer. Anal., 35:540–559, 1998.

88. P. Dostert, Y. Efendiev, and T.Y. Hou. *Multiscale finite element methods for stochastic porous media flow equations.* Comput. Meth. Appl. Math. Eng., 197(43-44):3445–3455, 2008.

89. P. Dostert, Y. Efendiev, T.Y. Hou, and W. Luo. *Coarse-gradient Langevin algorithms for dynamic data integration and uncertainty quantification.* J. Comp. Physics, 217(1):123–142, 2007.

90. J. Douglas, Jr. and T.F. Russell. *Numerical methods for convection-dominated diffusion problem based on combining the method of characteristics with finite element or finite difference procedures.* SIAM J. Numer. Anal., 19:871–885, 1982.

91. L.J. Durlofsky. *Numerical calculation of equivalent grid block permeability tensors for heterogeneous porous media.* Water Resour. Res., 27:699–708, 1991.

92. L.J. Durlofsky. *Coarse scale models of two-phase flow in heterogeneous reservoirs: Volume averaged equations and their relation to existing upscaling techniques.* Comp. Geosci. 2:73–92, 1998.

93. L.J. Durlofsky, Y. Efendiev, and V. Ginting. *An adaptive local-global multiscale finite volume element method for two-phase flow simulations.* Adv. Water Resour., 30:576–588, 2007.

94. L.J. Durlofsky, R.C. Jones, and W.J. Milliken. *A nonuniform coarsening approach for the scale-up of displacement processes in heterogeneous porous media.* Adv. Water Resour., 20:335–347, 1997.

95. B. Dykaar and P.K. Kitanidis. *Determination of the effective hydraulic conductivity for heterogeneous porous media using a numerical spectral approach: 1. Method.* Water Resour. Res., 28:1155–1166, 1992.

96. W. E. *Homogenization of linear and nonlinear transport equations.* Comm. Pure Appl. Math., XLV:301–326, 1992.

97. W. E and B. Engquist. *The heterogeneous multi-scale methods.* Comm. Math. Sci., 1(1):87-133, 2003.

98. W. E, P. Ming, and P. Zhang. *Analysis of the heterogeneous multiscale method for elliptic homogenization problems.* J. Amer. Math. Soc. 18(1):121–156, 2005.

99. Y. Efendiev. *Multiscale finite element method (MsFEM) and its applications.* Ph. D. Thesis, Applied Mathematics, Caltech, 1999.

100. Y. Efendiev and L. Durlofsky. *A generalized convection-diffusion model for subgrid transport in porous media.* SIAM MMS, vol.1(3):504–526, 2003.

101. Y. Efendiev and L.J. Durlofsky. *Numerical modeling of subgrid heterogeneity in two phase flow simulations.* Water Resour. Res., 38(8), pp. 1128, 2002.

102. Y. Efendiev, L.J. Durlofsky, and S. H. Lee. *Modeling of subgrid effects in coarse-scale simulations of transport in heterogeneous porous media.* Water Resour. Res., 36:2031–2041, 2000.

103. Y. Efendiev, V. Ginting, T.Y. Hou, and R. Ewing. *Accurate multiscale finite element methods for two-phase flow simulations.* J. Comp. Phys. 220(1):155–174, 2006.

104. Y. Efendiev, T.Y. Hou, and V. Ginting. *Multiscale finite element methods for nonlinear problems and their applications.* Comm. Math. Sci., 2:553–589, 2004.

105. Y. Efendiev, T.Y. Hou, and W. Luo. *Preconditioning Markov chain Monte Carlo simulations using coarse-scale models.* SIAM Sci., 28(2):776–803, 2006.

106. Y. Efendiev, T.Y. Hou, and T. Strinopoulos. *Multiscale simulations of porous media flows in flow-based coordinate system.* Comp. Geosci. DOI:10.1007/s10596-007-9073-7.

107. Y. Efendiev, T.Y. Hou, and X.H. Wu. *Convergence of a nonconforming multiscale finite element method.* SIAM J. Numer. Anal., 37:888–910, 2000.

108. Y. Efendiev, L. Jiang, and I. Mishev. *Multiscale finite element methods using partial upscaling.* In preparation.

109. Y. Efendiev, X. Ma, A. Datta-Gupta and B. Mallick. *Multi-stage MCMC using non-parameteric error estimators.* Submitted.

110. Y. Efendiev and A. Pankov. *Numerical homogenization of monotone elliptic operators.* SIAM Multiscale Model. Simul. 2(1):62–79, 2003.

111. Y. Efendiev and A. Pankov. *Homogenization of nonlinear random parabolic operators.* Adv. Differential Eq., 10(11):1235–1260, 2005.

112. Y. Efendiev and A. Pankov. *Numerical homogenization of nonlinear random parabolic operators.* SIAM Multiscale Model. and Simul., 2(2):237–268, 2004.

113. Y. Efendiev and A. Pankov. *Numerical homogenization and correctors for random elliptic equations.* SIAM J. Appl. Math., 65(1):43–68, 2004.

114. ———, *Meyers type estimates for approximate solutions of nonlinear elliptic equations and their applications.* DCDS-B, 6(3):481–492, 2006.

115. Y. Efendiev and B. Popov. *On homogenization of nonlinear hyperbolic equations.* Comm. Pure Appl. Anal., 4(2):295–309, 2005.

116. A. Ern and J.-L. Guermond. *Theory and practice of finite elements.* vol. 159 of Applied Mathematical Sciences, Springer-Verlag, New York, 2004.

117. R.E. Ewing and J. Wang. *Analysis of the Schwarz Algorithm for Mixed Finite Element Methods.* RAIRO Mathematical Modelling and Numerical Analysis, 26(6):739-756, 1992.

118. R. Eymard, T. Gallouët, and R. Herbin. *Finite volume methods.* In Handbook of numerical analysis, Vol. VII, Handb. Numer. Anal., VII, North-Holland, Amsterdam, 713–1020, 2000.

119. A. Fannjiang and G. Papanicolaou. *Convection enhanced diffusion for periodic flows.* SIAM J. Appl. Math., 54(2):333–408, 1994.

120. J. Fish and K.L. Shek. *Multiscale Analysis for Composite Materials and Structures.* Composites Sci. Technol.: Int. J., 60:2547–2556, 2000.

121. J. Fish and Z. Yuan. *Multiscale enrichment based on the partition of unity.* Int. J. Numer. Meth. Eng., 62:1341–1359, 2005.

122. J. Fish and A. Wagiman. *Multiscale finite element method for heterogeneous medium.* Comput. Mechan.: Int. J., 12:1–17, 1993.

123. D. Frias, M. Murad, and F. Pereira. *Stochastic computational modeling of highly heterogeneous poroelastic media with long-range correlations.* Int. J. Numer. Anal. Meth. Geomech., 27:1-32, 2003.

124. L. Franca, A. Madureira, and F. Valentin. *Towards multiscale functions: Enriching finite element spaces with local but not bubble-like functions.* Comput. Meth. Appl. Mech. Eng., 194(27–29):3006–3021, 2005.

125. F. Furtado and F. Pereira. *Scaling analysis for two-phase immiscble flow in heterogeneous porous media.* Comput. Appl. Math., 17(3):233–262, 1998.

126. F. Furtado and F. Pereira. *Crossover from nonlinearity controlled to heterogeneity controlled mixing in two-phase porous media flows.* Comput. Geosci., 7:115–135, 2003.

127. F. Furtado and F. Pereira. *On the scale-up problem for two-phase flow in petroleum reservoirs.* Cubo Math. J., 6:53–72, 2004.

128. B. Ganapathysubramanian and N. Zabaras. *Modelling diffusion in random heterogeneous media: Data-driven models, stochastic collocation and the variational multi-scale method.* J. of Comput. Phys., 226:326–353, 2007.

129. B. Ganapathysubramanian and N. Zabaras. *A stochastic multiscale framework for modeling flow through heterogeneous porous media.* Submitted.

130. A. Gelman and B. Rubin. *Inference from iterative simulation using multiple sequences.* Stat. Sci., 7:457–511, 1992.

131. M. T. van Genuchten. *A closed-form equation for predicting the hydraulic conductivity of unsaturated soils.* Soil. Sci. Soc. Am. J., 44:892–898, 1980.

132. D. Gilbarg and N.S. Trudinger. *Elliptic partial differential equations of second order.* Springer, Berlin, New York, 2001.

133. V. Ginting. *Computational upscaled modeling of heterogeneous porous media flows utilizing finite volume method.* PhD thesis, Texas A& M University, College Station, 2004.

134. J. Glimm, H. Kim, D. Sharp, and T. Wallstrom. *A stochastic analysis of the scale up problem for flow in porous media.* Comput. Appl. Math., 17:67–79, 1998.

135. J. Glimm and D. Sharp. *Multiscale science: A challenge for twenty-first century.*SIAM News 30(8):1–7, 1997.

136. A. Gloria. *An analytical framework for the numerical homogenization of monotone elliptic operators and quasiconvex energies.* Multiscale Model. Simul., 5:996–1043, 2006.

137. I.G. Graham, P. Lechner, and R. Scheichl. *Domain decomposition for multiscale PDEs.* Numer. Math. DOI 10.1007/s00211-007-0074-1 (2007) .

138. U. Grenander and M.I. Miller. *Representations of knowledge in complex systems (with discussion).* J. R. Statist. Soc. B, 56:549–603, 1994.

139. V.H. Hoang and C. Schwab. *High dimensional finite elements for elliptic problems with multiple scales.* SIAM MMS, 3(1):168–194, 2004/2005.

140. L. Holden and B.F. Nielsen. *Global upscaling of permeability in heterogeneous reservoirs: The output least squares (OLS) method.* Transport Porous Media, 40:115–143, 2000.

141. U. Hornung. *Homogenization and porous media.* Interdisciplinary Applied Mathematics. Springer, New York, 1997.

142. T.Y. Hou and D. Liang. *Multiscale analysis for convection dominated transport equations.* DCDS-A, accepted, 2008.

143. T.Y. Hou, X.H. Wu, and Y. Zhang. *Removing the cell resonance error in the multiscale finite element method via a Petrov-Galerkin formulation.* Comm. Math. Sci., 2(2):185–205, 2004.

144. T.Y. Hou, A. Westhead, and D.P. Yang. *A framework for modeling subgrid effects for two-phase flows in porous media.* SIAM Multiscale Model. and Simul., 5(4):1087–1127, 2006.

145. T.Y. Hou and X.H. Wu. *A Multiscale finite element method for elliptic problems in composite materials and porous media.* J. Comput. Phys., 134:169–189, 1997.
146. T.Y. Hou and X.H. Wu. *A multiscale finite element method for PDEs with oscillatory coefficients.* Proceedings of 13th GAMM-Seminar Kiel on Numerical Treatment of Multi-Scale Problems, Jan 24–26, 1997, Notes on Numerical Fluid Mechanics, Vol. 70, ed. by W. Hackbusch and G. Wittum, Vieweg-Verlag, pp. 58–69, 1999.
147. T.Y. Hou, X.H. Wu, and Z. Cai. *Convergence of a multiscale finite element method for elliptic problems With rapidly oscillating coefficients.* Math. Comput., 68:913–943, 1999.
148. T.Y. Hou, D.P. Yang, and K. Wang. *Homogenization of incompressible Euler equation.* J. Comput. Math., 22:220–229, 2004.
149. T.Y. Hou, D.P. Yang, and H. Ran. *Multiscale analysis in the Lagrangian formulation for the 2-D incompressible Euler equation.* Discrete Contin. Dynam. Syst., 13:1153–1186, 2005.
150. T.Y. Hou, D.P. Yang, and H. Ran. *Multiscale computation of isotropic homogeneous turbulent flow.* In Inverse Problems, Multi-Scale Analysis and Effective Medium Theory, Contemporary Mathematics, Vol. 408, pp. 111–135, 2006, ed. H. Ammari and H. Kang.
151. T. Y. Hou, D.-P. Yang, and H. Ran. *Multiscale analysis and computation for the 3-D incompressible Navier-Stokes equations.* SIAM Multiscale Model. and Simul., 6(4):1317–1346, 2008.
152. T.Y. Hou and X. Xin. *Homogenization of linear transport equations with oscillatory vector fields.* SIAM J. Appl. Math., 52:34–45, 1992.
153. Y. Huang and J. Xu. *A partition-of-unity finite element method for elliptic problems with highly oscillating coefficients.* Preprint.
154. T.J.R. Hughes. *Multiscale phenomena: Green's Functions, the Dirichlet-to-Neumann formulation, subgrid scale models, bubbles and the origins of stabilized methods.* Comput. Meth. Appl. Mech Eng., 127:387–401, 1995.
155. T.J.R. Hughes, G.R. Feijóo, L. Mazzei, and J.-B. Quincy. *The Variational multiscale method - A paradigm for computational mechanics.* Comput. Meth. Appl. Mech. Eng., 166:3–24, 1998.
156. M. Gerritsen and L.J. Durlofsky. *Modeling of fluid flow in oil reservoirs.* Ann. Rev. Fluid Mech., 37:211–238, 2005.
157. M.E. Gurtin. *An introduction to continuum mechanics.* Academic Press, San Diego, CA, 1981.
158. O. Iliev, A. Mikelic,and P. Popov. *On upscaling certain flows in deformable porous media.* SIAM MMS, 7(1):93–123, 2008.
159. P. Jenny, S.H. Lee, and H. Tchelepi. *Multi-scale finite volume method for elliptic problems in subsurface flow simulation.* J. Comput. Phys., 187:47–67, 2003.
160. P. Jenny, S.H. Lee, and H. Tchelepi. *Adaptive multi-scale finite volume method for multi-phase flow and transport in porous media.* SIAM MMS, 3:30–64, 2004.
161. P. Jenny, S.H. Lee, and H. Tchelepi. *An adaptive fully implicit multi-scale finite-volume algorithm for multi-phase flow in porous media.* J. Comp. Phys., 217:627–641, 2006.

162. L. Jiang, *Multiscale numerical methods for partial differential equations using limited global information and their applications*. Ph.D. thesis, Texas A& M University, 2008

163. L. Jiang, Y. Efendiev, and V. Ginting. *Global multiscale methods for wave equations*. Submitted.

164. V. Jikov, S. Kozlov, and O. Oleinik. *Homogenization of differential operators and integral functionals,* Springer, New York, 1994, Translated from Russian.

165. R. Juanes and T.W. Patzek. *A variational multiscale finite element method for multiphase flow in porous media*. Finite Elem. Anal. Des., 41(7–8):763–777, 2005

166. I.G. Kevrekidis, C.W. Gear, J.M. Hyman, P.G. Kevrekidis, O. Runborg, and C. Theodoropoulos. *Equation-free, coarse-grained multiscale computation: enabling microscopic simulators to perform system-level analysis*. Commun. Math. Sci., 1(4):715–762, 2003.

167. V. Kippe, J.E. Aarnes, and K.-A. Lie. *A comparison of multiscale methods for elliptic problems in porous media*. Comp. Geosci., DOI: 10.1007/s10596-007-9074-6

168. S. Knapek. *Matrix-dependent multigrid-homogenization for diffusion problems*. In the Proceedings of the Copper Mountain Conference on Iterative Methods, edited by T. Manteuffal and S. McCormick, volume I, SIAM Special Interest Group on Linear Algebra, Cray Research , 1996.

169. S. Kozlov. *The averaging of random operators*. Mat. Sb. (N.S.) 109(151)(2):188–202, 1979.

170. M.A. Krasnosel'skiĭ, P.P. Zabreĭko, E.I. Pustyl'nik, and P.E. Sobolevskiĭ, *Integral operators in spaces of summable functions*. Noordhoff International, Leiden, 1976. Translated from the Russian by T. Ando, Monographs and Textbooks on Mechanics of Solids and Fluids, Mechanics: Analysis.

171. J.R. Kyte and D.W. Berry. *New pseudofunctions to control numerical dispersion*. SPE 5105, 1975.

172. D.C. Lagoudas and E.L. Vandygriff. *Processing and characterization of NiTi porous SMA by elevated pressure sintering*. J. Intell. Mater. Sys. Struct., 13:837–850, 2002.

173. P. Langlo and M.S. Espedal. *Macrodispersion for two-phase, immiscible flow in porous media*. Adv. Water Resour., 17:297-316, 1994.

174. C. Lee and C.C. Mei. *Re-examination of the equations of poroelasticity*. Int. J. Eng. Sci., 35:329–352, 1997.

175. S.H. Lee, C. Wolfsteiner, and H. Tchelepi. *A multiscale finite-volume method for multiphase flow in porous media: Black oil formulation of compressible, three phase flow with gravity and capillary force*. Comp. Geosci.., doi:10.1007/s10596-007-9069-3, 2008.

176. J. Li, P. Kevrekidis, C.W. Gear, and I. Kevrekidis. *Deciding the nature of the coarse equation through microscopic simulations: the baby-bathwater scheme*. SIAM Rev., 49(3):469–487, 2007.

177. K. Lipnikov, F. Brezzi, and V. Simoncini. *A family of mimetic finite difference methods on polygonial and polyhedral meshes*. Math. Models Meth. Appl. Sci., 15:1533–1553, 2005.

178. R. Lipton. *Homogenization and field concentrations in heterogeneous media*. SIAM J. Math. Anal., 38(4):1048–1059, 2006.

179. J.S. Liu, *Monte Carlo strategies in scientific computing*. Springer, New-York, 2001.
180. W.K. Liu, H.S. Park, D. Qian, E.G. Karpov, H. Kadowaki, and G.J. Wagner. *Bridging scale methods for nanomechanics and materials*. Comput. Meth. Appl. Mech. Eng., 195(13–16):1407–1421, 2006.
181. W.K. Liu, Y.F. Zhang, and M.R. Ramirez. *Multi-scale finite element methods*. Int. J. Numer. Meth. Eng., 32:969–990, 1991.
182. M. Loève. *Probability theory*. 4th ed., Springer, Berlin, 1977.
183. B. Lu, T.M. Alshaalan, and M.F. Wheeler. *Iteratively coupled reservoir simulation for multiphase flow*. SPE 110114, 2007.
184. Z. Lu, D. Higdon, and D. Zhang. *A Markov chain Monte Carlo method for the groundwater inverse problem*. Proceedings of the 15th International Conference on Computational Methods in Water Resources, June 13–17, Chapel Hill, NC, 2004.
185. A. Lu, and D. Zhang. *Accurate, efficient quantification of uncertainty for flow in heterogeneous reservoirs using the KLME Approach*. SPE paper 93452, The 2005 SPE Reservoir Symposium, Houston, TX, Jan. 31-Feb. 2, 2005.
186. I. Lunati and P. Jenny. *Multi-scale finite-volume method for compressible multi-phase flow in porous media*. J. Comp. Phys., 216:616–636, 2006.
187. I. Lunati and P. Jenny. *Multi-scale finite-volume method for density-driven flow in porous media*. Comp. Geosci., doi:10.1007/s10596-007-9071-9, 2008.
188. I. Lunati and P. Jenny. *The multiscale finite volume method: A flexible tool to model physically complex flow in porous media*. 10th European Conference on the Mathematics of Oil Recovery Amsterdam, The Netherlands, 4–7 September, 2006.
189. I. Lunati and P. Jenny. *Multi-scale finite-volume method for three-phase flow influenced by gravity*. CMWR, 2006.
190. I. Lunati and P. Jenny. *Multi-scale finite-volume method for highly heterogeneous porous media with shale layers*. 9th European Conference on the Mathematics of Oil Recovery Cannes, France, 30 August–2 September, 2004.
191. X. Ma, M. Al-Harbi, A. Datta-Gupta, and Y. Efendiev. *A Multistage sampling approach to quantifying uncertainty during history matching geological models*. SPE 102476, accepted for publication, SPE J., 2008.
192. S. MacLachlan and J. Moulton. *Multilevel upscaling through variational coarsening*. Water Resour. Res., 42, W02418, doi:10.1029/2005WR003940, 2006.
193. Y. Maday, A.T. Patera, and G. Turinici. *Global a priori convergence theory for reduced-basis approximations of single-parameter symmetric coercive elliptic partial differential equations*. C. R. Acad. Sci. Paris Sér., 1 335:1–6, 2002.
194. Y. Maday, A.T. Patera, and G. Turinici. *A Priori convergence theory for reduced-basis approximations of single-parameter elliptic partial differential equations*. J. Sci. Comput., 17(1–4):437–446, 2002.
195. A. Madureira. *Multiscale numerical methods for partial differential equations posed in domains with rough boundaries*. Math. Comput., accepted.

196. M. Sarkis and H. Versieux. *Convergence analysis for the numerical boundary corrector for elliptic equations with rapidly oscillating coefficients.* SIAM J. Numer. Anal., 46(2):545–576, 2008.

197. A. Matache, I. Babuška, and C. Schwab. *Generalized p-FEM in homogenization.* Numer. Math., 6:319–375, 2000.

198. A. Matache and C. Schwab. *Homogenization via p-FEM for problems with microstructure.* Appl. Numer. Math., 33:43–59, 2000.

199. J. McCarthy. *Comparison of fast algorithms for estimating large-scale permeabilities of heterogeneous media.* Transport Porous Media, 19:123–137, 1995.

200. D. McLaughlin, G. Papanicolaou, and O. Pironneau. *Convection of microstructure and related problems.* SIAM J. Appl. Math., 45:780–797, 1985.

201. N.G. Meyers and A. Elcrat. *Some results on regularity for solutions of non-linear elliptic systems and quasi-regular functions.* Duke Math. J., 42:121–136, 1975.

202. G. Milton. *The theory of composites.* Cambridge University Press, Cambridge, UK, 2002.

203. P. Ming and X. Yue. *Numerical methods for multiscale elliptic problems.* J. Comput. Phys. 214(1):421–445, 2006.

204. S. Moskow and M. Vogelius. *First order corrections to the homogenized eigenvalues of a periodic composite medium: A convergence proof.* Proc. Roy. Soc. Edinburgh, A, 127:1263–1299, 1997.

205. M. Murad and J. Cushman. *Multiscale flow and deformation in hydrophilic swelling porous media.* Int. J. Eng. Sci., 34(3):313–336, 1996.

206. J. R. Natvig, K.-A. Lie, B. Eikemo, and I. Berre. *A discontinuous Galerkin method for single phase flow in porous media.* Adv. Water Resour., 30(12):2424–2438, 2007.

207. J. Nečas. *Introduction to the theory of nonlinear elliptic equations.* Wiley-Interscience, John Wiley & Sons, Chichester, 1986, preprint of the 1983 edition.

208. F. Nobile, R. Tempone, and C. G. Webster. *A sparse grid stochastic collocation method for elliptic partial differential equations with random input data.* SIAM J. Numer. Anal., 46(5):2309–2345, 2008.

209. J. Nolen, G. Papanicolaou, and O. Pironneau. *A framework for adaptive multiscale method for elliptic problems.* SIAM MMS, 7:171–196, 2008.

210. A. Novikov. *Eddy viscosity of cellular flows by upscaling.* J. Comput. Phys., 195(1):341–354, 2004.

211. A. Novikov, G. Papanicolaou, and L. Ryzhik. *Boundary layers for cellular flows at high Peclet numbers.* Comm. Pure Appl. Math., 58(7):867–922, 2005.

212. J. Obregon, M. Murad, and F. Rochina. Computational homogenization of nonlinear hydromechanical coupling in poroplasticity. Int. J. of Multiscale Comput. Eng., 4:693–732, 2007.

213. J.T. Oden, S. Prudhomme, A. Romkes and P.T. Bauman. *Multiscale modeling of physical phenomena: Adaptive control of models.* SIAM J. Sci. Comput., 28:2359–2389, 2006.

214. M. Ohlberger. *A posteriori error estimates for the heterogeneous multiscale finite element method for elliptic homogenization problems.* SIAM MMS, 4(1):88–114, 2005.

215. D. Oliver, L. Cunha, and A. Reynolds. *Markov chain Monte Carlo methods for conditioning a permeability field to pressure data.* Math. Geol., 29, 1997.

216. D. Oliver, N. He, and A. Reynolds. *Conditioning permeability fields to pressure data.* 5th European conference on the mathematics of oil recovery, Leoben, Austria, 3–6 September, 1996.

217. H. Owhadi and L. Zhang. *Homogenization of parabolic equations with a continuum of space and time scales.* SIAM J. Numer. Anal., 46(1):1–36, 2008.

218. ——, *Metric based up-scaling.* Comm. Pure Appl. Math., LX:675–723, 2007.

219. G. Panasenko. *Multi-scale modelling for structures and composites.* Springer, Dordrecht, 2005. xiv+398 pp.

220. A. Pankov. *G-convergence and homogenization of nonlinear partial differential operators.* Kluwer Academic, Dordrecht, 1997.

221. M. Park and J.H. Cushman. *On upscaling operator-stable Levy motions in fractal porous media.* J. Comp. Phys., 217:159–165, 2006.

222. H.S. Park, E.G. Karpov, W.K. Liu, and P.A. Klein. *The bridging scale for two-dimensional atomistic/continuum coupling.* Philosophi. Mag., 85(1):79–113, 2005.

223. H.S. Park and W.K. Liu. *An introduction and tutorial on multiple scale analysis in solids.* Comput. Meth. Appl. Mech. and Eng., 193:1733–1772, 2004.

224. A. Papavasiliou and I. Kevrekidis. *Variance reduction for the equation-free simulation of multiscale stochastic systems.* SIAM MMS, 6(1):70–89, 2007.

225. G. Pavliotis and A.M. Stuart, *Multiscale methods averaging and homogenization.* Series: Texts in Applied Mathematics , Vol. 53, Springer, New York, 2008.

226. M. Peszyńska. *Mortar adaptivity in mixed methods for flow in porous media.* Int. J. Numer. Anal. Model. 2(3):241–282, 2005.

227. M. Peszyńska and R. Showalter. *Multiscale elliptic-parabolic systems for flow and transport.* Electron. J. Differential Eq., 147, 2007.

228. M. Peszyńska, M.F. Wheeler, and I. Yotov. *Mortar upscaling for multiphase flow in porous media.* Comput. Geosci. 6(1):73–100, 2002.

229. W. V. Petryshyn. *On the approximation-solvability of equations involving A-proper and pseudo-A-proper mappings.* Bull. Amer. Math. Soc., 81:223–312, 1975.

230. O. Pironneau. *On the transport-diffusion algorithm and its application to the Navier-Stokes equations.* Numer. Math., 38:309–332, 1982.

231. D.K. PONTING. *Corner-point geometry in reservoir simulation.* In: King PR, editor. Proceedings of the first European Conference on Mathematics of Oil Recovery, Cambridge, 1989 (Oxford), Clarendon Press, pp. 45–65, 1989.

232. P. Popov, Y. Efendiev and Y. Gorb. *Multiscale finite element methods for fluid-structure interaction problem.* Submitted.

233. S. Prudhomme, P. T. Bauman and J. T. Oden. *Error control for molecular statics problems.* Int. J. Multiscale Comput. Eng., 4:647–662, 2006.

234. C. Prud'homme, D. Rovas, K. Veroy, Y. Maday, A.T. Patera, and G. Turinici. *Reliable real-time solution of parametrized partial differential equations: Reduced-basis output bound methods.* J. Fluids Eng., 172:70–80, 2002.

235. J.N. Reddy, *Introduction to the finite element method.* McGraw-Hill Science/Engineering/Math, New York, 1993.

236. L. Richards. *Capillary conduction of liquids through porous medium.* Physics, pp. 318–333, 1931.

237. C. Robert and G. Casella. *Monte Carlo statistical methods.* Springer-Verlag, New-York, 1999.

238. A.J. Roberts, I. Kevrekidis. *General tooth boundary conditions for equation free modeling.* SIAM J. Sci. Comput., 29(4):1495–1510, 2007.

239. R.E. Rudd and J.Q. Broughton. *Coarse-grained molecular dynamics and the atomic limit of finite elements .* Phys. Rev. B 58, R5898, 1998.

240. E. Sanchez-Palencia. *Non-homogeneous media and vibration theory.* Springer, New York, 1980.

241. G. Samaey, I.G. Kevrekidis, and D. Roose. *Patch dynamics with buffers for homogenization problems.* J. Comput. Phys., 213(1):264–287, 2006.

242. G. Samaey, D. Roose, and I.G. Kevrekidis. *The gap-tooth scheme for homogenization problems.* SIAM MMS, 4(1):278–306, 2005.

243. G. Sangalli. *Capturing small scales in elliptic problems using a residual-free bubbles finite element method.* SIAM MMS, 1(3):485–503, 2003.

244. R.E. Showalter. *Monotone operators in Banach space and nonlinear partial differential equations.* vol. 49 of Mathematical Surveys and Monographs, American Mathematical Society, Providence, RI, 1997.

245. I.V. Skrypnik. *Methods for analysis of nonlinear elliptic boundary value problems.* vol. 139 of Translations of Mathematical Monographs, American Mathematical Society, Providence, RI, 1994. Translated from the 1990 Russian original by Dan D. Pascali.

246. V.R. Stenerud, V. Kippe, A. Datta-Gupta, and K.A. Lie. *Adaptive multiscale streamline simulation and inversion for high-resolution geomodels.* SPE J., 13:1, pp. 99–111, 2008. DOI: 10.2118/106228-PA.

247. T. Strinopoulos. *Upscaling of immiscible two-phase flows in an adaptive frame.* PhD thesis, California Institute of Technology, Pasadena, 2005.

248. T. Strouboulis, I. Babuška, and K. Copps. *The design and analysis of the generalized finite element method.* Comput. Meth. Appl. Mech. Eng., 181:43–69, 2000.

249. A.M. Stuart, P Wiberg, and J. Voss. *Conditional path sampling of SDEs and the Langevin MCMC method.* Comm. Math. Sci., 199:279–316, 2004.

250. R. Sviercoski, B. Travis, and J.M. Hyman. *Analytical effective coefficient and a first-order approximation for linear flow through block permeability inclusions.* Comput. Math. Appl., 55(9):2118–2133, 2008.

251. E. B. Tadmor, R. Phillips and M. Ortiz. *Mixed atomistic and continuum models of deformation in solids.* Langmuir, 12:4529-4534, 1996.

252. E. B. Tadmor, R. Phillips, and M. Ortiz. *Quasicontinuum analysis of defects in solids.* Philosoph. Mag. A, 73:1529–1563, 1996.

253. L. Tartar. *Compensated compactness and applications to P.D.E..* Nonlinear Analysis and Mechanics, Heriot-Watt Symposium, Vol. IV, ed. by R. J. Knops, Research Notes in Mathematics 39:136-212, Pitman, Boston, 1979.

254. L. Tartar. *Solutions oscillantes des équations de Carleman*. Seminaire Goulaouic-Meyer-Schwartz (1980-1981), exp. XII. Ecole Polytechnique (Palaiseau), 1981.

255. L. Tartar. *Nonlocal effects induced by homogenization*. in PDE and Calculus of Variations, ed by F. Culumbini, et al, Birkhäuser, Boston, pp. 925-938, 1989.

256. H. Tchelepi, P. Jenny, S.H. Lee, and C. Wolfsteiner. *An adaptive multiphase multiscale finite volume simulator for heterogeneous reservoirs*. SPEJ, 12:185–195, 2007.

257. A. Toselli and O. Widlund. *Domain decomposition methods—algorithms and theory*. Springer Series in Computational Mathematics, 34. Springer-Verlag, Berlin, 2005. xvi+450.

258. H. Versieux and M. Sarkis. *Numerical boundary corrector for elliptic equations with rapidly oscillating periodic coefficients*. Comm. Numer. Meth. Eng., 22(6):577–589, 2006.

259. C. Wolfsteiner, S.H. Lee, and H. Tchelepi. *Modeling of wells in the multiscale finite volume method for subsurface flow simulation*. SIAM MMS, 5:900–917, 2006.

260. X.H. Wu, Y. Efendiev, and T.Y. Hou. *Analysis of upscaling absolute permeability*. Discrete Contin. Dynam. Syst., Ser. B, 2:185–204, 2002.

261. P.M. De Zeeuw. *Matrix-dependent prolongation and restrictions in a blackbox multigrid solver*. J. Comput. Appl. Math., 33:1–27, 1990.

262. S. Verdiere and M.H. Vignal. *Numerical and theoretical study of a dual mesh method using finite volume schemes for two-phase flow problems in Porous Media*. Numer. Math., 80:601–639, 1998.

263. T.C. Wallstrom, M. Christie, L.J. Durlofsky, and D.H. Sharp. *Effective flux boundary conditions for upscaling porous media equations*. Transport Porous Media, 46:139–153, 2002.

264. T.C. Wallstrom, M. Christie, L.J. Durlofsky, and D. H. Sharp. *Application of effective flux boundary conditions to two-phase upscaling in porous media*. Transport Porous Media, 46:155–178, 2002.

265. T.C. Wallstrom, S.L. Hou, M. Christie, L.J. Durlofsky, and D.H. Sharp. *Accurate scale-up of two phase flow using renormalization and nonuniform coarsening*. Comput. Geosci, 3:69–87, 1999.

266. W.L. Wan, T. Chan, and B. Smith. *An energy-minimizing interpolation for robust multigrid methods*. SIAM J. Sci. Comput., 21(4):1632–1649, 2000.

267. X. Wan and G. Karniadakis. *Multi-element generalized polynomial chaos for arbitrary probability measures*. SIAM J. Sci. Comput., 28:901–928, 2006.

268. A.W. Warrick. *Time-dependent linearized infiltration: III. Strip and disc sources*. Soil. Sci. Soc. Am. J., 40:639–643, 1976.

269. A. Westhead. *Upscaling two-phase flows in porous media*. PhD thesis, California Institute of Technology, Pasadena, 2005.

270. C. Wolfsteiner, S.H. Lee, H. Tchelepi, P. Jenny, and W.H. Chen. *Unmatched multiblock grids for simulation of geometrically complex reservoirs*. 9th European Conference on the Mathematics of Oil Recovery Cannes, France, 30 August–2 September, 2004

271. E. Wong. *Stochastic processes in information and dynamical systems*. McGraw-Hill, New York, 1971.

272. D. Xiu and J. Hesthaven. *High-order collocation methods for differential equations with random inputs.* SIAM J. Sci. Comput., 27(3):1118–1139, 2007.

273. J. Xu and L. Zikatanov. *On an energy minimizing basis for algebraic multigrid methods.* Comput. Vis. Sci. 7:3-4, 121–127, 2004.

274. E. Zeidler. *Nonlinear functional analysis and its applications. II/B.* Springer-Verlag, New York, 1990. Nonlinear monotone operators, Translated from the German by the author and Leo F. Boron.

275. D. Zhang. *Stochastic methods for flow in porous media: Coping with uncertainties.* Academic Press, San Diego, CA, pp. 350, 2002.

276. D. Zhang and Z. Lu, *Monte Carlo simulations of solute transport in bimodal randomly heterogeneous porous media.* Proceeding of World Water & Environmental Resources Congress 2003 and Symposium of Probabilistic Approaches and Groundwater Modeling, June 22–26, 2003, Philadelphia.

277. D. Zhang and Z. Lu, *An efficient, high-order perturbation approach for flow in random porous media via Karhunen-Loeve and polynomial expansions.* J. of Comput. Phys., 194(2):773–794, 2004.

278. V. Zikov, A. Kozlov, O. Oleinik, and H.T. Ngoan. *Averaging and G-convergence of differential operators.* Uspekhi Mat. Nauk 34, 5(209):65–133, 1979

# Index